Urban Disasters and Resilience in Asia

Urban Disasters and Resilience in Asia

Edited by

Rajib Shaw
Professor,
Graduate School of Global Environmental Studies,
Kyoto University, Kyoto, Japan

Atta-ur-Rahman
Associate Professor,
Institute of Geography, Urban and Regional Planning,
University of Peshawar, Peshawar, Pakistan

Akhilesh Surjan
Associate Professor,
Faculty of Engineering, Health, Science
and the Environment,
Charles Darwin University, Darwin, Australia

Gulsan Ara Parvin
Researcher,
Graduate School of Global Environmental Studies,
Kyoto University, Kyoto, Japan

ELSEVIER

AMSTERDAM • BOSTON • HEIDELBERG • LONDON
NEW YORK • OXFORD • PARIS • SAN DIEGO
SAN FRANCISCO • SINGAPORE • SYDNEY • TOKYO
Butterworth-Heinemann is an imprint of Elsevier

B H

Butterworth-Heinemann is an imprint of Elsevier
The Boulevard, Langford Lane, Kidlington, Oxford, OX5 1GB, UK
225 Wyman Street, Waltham, MA 02451, USA

Notices
Knowledge and best practice in this field are constantly changing. As new research and experience broaden our understanding, changes in research methods, professional practices, or medical treatment may become necessary.

Practitioners and researchers must always rely on their own experience and knowledge in evaluating and using any information, methods, compounds, or experiments described herein. In using such information or methods they should be mindful of their own safety and the safety of others, including parties for whom they have a professional responsibility.

To the fullest extent of the law, neither the Publisher nor the authors, contributors, or editors, assume any liability for any injury and/or damage to persons or property as a matter of products liability, negligence or otherwise, or from any use or operation of any methods, products, instructions, or ideas contained in the material herein.

ISBN: 978-0-12-802169-9

Library of Congress Cataloging-in-Publication Data
A catalogue record for this book is available from the Library of Congress

British Library Cataloguing-in-Publication Data
A catalogue record for this book is available from the British Library

For information on all Butterworth-Heinemann publications visit our website at http://store.elsevier.com/

Contents

List of Contributors

Vinayak Adane Professor, Department of Architecture and Planning, Visvesvaraya National Institute of Technology, Nagpur, India

Reazul Ahsan Research Fellow, MIT-UTM Sustainable Cities Program, Massachusetts Institute of Technology, Cambridge, MA, USA

Mohammad Shakil Akther Professor, Department of Urban and Regional Planning, Bangladesh University of Engineering and Technology, Dhaka, Bangladesh

Atta-ur-Rahman Associate Professor, Institute of Geography, Urban and Regional Planning, University of Peshawar, Peshawar, Pakistan

Michiko Banba Education and Research Center for Disaster Education, University of Hyogo, Japan

Uttama Barua Graduate Student, Department of Urban and Regional Planning, Bangladesh University of Engineering and Technology and Research Planner, Bangladesh Network Office for Urban Safety (BNUS), Bangladesh University of Engineering and Technology, Dhaka, Bangladesh

Ranit Chatterjee Graduate School of Global Environmental Studies, Kyoto University, Kyoto, Japan

Priya Choudhary Smt. Manoramabai Mundle College of Architecture, Seminary Hills, Maharashtra, India

Sameer Deshkar Assistant Professor, Department of Architecture and Planning, Visvesvaraya National Institute of Technology, Nagpur, India

Glenn Fernandez Disaster Risk Management Systems, Asian Disaster Preparedness Center, Bangkok, Thailand

Ishrat Islam Professor, Department of Urban and Regional Planning, Bangladesh University of Engineering and Technology, Dhaka, Bangladesh

Nafesa Ismail Graduate School of Global Environmental Studies, Kyoto University, Kyoto, Japan

Sadasivam Karuppannan University of South Australia, Australia

Jon Kellett University of Adelaide, Adelaide, Australia

Huy Nguyen Institute for Social and Environmental Transition, Vietnam

Kenji Okazaki Professor, Kyoto University, Japan

Gulsan Ara Parvin Researcher, Graduate School of Global Environmental Studies, Kyoto University, Kyoto, Japan

Krishna S. Pribadi Professor, Bandung Institute of Technology, Indonesia

Bhaswati Ray Assistant Professor, Sivanath Sastri College, Kolkata, India

Rajib Shaw Professor, Graduate School of Global Environmental Studies, Kyoto University, Kyoto, Japan

Kazi Farzana Shumi Department of Business Administration (DBA), International Islamic University Chittagong (IIUC), Dhaka, Bangladesh

Nitin Srivastava Graduate School of Global Environmental Studies, Kyoto University, Kyoto, Japan

Akhilesh Surjan Associate Professor, Faculty of Engineering, Health, Science and the Environment, Charles Darwin University, Darwin, Australia

Phong Tran Institute for Social and Environmental Transition, Vietnam

Tho Tran Graduate School of Global Environmental Studies, Kyoto University, Kyoto, Japan

About the Editors

Rajib Shaw

Professor, Graduate School of Global Environmental Studies, Kyoto University, Kyoto, Japan

Rajib Shaw is a professor at the Graduate School of Global Environmental Studies of Kyoto University, Kyoto, Japan. He has worked closely with local communities, nongovernmental organizations, governments, and international organizations, including the United Nations (UN), especially in Asian countries. His research interests include community-based disaster risk management, climate change adaptation, urban risk management, and disaster and environmental education. He is currently the president of the Asian University Network of Environment and Disaster Management, and the co-chair of the UN International Strategy of Disaster Reduction (ISDR) Asia Science Technology Academia Advisory Group. Professor Shaw has been published extensively in different journals, books and edited volumes.

Atta-ur-Rahman

Associate Professor, Institute of Geography, Urban and Regional Planning, University of Peshawar, Peshawar, Pakistan

Atta-ur-Rahman is an associate professor at the Institute of Geography, Urban and Regional Planning, University of Peshawar, Pakistan. He recently completed his postdoctoral studies on modeling disaster risk at Kyoto University, Kyoto, Japan. He is also a visiting faculty member at the Centre for Disaster Preparedness and Management, University of Peshawar, Pakistan. His specialties are disaster risk reduction (DRR) and environmental impact assessment. He is currently working with a number of international organizations on various aspects of

DRR and supervising research students in the field of disaster risk management. He is a member of the editorial boards of several prestigious journals and has written numerous books and research articles.

Akhilesh Surjan

Associate Professor, Faculty of Engineering, Health, Science and the Environment, Charles Darwin University, Darwin, Australia

Akhilesh Surjan has successfully dealt with issues of climate and disaster risk reduction (DRR) and urban environmental management. He served as a lead author for the Fifth Assessment Report (AR5) of the Intergovernmental Panel on Climate Change (IPCC). He also served as a contributing author for the United Nation's Global Assessment Report on Disaster Risk Reduction, published in May 2011. In particular, he focuses on issues surrounding the sustainability of development and adaptation concerns of emerging cities. Dr. Surjan trained at Kyoto University in Kyoto, Japan, where he successfully completed his doctoral study, focusing on resilience to environment and DRR in the Asia-Pacific region.

Gulsan Ara Parvin

Researcher, Graduate School of Global Environmental Studies, Kyoto University, Kyoto, Japan

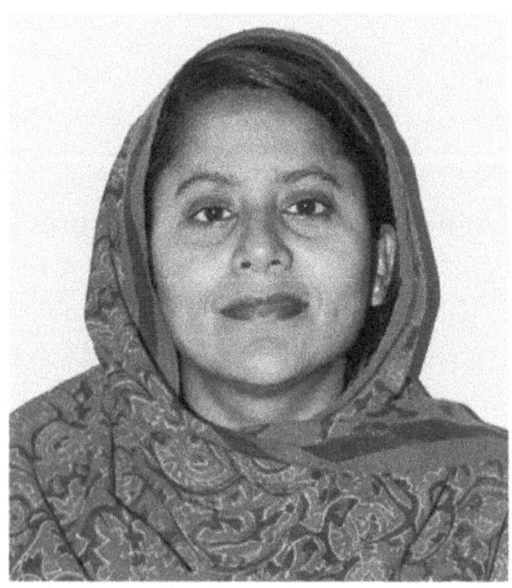

Gulsan Ara Parvin is a researcher at the Graduate School of Global Environmental Studies, at Kyoto University, Kyoto, Japan. She has 12 years of teaching and research experience at the Bangladesh University of Engineering and Technology (BUET). She was an associate professor in the Urban and Regional Planning Department of BUET. Dr. Parvin was awarded a postdoctoral fellowship from the Japan Society for the Promotion of Science to conduct research at Kyoto University from 2007 to 2010. She obtained a Ph.D degree from the Urban Engineering Department of the University of Tokyo in 2003 and an M.Sc degree from the Agriculture Conservation and Rural Development Program of the Asian Institute of Technology in 1999. Her research interests mostly concentrate on community development, vulnerability and disaster management for the poor, climate change impact and adaptation, female empowerment, and the role of nongovernmental organizations.

Preface

Globally, the intensity and frequency of disasters are on the rise. In urban areas, more than 50% of the world's population is living on just 2% of the land surface. Most of these cities are located in Asia, which consists mostly of developing nations. In these cities, more than 30% of the residents are living in slums and squatters. Unforeseen disaster events hit cities and communities in both the developing and developed world, but developing nations are more vulnerable and suffer more intensely. Numerous cities have been affected by natural and human-created disasters, with thousands of the inhabitants either buried under debris or washed away by gushing water. Over time, urban disasters put unprecedented pressure on city budgets, which must accommodate emergency response and recovery. Urban centers are the hub of industrial and commercial activity, so cities function to empower their societies. Wherever disasters hit urban areas, their severity brought widespread devastation in terms of human losses and adverse economic consequences and setbacks.

Urban resilience is largely a function of resourceful citizens and governments. The strong and committed involvement of citizens at the grassroots level can lead to a resilient city. There are many factors that affect urban resilience, and no two cities are alike in their inherent capacities. Therefore, building cities that are more resilient to both external and internal negative factors may bring about more productive economic returns. It is important to note that mitigation prior to the occurrence of a disaster is much more effective than picking up the pieces afterward.

This is a very demanding area, which deserves special attention from academia, government institutions, disaster managers, urban authorities, international governmental organizations, scientific community, practitioners, and other experts. It should and will be the focus of policy makers, practitioners working in both the public and private sectors, students, academia, and government officials.

This book particularly highlights several key areas: urban risk, disaster and resilience, building code and land-use planning, coastal cities, urban expansion, urban rural linkages, urban microfinancing, food security, risk communication, and private sector involvement. The text also illustrates some case studies from Asian countries, highlighting experiences from a number of cities.

This book is geared toward a wide audience. Initially, the target group is students, teachers, and researchers working in the fields of urban planning, architecture, disaster preparedness and management, social sciences, and earth and structural sciences. The chapters draw on evidence-based data from the scientific integration of databases and feedback on conceptualization, idea generation, field surveying on urban disasters, and building city resilience. Similarly, disaster managers, field practitioners, decision makers, disaster-related authorities, and city government staffers are other target readers of this valuable reference.

Rajib Shaw
Atta-ur-Rahman
Akhilesh Surjan
Gulsan Ara Parvin

About the Book

Urbanization is on the rise and disasters are increasing throughout the world. Asia, where significant numbers of people still live in informal settlements, bears the brunt of these developments. Numerous Asian cities have been affected by natural and human-created disasters in recent years, and there are many examples of innovative risk reduction approaches designed to improve the resilience of urban areas on the continent. This book is an attempt to explain some of the key lessons of urban resilience based on Asian examples. The book, the culmination of years of effort from renowned urban scientists and other experts, has 20 chapters, some of them highlighting common issues like coastal management, building vulnerability, urban rural linkage, and community engagement, and others presenting case studies of different cities highlighting and linking these issues. This topic is a demanding one that definitely merits special attention from academia, government institutions, disaster managers, urban authorities, international nongovernmental organizations, scientific community, and practitioners. The book will be useful to policy makers, practitioners working in both the public and private sectors, students, academia, and government officials.

Urban Disasters and Approaches to Resilience

1

Atta-ur-Rahman[1], Rajib Shaw[2], Akhilesh Surjan[3], Gulsan Ara Parvin[4]
[1]Associate Professor, Institute of Geography, Urban and Regional Planning, University of Peshawar, Peshawar, Pakistan; [2]Professor, Graduate School of Global Environmental Studies, Kyoto University, Kyoto, Japan; [3]Associate Professor, Faculty of Engineering, Health, Science and the Environment, Charles Darwin University, Darwin, Australia; [4]Researcher, Graduate School of Global Environmental Studies, Kyoto University, Kyoto, Japan

1.1 Introduction

More than half of the world population is now living in urban areas (UN, 2014). The urban population is increasing at a rapid rate, and it is projected that by the year 2030, 65% of the world's population will be living in cities, mostly in the developing world (Sharma et al., 2011). Most of the top 20 cities in the world are in Asia, mainly located in the developing world. The data reveals that in the developing world, urban populations are increasing at a rapid pace that poses a series of threats to them. It has been estimated that in Asia, over 40% of its urban dwellers are living in slums and squatter settlements. Large cities are particularly vulnerable to a wide variety of hazards, with the majority of these populations living in high- to moderate-risk zones.

The so-called super cities, including Tokyo, Delhi, Shanghai, Dhaka, Mumbai, Karachi, Manila, Jakarta, Bangkok, and Calcutta, have experienced serious incidents of flooding, cyclone surges, and earthquakes in the past decade (Douglass, 2013). Meanwhile, several other Asian cities have faced heat waves, droughts, urban flooding, and intense rainfall. The effects of such incidents have been intensified by climate change. Cities are the hub of educational and cultural innovation and provide industrial, commercial, and infrastructure services (Shaw et al., 2009). Such links have positive implications to accelerate both the economic and political situations. Cities are certainly strong, but they are also vulnerable to wide range of disasters. This is why the urban authorities are called upon to develop city disaster risk reduction (DRR) plan(s) to cope, adapt to, or withstand shock, stress, and disturbances with minimum human casualties and damage (Rahman & Shaw 2015).

The continent of Asia is where the world's least-urbanized countries are located. In Sri Lanka, Afghanistan, Tajikistan, Bangladesh, Vietnam, Nepal, India, Thailand, and Laos, less than 30% of the population lives in urban areas (UN, 2014). Singapore, Hong Kong, Qatar, Kuwait, Israel, South Korea, and Japan are among the most urbanized countries, with over 90% of the total population residing in cities. As a whole, the urban population in Asia is rapidly increasing compared to other continents. In Asia, in terms of degree of urbanization, 27 countries have more than 50% of their population living in urban areas. Of the top 20 megacities in the world, 13 are in Asia—namely, Tokyo, Delhi, Shanghai,

Urban Disasters and Resilience in Asia.

Mumbai, Beijing, Osaka, Dhaka, Karachi, Calcutta, Istanbul, Chongqing, Manila, and Guangzhou, with populations of over 10 million (UN, 2014). Out of these cities, four are in China, three in India, two in Japan, and one each from Bangladesh, Pakistan, Turkey, and the Philippines. It is urban centers that accelerate the economic growth rate of high-income countries (i.e., Japan, South Korea, China, and Singapore), middle-income countries (i.e., Azerbaijan, India, Iran, and Pakistan) and low-income nations (i.e., Bangladesh, Cambodia, and Kyrgyzstan) (UNHABITAT, 2010).

Figure 1.1 depicts the spatial distribution of the major urban centers in Asia in the year 2000. These cities are growing at an alarming rate, and between 2000–2010, cities in the developing world accounted for a two-thirds increase. During the same period, the population of Karachi, Pakistan, has grown remarkably, by 80%. In these cities, over 37% of urban citizens are living in slums and squatter developments. As a consequence, the intensity and occurrences of urban disasters has increased, and as a result the urban authorities have been hard-pressed to cope with and build urban resilience to these events. The analysis presented here shows that urban resilience is largely a function of resilient and resourceful citizens. The strong and committed involvement of citizens at the grass-roots level may lead to cities that can withstand and react well to disasters.

In the scientific research that is currently available, city resilience is considered as the capability of an established system to cope with and withstand the impact of a major disaster and recover quickly to normal city functioning. However, resilience largely varies from city to city and study to study, depending on the use and application of resilience methods. Similarly, vulnerability and exposure to such events also vary from city to city. Some cities are extremely vulnerable to coastal hazards, like Mumbai, Shanghai, Karachi, Chennai, Chittagong, Yangon, Ho Chi Minh City, Osaka, Singapore, and Semarang. The urban agglomerations in the Bohai Bay area (China), the Ganges-Brahmaputra deltaic region (Bangladesh), the Indus river delta (Pakistan), the Yangtze River delta (China) and the Pearl River delta region (China) are exposed to various coastal hazards. Some Asian cities are exposed to river flooding, like Dhaka, Delhi, Bangkok, Lahore, and Bandung. Several cities in India, the Philippines, Vietnam, Bangladesh, Pakistan, Myanmar, and China are frequently exposed to violent storms. Earthquakes are another type of devastating event, to which many cities in Japan, Indonesia, China, India, Pakistan, Iran, Bangladesh, Vietnam, and Thailand are comparatively more vulnerable.

It has been observed that developing cities generally spend only a small fraction of their budgets on disaster preparedness. Such limited investment in urban resilience can lead to massive damage after catastrophic events occur. Experience has shown that even a small investment in urban risk reduction is much more effective than picking up the pieces after a disaster (Rahman & Shaw, 2015). As cities are the hubs of commercial, industrial, and social activities, they contain large numbers of people in zones of great population density. They also act as engines for national economic growth and prosperity. It is cities that empower societies, and hence it is important to give them the attention they need in order to withstand disastrous events. The resilient capability of a city varies from location to location, and for this reason, increasing resilience is mainly a function of a city's resilient and resourceful citizens. The committed and effective participation of city dwellers at the community level, and effectively addressing both internal and external negative factors, may yield productive and resilient cities.

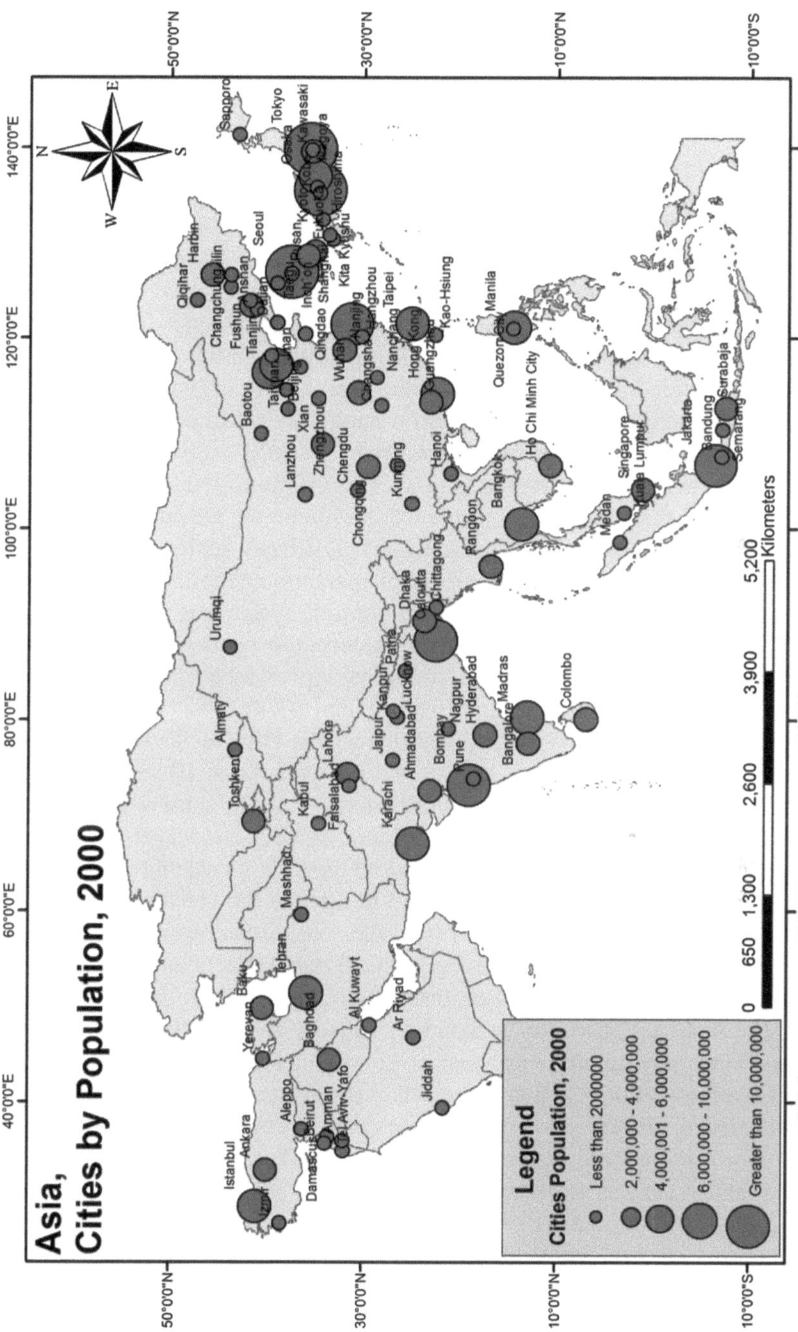

Figure 1.1 Distribution of Asian cities, 2000.

1.2 Resilience in a Global Context

An earthquake occurring in the Indian Ocean in 2004, followed soon after by a tsunami, was a turning point in the history of global disaster risk management systems. After the Indian Ocean tsunami, the United Nations World Conference on Disaster Reduction (UNWCDR) was held on January 18–22, 2005, in Kobe, Japan. The UNWCDR provided a platform to bring together the scientific community, government stakeholders, and practitioners under a single but comprehensive agenda of reducing disaster vulnerabilities. The Hyogo Framework for Action: Building the Resilience of Nations and Communities to Disaster (HFA 2005–2015) was the outcome of this conference, which insisted that nations explicitly work on five priority areas (GoP, 2012).

HFA 2005–2015 is the agreed structure for making the world safer from extreme events and enhancing community resilience against disasters. In this agreement, 168 UN member-states decided on five action priorities, and a 10-year plan was set up to achieve a sizable lessening of disaster impacts on human lives and economic, social, and environmental assets of communities and nations.

Overall, the HFA has provided critical guidance in efforts to reduce disaster risk and contributed toward the achievement of the UN's Millennium Development Goals (MDGs). The HFA priorities included ensuring that DRR is a national and a local priority with a strong institutional basis for its implementation; identifying, assessing, and monitoring disaster risks and enhancing early-warning systems; using knowledge, innovations, and education to build a culture of safety and resilience at all levels; reducing the underlying risk factors; and strengthening disaster preparedness for effective response at all levels (Queensland Government, 2014).

During the third United Nations World Conference on Disaster Risk Reduction, held in Sendai, Japan, from March 14–18, 2015, it was stated that 10 years after the adoption of the HFA, disasters continue to undermine efforts to achieve sustainable development in the developing world. Because of this, an HFA 2015-2030 agreement was reached with the aim of achieving, in the next 15 years, a substantial reduction of disaster risk and damage to lives, livelihoods, and health and the economic, physical, social, cultural, and environmental assets of people, businesses, communities, and countries as a whole. This will require strong political commitment and involvement in each country at all levels. Governments throughout the world are fully dedicated to enhancing communities' capacity to handle disasters and building nations and community resilience against extreme events. Almost all UN member states have taken legislative and constitutional actions to establish disaster management agencies to mitigate, prepare for, prevent, and effectively respond to disasters and recover from emergency situations. In Asia, almost all the UN member states have approved legislation to establish disaster management authorities.

The Queensland Government, 2014 defined resilience as the capacity to prepare for, withstand, respond to, and recover from disasters. From this perspective, the basic idea is to build cities that are stronger and more resilient. UNISDR (2009) defined resilience as the ability of a system, community, or society to absorb, resist, accommodate to, and recover from disaster impacts in a timely and efficient manner, including through the restoration of its essential basic functions and structures. The condition of resilience has strengthened with time, which enhances the ability of many cities to minimize the effects of disasters in the future.

1.3 Impact of Disasters and Extent of Resilience

Of the 10 most damaging natural disasters throughout the world in 2013, 8 were reported in Asia (Caulderwood, 2014). The Philippines, China, and Vietnam suffered the most from Typhoon Haiyan in November 2013. Similarly, India, Nepal, and Pakistan were hit by flooding, which resulted in 7194 deaths, while an earthquake in Pakistan killed 825 people in September. In 2013, the most economically expensive disasters were that of flooding in central Europe, which cost $22 billion; an earthquake that occurred in Sichuan province, China, on April 20, which cost $14 billion; Super Typhoon Haiyan, which cost $13 billion; Typhoon Fitow in October in China and Japan, which cost $10 billion;droughts in China, which cost $10 billion; a series of droughts in Brazil, which cost $8 billion;flooding in Alberta, Canada, in July, which cost $5.2 billion; floods in north China in August–September, which cost $5 billion; another flood in southwest China, which cost $4.5 billion; and Hurricane Manuel in Mexico, which cost $4.2 billion(Caulderwood, 2014).

Disastrous events have occurred in both developing and developed nations, but developing nations are more vulnerable and experience such incidents more intensely (Rahman & Shaw, 2015). In the past decade, numerous cities have been affected by natural and human-induced disasters, where thousands of inhabitants either buried under debris or swept away by gushing water. Eventually, such urban disasters have extraordinary impacts on city budgets (Rahman & Shaw, 2015). Whenever any extreme event strikes an urban center, it seriously affects the residents in terms of both human casualties and physical and economic losses. Historically, urban centers are designed to empower the societies that contain them, as cities are the hubs of commercial and industrial activities. When disaster strikes, this pressures urban authorities to shift budget monies from urban development to emergency response, evacuation, rehabilitation, and early recovery (Rahman & Shaw, 2015).

Historically, many cities were severely affected either by human-induced or natural extreme events and incurred numerous economic losses, disruption of the urban systems, and human casualties. In the predisaster phase, city governments were not prepared, so these communities suffered great devastation. A government needs to formulate and develop a preparedness plan so that its cities can effectively withstand unforeseen catastrophes. In Asian cities, the existing emergency response systems should be planned in a way that employs the existing capability of organizations, communities, technical experts, and resourceful citizens to respond efficiently. Due to the lack of such effective disaster management plans, many city budgets were consumed with emergency response, rehabilitation, early recovery, and reconstruction duties after disaster strikes.

The concept of urban resilience means the capacity of a city to bounce back effectively and quickly from the impacts of a disastrous event. Likewise, it is considered a joint function of the resilience and high capacity of citizens (Campanella, 2006).

It has been observed that the committed and powerful involvement of residents at the community and local government levels may lead to a resilient city. However, there are numerous determining factors that influence city resilience. The research literature reveals that no two cities are alike in their inherent resilience capacities and coping mechanisms. It is because of this fact that enhancing city resilience to internal and external adverse factors may lead to more productive economic returns (Rahman & Shaw, 2015).

1.4 Why Inclusive Urban Resilience?

UNISDR (2007) stated that the role of women in disasters is not merely as victims, but rather as active stakeholders in DRR planning. The one UN program that concerns disaster preparation has focused on the need for increasing community resilience in response to disasters, climate change, temporary relocated persons, humanitarian assistance, disease transmission, and gender equality (UN, 2009). Recent DRR strategies have stressed inclusive and sensitive sustainable planning and development. The vulnerable group has low resilience capacity and needs more attention at all levels. In developing countries, majority of women are housewives, who are primarily responsible for household activities including cooking, washing, cleaning, child care, and care of elderly or disabled persons (Samiullah et al., 2015) and have limited access to disaster response. Sudden misfortune increases women's vulnerabilities and complicates their duties as heads of household. Scientific studies reveal that in most of these countries, women generally are less educated, low capacity, and have limited access to resources and knowledge about disasters and climate change, low awareness of their rights, limited exposure to external environment, and few life skills. These are the determining factors that make women extremely vulnerable during any extreme event.

In gender-sensitive risk reduction strategies, women have the capacity to cope with disaster events and work as activists in search and rescue activity. Similarly, most often immediately during disasters, women are devoted to taking care of their household belongings and the preservation of valuable properties. Disabled people, children, the elderly, and pregnant women need special care and sometimes have difficulty finding and getting to shelters. In the developing countries of Asia, women have very low disaster and climate change education, and therefore, they cannot take a proactive approach to functioning in all three (pre, during and post) phases of disaster. For instance, during the earthquake that hit Kashmir, Pakistan, in 2005, numerous cases of harassment of women, child kidnapping, and sexual abuse were the result of negligence on the part of both government functions and the local community (Samiullah et al., 2015).

It is very important to realize that urban authorities need to prepare city disaster risk management plans to enhance disaster resilience and multihazard early warning and mapping. Men and women must work together to recover from the impact of a disaster. Generally, women are predominantly seen as helpless in these societies, and recently, attempts have been made to involve vulnerable groups (including women) in the decision-making process. The empowerment of women is an important element of increasing their role in mentoring, management, leadership, and policy formulation (Samiullah et al., 2015). Mainstreaming vulnerable group in risk assessment process and disaster management is very effective, and leads to better decision making and implementing programs at the institutional, community and regional levels. In gender-specific planning, the emphasis remains on taking particular care of marginalized and vulnerable groups, such as women, children, the disabled, and the elderly, to ensure that they receive sufficient attention during emergencies (Samiullah et al., 2015).

1.5 Approaches to Urban Resilience

Globally, in DRR efforts, a paradigm shift was noted after the Indian Ocean tsunami in 2004. The UNISDR, international organizations, and the countries in this area of the world quickly adopted initiatives for legislative and policy actions at the national, regional, and community levels for DRR. The same landmark development was addressed in the HFA 2005–2015 declaration with five priority areas for DRR. The gradual introduction of innovative techniques and approaches was involved, including an attempt to shift from reactive to proactive approaches, with specific attention paid to hazard identification, preparedness, mitigation, enhancing community resilience, capacity development, early warning, emergency response, and early recovery. As a result, disaster awareness, coping mechanisms, and endorsement of urban risk reduction strategies in planning and development have increased.

The HFA insisted that nations incorporate DRR into their urban risk reduction plans. In Asia, almost all nations have taken initiatives toward policy formulation, legislation, establishment of disaster management mechanisms, institutionalization of organizations, and disaster risk management planning at the national, regional, and community levels, and the focus has changed from reactive to proactive DRR approaches. The governments of Japan, China, South Korea, the Philippines, Vietnam, Indonesia, Malaysia, Thailand, Sri Lanka, Bangladesh, India, Pakistan, Nepal, and other nations have strengthened their resilience to disasters by implementing risk management legislation, establishing appropriate institutions, implementing mitigation, risk assessment, preparedness, forecasting and early warning systems, mechanisms for emergency response, early recovery, and disaster and climate change education in order to minimize the underlying risk factors. Nevertheless, the key urban risk reduction strategies include land-use planning, protecting critical structures, prohibiting specific urban functions in incompatible areas, reducing pollution, protecting the environment, controlling emissions, enforcing building bylaws, and creating early warning and emergency response systems (Rahman & Shaw, 2015). Such innovative and proactive urban risk reduction strategies have been shown to be effective in handling urban risks. Climate-related urban disasters are gaining attention due to their increasing occurrence, and urban authorities need to properly endorse DRR in their city planning and development processes.

In Asia, almost every country has only a few major cities and many medium-sized and small urban centers. In addition to the megacities, governments should take into account medium-sized and small cities when mainstreaming DRR into their urban policy planning and development. These administrations are attempting to reduce the impacts of multiple hazards and make cities more resilient to the hazards presented by cyclones, regional storm surges, heat waves, seismic shocks, urban floods, tsunamis, urban fires, intense rainfall, unhygienic conditions, poor drainage systems, and lack of effective solid waste collection and disposal. Similar hazards are regularly reported from almost all Asian cities located in the developing world. It is, however, essential to anticipate and formulate urban risk reduction plans to effectively protect people, communities, livelihoods, cultural heritage, ecosystems, education, health, and other vital assets.

In the domains of urban disasters, capacity mechanisms, and resilience, there are several critical issues that need to be addressed well in advance of actual crises; otherwise, there will be far-reaching and irreversible negative impacts on the city dwellers, urban systems, and city authorities. Some of the key urban challenges include low structural resilience, lack of urban regulation and enforcement, weak urban-rural linkages, farmland conversion into built-up area, threats to urban food security, poor risk communication to urban citizens, lack of procedure for urban recovery, and lack of public-private partnerships in handling urban disasters. These are among the wide range of urban issues dealt with by infrastructural scientists, social scientists, architects, geologists, and experts in urban planning and disaster preparedness and management.

In city planning, mainstreaming DRR into policies, plans, legislation, land-use regulations, and building codes may help in minimizing the impact of disasters on human lives and property. In urban risk reduction planning, prevention is less costly than postdisaster reactions, and early recovery and resilience may reduce the extent of any damage from a crisis. City governments are hard-pressed to provide basic services to their citizens during emergencies, mainly because of financial and technical capacities to withstand shocks and extreme events (Rahman & Shaw, 2015). The gap between the urban growth rate and the provision of services is increasing at a faster pace than city administrators can cope with. These days, it is vital for urban authorities to develop effective institutional frameworks for increasing city resilience and climate change adaptation.

1.5.1 Legislation and Urban Risk Reduction

Following the 2004 tsunami, the countries in the region were urged to institutionalize DRR through disaster management legislation and policy. Almost all the countries in Asia have mainstreamed DRR into their policy and planning. For example, after the 2005 Kashmir earthquake, the government of Pakistan established a disaster management commission and a disaster management ordinance was issued by President Pervez Musharraf in 2006. This in turn paved the way for the establishment of the National Disaster Management Authority (NDMA). Similarly, the Provincial Disaster Management Authority (PDMA) and district disaster management authorities (DDMAs) were established. The federal government was fully dedicated to further strengthening its legislative structures and disaster management framework, the ministry of climate change, and Pakistan disaster management policy. Recently, with the technical support of Japan International Cooperation Agency (JICA), the National Disaster Management Plan (NDMP) 2012–2022 was developed to increase disaster resilience and minimize risks through various risk reduction efforts. The NDMA and PDMA frequently work to implement the strategies, programs, and recommendations of NDMP so that Pakistani communities can effectively withstand the impact of extreme natural disasters.

1.5.2 Bylaws and Reinforcement

In Asian cities, structure resilience is very poor and in great need of reinforcement. There are numerous cities for which building bylaws have been prepared in response to this issue. Following a devastating earthquake that hit Quetta (in what is now Pakistan), in 1935, building codes were developed and subsequently enforced to minimize losses to such events in the future (Khan, 2003). Building bylaws provide guidelines, procedures, and recommendations for structural design in various conditions, as well as the selection of building materials. Building codes are mainly developed to specify the minimum acceptable level of safety for structures. These are enforced to protect public health, safety, and welfare pertaining to occupancy of buildings and structures (Rahman & Shaw, 2015). Keeping this challenging scenario in mind, the Ministry of Housing and Works in the Pakistani government recently asked the National Engineering Services Pakistan (NESPAK) to prepare building codes (MOHW, 2007). Citywise enforcement of building codes and regulations, particularly in the urban areas, is a current concern (Rahman & Shaw, 2015). For community safety, regular updating of building bylaws according to the latest techniques is crucial.

At strategic locations, every city authority can establish a robust communication system and an efficient transport corridor and logistics system to be used during emergency situations. Similarly, it can devise and implement emergency response plans in relevant ministries and departments at the federal, state, district, and community level. Likewise, a national disaster management fund was established to enable the federal government to manage its emergency response effectively.

1.5.3 Preparedness, Early Warning Systems, and Emergency Response Systems

When rapid onset disasters occur, effective disaster preparedness, early warning systems, and emergency response systems are key components of urban risk reduction strategies (Parvin et al., 2013). In Asia, very few cities to date have incorporated these DRR components into urban planning processes; there remains a need to make it compulsory to implement them across the board. It has been observed that community involvement in the principal components, including disaster preparedness, early warning, and emergency response systems has always played a role in reducing the impacts of disasters. In addition, capacity building within communities can further enhance the functional performance, roles, and responsibilities of individuals and organizations. Regular arranging of simulation exercises and drills is an effective way to raise awareness and response capacity. As a disaster preparedness strategy, the establishment of a stock repository may further improve a city's ability to recover quickly from a disaster and will help its government have sufficient supplies for emergency relief and rehabilitation. Therefore, urban authorities must create and strengthen warehouses and stockpiling mechanisms for storing food, medicine, relief items, shelters, and rescue equipment (Rahman & Shaw, 2015).

In Asian cities, the frequency and intensity of disastrous events and climate change exacerbations are on the rise, both in terms of fatalities and property loss. In these urban centers, effective disaster forecasting and timely dissemination of warnings are either entirely lacking or minimal (done by only a few major cities). In disaster preparedness, forecasting and early dissemination of warnings to a community about to experience a natural disaster is the primary responsibility of concerned government agency in conjunction with the city government. All urban governments that lack forecasting and early warning systems, need to formulate strategies to establish effective forecasting and timely warning of citizens in high-risk areas and prepare emergency response plans and early recovery for all urban sectors.

In general, emergency response systems in Asian cities are either completely lacking or have insufficient capacity to withstand and respond effectively to disaster situations. Therefore, the establishment of rapid response teams and mechanisms can help in minimizing the impacts of unforeseen events in these communities. In Pakistan, the National Disaster Management Act (NDMA) of 2010 has especially endorsed the importance of having emergency response systems at both the city and community levels. In this regard, the NDMA has formulated a national disaster response plan to respond effectively to disaster situations (Rahman & Shaw, 2015). However, the capacity of existing response systems is very limited. The national disaster response mechanism is required for undertaking search and rescue efforts in urban areas. This calls on city governments to build city resilience through the establishment of emergency response systems at all levels. However, priority needs to be given to creating emergency operation centers, civil defense divisions, and urban search-and-rescue teams in all urban centers (Rahman & Shaw, 2015).

1.5.4 Urban Floodproofing

Worldwide, there is an increasing trend of urban flooding, with resultant damage, due to climate change. In Asia, however, very few cities thus far have developed urban floodproofing and flood management plans, and much still has to be done by city governments. In intense rainfall conditions, the existing drainage system fails to accommodate access water, and as a result, the water overflows, inundating urban services and structures. Similarly, urban floodproofing and management through guided head spurs, marginal embankments, levees, the prohibition of encroachment onto the flood channels, chocking of drainage systems, flood forecasting, and early warning and flood defense systems need to be properly incorporated into urban development planning.

1.5.5 Earthquake Risk Reduction

In an urban environment, effective mainstreaming of DRR and understanding urban scenarios, risk profiles, and trends of industrialization and urbanization are essential (Sharma et al., 2011). Asia is one of the world's most seismically active regions, and almost all of its megacities are vulnerable to the risk of earthquakes and placed from high to minor area. Over the past two decades, the region has experienced several episodes of high-intensity earthquakes, with thousands of human casualties. The urban areas have incurred the heaviest amount of casualties and property damage. The past decade has seen a number of such tragedies, including the 2005 Pakistan earthquake, with over 70,000 fatalities; an earthquake in Lushan, China, in 2013, which caused economic losses of $14 billion and 196 deaths; and an earthquake in Pakistan, which killed 825 (Caulderwood, 2014). The recent April and May 2015 Nepal earthquakes was devastating with thousands of fatalities. It is, therefore, important to devise strategies for reducing urban seismic risk and build resilience against shocks.

1.5.6 Urban Disaster Management Plans

Keeping in mind the facts mentioned in this chapter up to now, there has been a specific focus on making cities more resilient against unforeseen extreme events. The World Bank (2011) specifically highlighted the importance of including proactive approaches for urban risk reduction while devising urban development plans. In addition to increasing urban populations, the situation is further exacerbated by daytime commuters in the urban centers of developing countries. The increasing urbanization poses challenges to urban authorities in extending urban services. Therefore, urban administrations, in consultation with regional disaster management authorities, should prepare disaster management plans that clearly reflect all the stages, including forecasting, creating an early warning system and emergency response mechanisms, capacity building, disaster and climate change education, mitigation, making evacuation plans and early recovery systems, and coordination with government agencies. This will ultimately help in building city resilience and to help them cope with disturbances with minimum human losses and property damage (Campanella, 2006). Such processes, as well as targeted investment in urban risk reduction efforts, can minimize the vulnerabilities of city dwellers. Godschalk (2005) stated emphatically that investment in urban mitigation prior to the fall of a disaster is better than exerting the effort needed to react to a crisis, say, in searching for human bodies and combing through debris to find precious property.

1.5.7 Recovery Planning

Currently, these cities are more resilient than they have been in the past. Cities with a strong economic and political base may recover faster than cities with weak systems and networking. For any urban setup, a fully equipped preparedness plan can enable a city to bounce back after a crisis, with minimal casualties and economic losses.

1.5.8 Cities' Resilience and Climate Change

Historically, cities have relied on ecological resilience but with the passage of time, the extent of resilience expanded. Leichenko (2011) explained urban resilience as the potential to absorb disturbance by enhancing both structural and non-structural resilience.

At the third UNWCDR in 2015, it was reported that viable progress has been achieved in building city resilience and minimizing the impact of disasters, but much still needs to be done. Cities should have self-organizing environmental, physical, ecological, and socioeconomic systems. The Sendai conference set four priorities for action. Taking into account the experience gained through the implementation of the HFA, and in pursuance of the expected outcomes and goals, there is a need for focused action within and across sectors by states on the local, national, regional, and global levels in the four priority areas, which include understanding disaster risk, strengthening disaster risk governance, investing in DRR for resilience, enhancing disaster preparedness for effective response to extreme events, and to Build Back Better in recovery, rehabilitation, and reconstruction (UN, 2015). In recent years, with climate change scenarios becoming more prevalent, the negative consequences of hydrometeorological hazards to cities are multiplying fast. It is essential for cities to enhance their capacities using innovative tools and techniques to resist stresses and shocks and cope with uncertainty in the face of changing climate.

1.6 Why Make a City Resilient?

In Asia, cities are growing are a rapid pace, and there is no sign that this haphazard urbanization will be checked in the future. Similarly, the increasing industrial nature of these cities further contributes to global warming. The World Bank (2011) reported that cities are the first respondents to climate change impact. Urban disasters disrupt a city's lifeline and economic activities and put unprecedented pressure on its budget (IPCC, 2011). After disaster strikes, roads and drains are blocked, and other essential urban services need rehabilitation and fast recovery. After a devastating event, huge amounts of money are spent on the response and recovery. It is cities that have the potential to bring about changes in a country's economic growth and switch from a reactive approach to a proactive one (World Bank, 2011).

Parallel to their other responsibilities, the city governments also must work on urban risk reduction and making their communities best able to respond effectively to disasters. In each city, government officials are particularly accountable for their decisions to city residents who can vote them out (Rahman & Shaw, 2015). In urban environments, the presence of technical personnel is the strength to prepare master plans, structure plan and local plan for sustainable city development and mainstreaming DRR. City officials should have a missionary zeal to meet the challenges of disasters and climate change. The scientific literature shows that cities are the future of society; and in the next decade, almost half of all Asians will be residing in cities. This calls for urban innovative plans for building city resilience via urban risk reduction and climate change adaptation (Rahman & Shaw, 2015).

1.7 About the Book

This book, entitled *Urban Disasters and Resilience in Asia*, is a pioneering regional work of its kind; it provides a balanced approach of practice and theory in urban disaster and resilience. The text analyzes the status of urban disasters and resilience in Asia, drawing examples and lessons from the output of regional-, national-, and community-level projects and programs and experiences in the developed countries.

The book covers different types of urban disasters that face Asia and the extent of available resilience and coping mechanisms. It attempts to incorporate and describe some of the key lessons learned from the predisaster phase, through the disaster phase, and finally to the postdisaster phase; and provides insight for developing effective frameworks. The content is rich and based on a selection of available documents, a consultative workshop with academicians from different universities undertaking DRR higher education programs, and the editors' own knowledge and experience in the field. Special emphasis is given to analyzing field experiences from an academic perspective and pinpointing key issues and the policy relevance of urban disaster and resilience.

It is expected that this book will have a wide audience. However, the main target group includes students, teachers, and researchers working in the field of urban planning, architecture, disaster preparedness and management, social sciences, and earth and structural sciences. They can take evidence-based data from this scientific integration of databases and can get feedback for conceptualization, idea generation, and field surveying on urban disasters and building city resilience. Similarly, disaster managers, field practitioners, decision makers, disaster-related authorities, and key stakeholders of city government are also part of the desired readership.

This reference particularly highlights a number of key areas, including urban disasters, urban risk, urban resilience and risk reduction approaches, city government and resilience, vulnerability and climate change, climate change and urban adaptation strategies, urban floods, and changing climate. There is no book on this major topic, so this is an attempt to fill that void. The discussion especially focuses on Asian cities, where more than 60% of the world population currently lives—This was the major reason behind its dire need especially in the world market.

1.8 Structure of the Book

Chapter 1 deals with urban disaster and risk reduction approaches. Worldwide, the frequency and intensity of extreme events are increasing, and cities, where half of the world population is living, have been exposed to a series of disasters. Urban disasters have hit both the developed and developing world, but it is particularly the cities in the developing world that have low resilience and high vulnerability. In history, numerous cities have been buried by disasters and thousands of inhabitants either buried under debris or washed away by gushing water. Such disastrous events had put unprecedented pressure on city budgets, as cities are the hub of commercial and industrial activities. Whenever any disaster hits a city, its severity remains widespread and the focus of the budget shifts from development to response, recovery, and reconstruction. There is an increasing trend of urban disasters in recent years, and city governments must put their attention on building city resilience to effectively withstand these extreme events.

Chapter 2 focuses on urban risk, the role and responsibility of city government, and the extent of resilience in Asian cities. In the face of uncontrolled and rapid growth, urbanization is considered as one of the major risk factors. It is suspected that MDGs may not be effectively achieved if a city government fails to build a city's resilience by prioritizing DRR. In this regard, urban scientists frequently emphasize the status of poor urban governance in an attempt to build safer cities and thus to achieve MDGs. Since Asia is the epicenter of the current urbanization surge and highly vulnerable to natural disasters, these risks are particularly highlighted. The chapter also addresses the urban risk and role of city governments to build resilient communities by presenting examples from Dhaka.

Chapter 3 highlights the vulnerabilities of Asian cities in the face of changing climate. Globally, climate change is increasingly being recognized as one of the most serious threats to humanity. Cities occupy the center stage of a dialogue about mitigating the effects of greenhouse gases and adapting to the expected impacts of climate change. Cities in Asia have already witnessed the impacts of climate change–induced disasters such as flooding, water stresses in summer, storm surges and cyclones, heat waves, intense rainfall, and strong typhoons. The chapter also discusses the fact that Asian cities have strong potential to incorporate strategies for reducing impacts of urban disasters by building the resilience of their cities, and elaborates on various strategies to minimize the impacts of urban disasters and combat increasing climate change. These cities should devise innovative risk reduction strategies and create smart and sustainable environments.

Chapter 4 is devoted to discussing how resilient houses makes cities resilient. Globally, the recurrence and devastating impacts of earthquakes are on the rise, and people and houses are densely concentrated in urban areas, where these events can hit the hardest. In earthquake disasters, most deaths and injuries are caused by the total or partial collapse of buildings, particularly houses, most of which are not built by engineers. Further, vulnerable houses block streets when they collapse, which hampers evacuation, relief, and firefighting activities. Thus, resilient houses make their cities resilient as well. The field surveys described in this chapter were conducted in several developing countries, including India, Indonesia, Nepal, and Pakistan. The authors also briefly introduce efforts currently underway in Indonesia to facilitate housing safety in new construction through the appropriate implementation of building codes and dissemination of technical guidebooks.

Chapter 5 describes urban regulations and enforcement, which present a challenge in cities in developing nations. Historically, cities have been built following certain norms and principles. In Asian cities, however, this process is unique because most of the cities have urban development–related rules and regulations that have not been truly enforced. Institutional weaknesses are often regarded as the main cause of the poor implementation of urban regulations. However, lack of awareness about the existing urban regulations, lack of transparency in the system, poor monitoring mechanisms, financial constraints, and other factors could also be regarded as root causes for poor implementation of urban development norms.

Chapter 6 deals with expanding coastal cities, which pose urban risk. Coastal cities are where a significant percentage of people in Asia live. Historically, coasts have attracted large scale population, and this has been even more the case over the past few decades. Extremely sophisticated and vast infrastructure is being developed in coastal areas of several Asian cities to expand production and trade. This chapter discusses a number of examples concerning coastal cities in Asia, including demographic expansion, trend in urban disasters, prevailing socioeconomic conditions, structure of local governance and its priorities in risk reduction activities, and community engagement in risk reduction. The chapter also discusses recent initiatives in addressing risk in coastal cities.

Chapter 7 analyzes the impact of urban expansion on farmland. Globally, with increasing urbanization and economic growth, the cities are expanding at an alarming pace and multiplying built-up areas in the large urban centers of developing countries. This expansion has modified the way land is used, and in most cases, it has resulted in the reduction of productive agricultural land around the cities, which was once a major source of food for urban residents. The study discussed in this chapter focuses on the impact of urban expansion on farmland in Asian cities, particularly on the city of Peshawar, Pakistan, as a case study. The analysis shows that urban areas have been growing at a very rapid pace in recent years, and in the absence of planning control, there has been tremendous unplanned urban expansion in a leapfrog fashion, encroaching upon the best agricultural land and posing a serious threat to food security and urban air quality as well. However, there is no clear agreement on how to regulate expansion and conserve farmland.

Chapter 8 deals with enhancing city resilience and promoting urban-rural links in Asian cities. Urban communities in developing countries are failing to gather resources to allow them to withstand a shock, while stresses erode resilience and slowly increase the vulnerability of the population. Additionally, a city is resilient only if all its constituents can withstand and recover from the effects of a disaster. Similarly, a region is resilient if it can face the negative consequences of disasters and recover in a minimum period of time. This chapter discusses the interdependency of cities over villages and vice versa and how these urban-rural links can be utilized to build the resilience of cities.

Chapter 9 deals with urban DRR in Vietnam, particularly the nation's gaps, challenges, and approaches. With a combination of urban issues, many communities in cities and peri-urban areas have become increasingly vulnerable to extreme events. DRR efforts in Vietnam to date have primarily focused on rural areas and often employ effective community-based disaster risk management (CBDRM) methods, and yet there is an intensifying need for better DRR approaches in urban contexts. Current disaster management policies and efforts in this country have focused largely on emergency response and disaster recovery rather than on risk reduction, prevention, and adaptation. This chapter analyzes those gaps and challenges and introduces resilient approaches by describing some good practices from the Asian Cities Climate Resilience Network (ACCCRN).

Chapter 10 discusses the urban disasters and microfinancing in Asian cities. There is no doubt that disasters have a disproportionate impact on the poor. In particular, the urban poor living in slums (now estimated at approximately 1 billion people) are at risk. But financial support and microfinancing can help rebuild livelihoods, strengthen community bonds, and protect the urban poor from income shortfalls. Nonetheless, it has been proven that microfinancing can support economic and social rejuvenation after disasters occur. Focusing on the practices of a number of cities in Bangladesh, India, Indonesia, and the Philippines, this chapter gives an overview of the availability, accessibility, and role of microfinancing in achieving urban risk reduction and recovery of urban poor.

Chapter 11 assesses the urban food security issue as a growing threat in Asian cities. Urban food security is emerging as a key area of developmental policy deliberations around the world. By 2020, more than half of the Asian population will be in urban centers, and as a consequence, the demand for food will be high. Characteristically, these urban areas are dependent on the neighboring peri-urban and rural areas for agriculture-based food. As these cities seamlessly merge with the peripheral areas, the agriculture lands are getting built up to meet the increasing demand for residential areas. This trend negatively affects the local food production, resulting in increasing dependency on national and global supply chains, which can be more expensive and difficult. The chapter also considers Asia's high vulnerability to natural hazards and need to strengthen the supply chain and food storage and enhance urban resilience.

Chapter 12 discusses urban disasters and public-private partnerships in Asia. In most Asian economies, small and medium-scale enterprises (SMEs) make up a major part of private-sector investment, and they are among the highest employment generators. At this point, it can be said that involving SMEs in disaster management is an effective measure to reduce the impact of disasters in urban areas. This chapter looks at the possible involvement of SMEs in urban resilience as being essential for their own survival, drawing heavily on the good practices of Japan and Thailand in the wake of recent disasters.

Chapter 13 describes urban disasters and risk communication in the context of youth groups in Makati, a city in the Philippines. It has been shown that traditional mass media (i.e., print, recordings, cinema, radio, and television) cannot meet all of the information-sharing and communication needs before, during, and after disasters. The Internet and mobile communication avenues has opened many new opportunities for coordination among affected communities, government, and other stakeholders. In recent years, the use of social media in DRR has increased around the world. Twitter, Facebook, and new types of social media are being used in collecting information on needs assessment and building city resilience.

Chapter 14 analyzes the urban resilience and DRR strategies through the process of urbanization in Dhaka, Bangladesh, whereas spatial planning has been launched in the city with the intent to increase city resilience and offer healthy living environments. Nevertheless, in addition to planned urban growth, cities are also haphazardly expanding at an alarming pace. The frequency and intensity of hydrometrological disasters are becoming more prevalent. Disasters set back development prospects, and poor areas of the society are hit the hardest. The chapter analyzes historic growth of Dhaka in particular and analyzes the city's policies and risk reduction measures.

Chapter 15 analyzes the urban recovery of Kobe, Japan, 20 years after it experienced the Hanshin Awaji earthquake in 1995. At this point, reconstruction is almost complete. The intention was to rebuild Kobe as a leading 21st-century city. It is the Kobe city organization to determine the direction of the recovery of city through consensus building among citizens. Measures for new industrial development and special zones for structural reform have been taken to support a wide range of businesses and enable the city to revitalize. Thus, the recovery of Kobe has been achieved through public involvement in the process of recovery and concrete measures have been taken for sustainable development of the city.

Chapter 16 discusses the community resilience approach for prioritizing infrastructure development in urban areas. Concentration of populations in risk-prone locations has become a typical feature of today's megacities. When populations expand faster than the capacity of local governing authorities to accommodate them, there are always nonresilient structures. A paradigm shift in urban planning practice is needed to achieve resilient development by integrating the principles of disaster risk management. This chapter emphasizes the importance of a community's access to life support services and the availability of urban infrastructures as the key measures for resiliency. It explores the role of urban infrastructures in achieving community resilience and addresses critical scenarios arising out of inadequate and ineffective service provided in several Indian cities.

Chapter 17 discusses the vernacular architecture in India is an indigenous approach for resilience. India is a vast country, with diverse climatic, topographic, and sociocultural conditions. Yet each region has an immense indigenous knowledge in the form of its splendid yet mostly unmapped vernacular architecture. The vernacular built environments in India are centuries old and thus have proved to be resilient. In this chapter, the vernacular architecture of two different regions with different social, climatic, and topographic conditions are discussed with respect to the inherent resilient aspects of spatial planning, materials, and technology. Vernacular architecture in the cold Himalayan region and in the central hot dry Indian plain was studied. In the Himalayan region, earthquakes, landslides, intense rains, and floods are prominent hazards, whereas the central Indian plains experience extremes in temperature and drought. The chapter concludes that the vernacular structures are more resilient and culturally sustainable.

Chapter 18 deals with building community resiliency and linkages between individuals, communities, and local governments. Community resilience in an urban area is a growing challenge. Community-based risk reduction has been commonly practiced in rural areas, but nevertheless, some megacities have documented innovative approaches of enhancing community resilience. The chapter describes the link between individual and community actions for urban risk reduction. One of the key aspects of community resilience is sustainability through proper institutional linkages. The text also highlights the local governance linkage of community and individual approaches of risk reduction.

Chapter 19 discusses the climate migration and urban changes in Bangladesh. The impact on human settlements by extreme climate events has been significant. But the results of such climate-induced events are not just the immediate effects; they ripple through societies over time. In Asia, millions of people, mostly in low-lying coastal zones, face forced displacement, and they are searching for alternative livelihoods. In most cases, the destinations of such displaced populations are the nearest urban area. Bangladesh is a region where a complex set of climate-driven outcomes is already evident. Using the region as an example, this chapter covers the level of climate change's impact on natural systems through secondary population displacement. Tertiary impacts are also discussed in terms of the socioeconomic impacts on urban systems and the links between climate change, migration, and urbanization.

Finally, Chapter 20 discusses water scarcity in the megacity of Kolkata, India. It analyzes the historic perspective of the water supply system and traces the evolution of the city's water supply system over a period of 200 years. Water demand and usage has increased significantly in the city, mainly due to population growth and lifestyle changes in the population. Therefore, the city's current water supply system cannot accommodate these needs, and so the area is struggling with severe water scarcity, especially in the dry season. The chapter finally provides a set of recommendations for developing a resilient water supply system in Kolkata.

References

Campanella, T. J. (2006). Urban resilience and the recovery of New Orleans. *Journal of American Planning Association*, *72*(2), 141–146.

Caulderwood, K. (2014). Report: The Ten Most Expensive Natural Disasters in 2013. *International Business Times*. January 15, 2014. Accessed on March 20, 2014, from http://www.ibtimes.com/report-ten-most-expensive-natural-disasters-2013-1540058.

Douglass, M. (2013). *The Urban Transition of Environmental Disaster Governance in Asia. Working Paper Series No. 210*. Singapore: Asia Research Institute. National University of Singapore. Available at http://www.ari.nus.edu.sg/publication_details.asp?pubtypeid=WP&pubid=2334 Accessed on June 15, 2015.

Godschalk, D. (2005). Viewpoint: Mitigate, mitigate, mitigate. *Planning*, *71*(10), 58.

Government of Pakistan (GoP). (2012). *National Climate Change Policy. Ministry of Climate Change*. Islamabad: Government of Pakistan.

Intergovernmental Panel on Climate Change (IPCC). (2011). Summary for policymakers. In C. Field, et al. (Ed.), *Intergovernmental Panel on Climate Change Special Report on Managing the Risks of Extreme Events and Disasters to Advance Climate Change Adaptation*. Cambridge, UK: Cambridge University Press, pp. 1–20.

Khan, F. K. (2003). *Geography of Pakistan: Population, economy, and environment*. Karachi, Pakistan: Oxford University Press.

Leichenko, R. (2011). Climate change and urban resilience. *Current Opinion in Environmental Sustainability*, *3*, 164–168.

Ministry of Housing and Works (MOHW). (2007). *Building codes of Pakistan; seismic provision 2007*. Ministry of Housing and Works, Government of Pakistan.

Parvin, G. A., Ahsan, S. M. R., & Shaw, R. (2013). Urban risk reduction approaches in Bangladesh. In R. Shaw, F. Mallick, & A. Islam (Eds.), *Disaster risk reduction approaches in Bangladesh* (pp. 235–257). Tokyo: Springer.

Government, Queensland (2014). *Rebuilding a stronger, more resilient Queensland*. Queensland: Reconstruction Authority, Queensland. Accessed on April 14, 2014, from http://www.qldreconstruction.org.au/u/lib/cms2/rebuilding-resilient-qld-full.pdf.

Rahman, A., & Shaw, R. (2015). Urban risk and reduction approaches in Pakistan. In A. Rahman, A. N. Khan, & R. Shaw (Eds.), *Disaster risk reduction approaches in Pakistan* (pp. 295–314). Tokyo: Springer.

Samiullah, Rahman A., & Shaw, R. (2015). Gender and disaster risk reduction in Pakistan. In A. Rahman, A. N. Khan, & R. Shaw (Eds.), *Disaster risk reduction approaches in Pakistan* (pp. 379–394). Tokyo: Springer.

Sharma, A., Surjan, A., & Shaw, R. (2011). Overview of urban development and associated risks. In Shaw, et al. (Ed.), *Climate and disaster resilience in cities: Community, environment, and disaster risk reduction* (pp. 1–16). Bingley, UK: Emerald Group Publishing Ltd.

Shaw, R., Srinivas, H., & Sharma, A. (2009). *Urban risk reduction: An Asian perspective*. Bingley, UK: Emerald Publisher.

UNHABITAT. (2010). *The State of Asian Cities 2010/11. United Nations Human Settlements Programme (UNHABITAT), Regional Office for Asia and the Pacific*. Japan: ACROS Fukuoka Building.

United Nations (UN). (2009). *One UN Program in Pakistan*. Improving lives and helping people. United Nations Reform in Pakistan.

United Nations (UN). (2014). Department of Economics and Social Affairs Division, Population Division. *World Urbanization Prospects; 2014 Revisions*. Accessed on July 8, 2015, from http://esa.un.org/unpd/wup/CD-ROM/.

United Nations (UN). (2015). *Third United Nations World Conference on Disaster Risk Reduction*. Japan: Sendai. March 14–18, 2015, Agenda item 11 Adoption of the final outcomes of the conference, Sendai, Japan.

UNISDR. (2007). *Gender Perspective: Working Together for Disaster Risk Reduction: Good Practices and Lessons Learned Geneva, June 2007. United Nations International Strategy for Disaster Reductiona*, Geneva.

United Nations Office for Disaster Reduction (UNISDR). (2009). *UNISDR Terminology on Disaster Risk Reduction*. Geneva. http://www.unisdr.org/we/inform/terminology.

World Bank. (2011). *Guide to Climate Change Adaptation in Cities*. Accessed October 16, 2013, from http://siteresources.worldbank.org/INTURBANDEVELOPMENT/Resources/336387-1318995974398/GuideClimChangeAdaptCities.pdf#page=5.

World Economic Forum (WEF). (2014). Global Risks 2014. *World Economic Forum, Geneva*. Available at www.weforum.org/risks.

Urban Risk, City Government, and Resilience

2

Gulsan Ara Parvin[1], Akhilesh Surjan[2], Atta-ur-Rahman[3], Rajib Shaw[4]
[1]Researcher, Graduate School of Global Environmental Studies, Kyoto University, Kyoto, Japan; [2]Associate Professor, Faculty of Engineering, Health, Science and the Environment, Charles Darwin University, Darwin, Australia; [3]Associate Professor, Institute of Geography, Urban and Regional Planning, University of Peshawar, Peshawar, Pakistan; [4]Professor, Graduate School of Global Environmental Studies, Kyoto University, Kyoto, Japan

2.1 Introduction

Cities, which contain half of the world's population and much of the world's assets, are becoming the focal point of the destruction and loss caused by natural disasters. It has been argued that due to rapid urbanization and global environmental change, a number of human settlements are turning into potential hot spots for disaster risk (UNHABITAT, 2007) United Nations (UN) estimates show that the number of recorded disasters is increasing as the number of people living in cities increases. For instance, the past decade has seen a growing number of natural disasters, including an earthquake in Nepal, flooding in Sydney, Australia, in areas of Texas and Oklahoma, and in Tbilisi, Georgia in 2015; the earthquake and tsunami that struck northern Japan in 2011; earthquakes in Haiti, Chile, and China and flooding in China, Pakistan, and Brazil in 2010; the South Asian earthquake in 2005; and the Indian Ocean tsunami in 2004. All these incidents have unveiled the extent of disaster risk and vulnerability of towns and cities in both developing and developed countries. Urban disasters are becoming more common and increasingly expensive and their economic impact has magnified (Basu et al., 2013). An assessment in 2008 by the Asian Development Bank (ADB) claimed that 8 of the 10 most populous cities in the world are facing moderate to high earthquake hazard. Similarly, 8 out of 10 of the most populous cities are located on coastlines and are vulnerable to storm surge and tsunami waves (ADB, 2008).

Scholars claim that underlying reasons for high risks are rooted in every process of rapid urbanization (Sharma et al., 2011). Over the past few decades, the whole world has witnessed increasing threats and risks to cities. Unplanned, uncontrolled, and rapid urbanization is considered the principal cause of these increasing risks. Furthermore, since cities (especially those in developing countries) have high population density, high concentration of activities, a great reliance on money-based economy, concentration of poverty, and political competition, they are highly vulnerable to disasters (UNHSP, 2004, cited in Tran et al., 2009).

Unfortunately, in most cases, this uncontrolled and rapid increase in urban populations is going on in middle- and lower-income countries, which have a more limited

Urban Disasters and Resilience in Asia.

capacity to manage the new risks being created and existing risks being exacerbated by global urban transitioning (World Economic Forum, 2014). Baker, (2012) also warns that during the next 20 years, over 95% of the population growth in developing countries will take place in urban areas. It also has been argued that urbanization is an irreversible process and therefore, urban vulnerabilities become reality (Surjan et al., 2011a). However, according to the UN Human Settlement Program (UNHSP, 2004), vulnerabilities do not spread due to urbanization alone. Rather, disaster risks and vulnerabilities increase due to poor governance, bad planning, and weak institutions. Therefore, the role of city governments in dealing with risk reduction and enhancing resilience have been recognized as one of the key concerns throughout the world. It is suspected that sustainable development goals may not be achieved if a city government fails to enhance the city's resilience by prioritizing disaster risk reduction (DRR). Since Asia is the epicenter of the current urbanization surge, and this part of the world is highly vulnerable to natural disasters and risks, this chapter highlights urban risks in Asian cities. It aims to address urban risk reduction approaches and the role that city governments can play in building resilient communities, with a special focus on the urban risks of Bangladesh and the resilience of its capital, Dhaka, which is one of the most rapidly urbanized and disaster-vulnerable cities in the developing world.

2.2 Role of City Government in Enhancing Resilience Through Risk Reduction

In the arena of disaster management (DM), due to paradigm shift, compare to disaster response and recovery, DRR has attained more concern of development initiators, designers, and scholars. According to ISDR (2004), DRR embraces all efforts aimed at minimizing human vulnerability and disaster risk to prevent or limit the adverse impacts of hazards within the broad context of sustainable development, thus enhancing resilience. Here, it should be noted that resilience is the capacity of a system, community, or society potentially exposed to hazards to change by coping or adapting in order to reach and maintain an acceptable level of functioning and structure. DRR for enhancing resilience has been considered as the three prioritized sectors of works (priorities for actions: priority 1 to priority 3) by the Sendai Framework for Disaster Risk Reduction 2015–2030 (SFDRR). This framework includes SFDRR Priority 2: Strengthening Disaster Risk Governance to Manage Disaster Risk, and SFDRR Priority 3: Investing in Disaster Risk Reduction for Resilience. In this approach of DRR and resilience enhancement, urban local governments have the primary responsibility for implementing all the necessary measures. Since city governments have the power and responsibility to protect residents and their assets, DRR and enhancing the city's disaster resilience must be the integral part of planning and investment (Surjan et al., 2011b).

Emphasizing the increasing risk of disasters in the urbanized world, the United Nations International Strategy for Disaster Reduction (UNISDR) launched a campaign called "Making Cities Resilient—My City Is Getting Ready" in 2010. The target of this program was to achieve at least 25 cities or local governments that could serve as role models for resilience, and 50 additional participating ones as well. By 2011, the targets had not only been met, but far exceeded. In less than 2 years, almost a thousand cities and local governments had signed up for the campaign, with many hundreds more in the process of joining (Basu et al., 2013). These initiatives have encouraged cities all over the world to start to incorporate disaster and climate-change risk into their planning and investment, achieving positive outcomes and imparting important lessons. These cities include Boston, Massachusetts; Cape Town, South Africa; Ho Chi Minh City, Vietnam; Quito, Ecuador; Rotterdam, the Netherlands; Toronto, Canada; and London. At a minimum, these cities have identified risk-prone areas and, through urban planning, discouraged new construction in these areas (Baker, 2012). Furthermore, it is worthy to note that the Cornwall County Council in the United Kingdom has developed an action plan for developing a climate change strategic framework, and the city of Pasadena, California, has launched a comprehensive environmental action plan (Fernandez et al., 2011) Similar approaches and initiatives have been started in many other cities as well, both in developed and developing countries.

In Asia, a number of city governments are taking significant steps toward enhancing their city's resilience and reducing disaster risks. The speed of urbanization is highest in Asia (Shaw et al., 2009), and so are the disaster risks in Asian cities. UNHABITAT, (2007) states that when considering vulnerability in terms of the different types of natural hazard exposure, Manila, the Philippines; Kolkata, India; Osaka, Kobe, Kyoto, and Tokyo in Japan; Jakarta, Indonesia; and Dhaka, Bangladesh are all high-risk metropolises. All are Asian cities with more than 10 million residents and have high exposure to at least two different kinds of natural hazard. In this chapter, snapshots of the role of city government in a few of these Asian cities are featured in boxes. Box 2.1 describes the DRR efforts in Jakarta.

It is well known to the world that the Great Hanshin-Awaji Earthquake on January 17, 1995, caused devastating loss and damages in Kobe, Japan. But the city's government learned several important lessons from having to recover from the disaster; it reacted by initiating several efforts designed to enhance the city's DRR and resilience. The encouragement of voluntary organizations for disaster prevention is one of the

Box 2.1 Jakarta Risk Planning

Jakarta's DRR plan for 2010–2030 calls for incorporating risk reduction into long-term spatial planning for the city. The elements include restoring the mangrove forests, improving public facilities and mass transit, refining building and environmental regulations that affect hazard risk, redesigning technology and engineering in disaster areas, and improving the provision of open space for anticipated increases in intense rainfall. The city also created a dedicated body for DRR in 2010, which is intended to help with planning and mainstreaming at the neighborhood level. A Kelurahan Empowerment Initiative is under way in five neighborhoods to develop Local Resilience Action Plans (LRAPs) with local and community leaders in Jakarta (Baker, 2012).

Box 2.2 The BOKOMI Disaster Management Organization

In Kobe, community-based disaster management organizations are considered a major element of disaster prevention throughout the city. In Japanese, it is called as Bosai Fukushi Community (BOKOMI for short). The BOKOMI concept targets elementary school areas; and at present, 19 BOKOMI cover all households in Kobe. Each BOKOMI consists of its headquarters and an information group, a firefighting group, a rescue group, an evacuation guidance group, and a livelihood group. The leader and deputy leader belong to the headquarters. The Kobe City Office is conducting the following for supporting BOKOMI. Fire, Health and Welfare Department of Kobe City support BOKOMI. It supports BOKOMI by providing disaster management equipment, funds and training for disaster management leaders.

Source: Matsuoka et al. (2012).

most remarkable of these developments. BOKOMI is such a community-based disaster management organization, as described in Box 2.2.

Tokyo, which is one of the most urbanized and disaster-prone cities in the world, has a very active and efficient government when it comes to disaster management. In particular, the Tokyo Disaster Prevention Plan should be able to offer tremendous support for the city in times of crisis. The Tokyo metropolitan government has the goal of having this plan in place for 2020; it will provide a list of disaster-prevention measures in response to earthquakes, floods, and other natural disasters to be taken in advance by citizens, regions, businesses, and government. The plan (described in Box 2.3) will feature a well-organized webpage, which includes information on disaster prevention and response, action manuals for earthquakes and floods, and special disaster preparedness guidance for children and the elderly.

The following section focuses in particular on the risks faced in Bangladesh, especially by Dhaka, which is one of the world's most vulnerable cities in terms of climate change and natural and other disasters. At the same time, this city has been experiencing one of the highest rates of urbanization over the last several decades.

2.3 Urban Risks of Bangladesh

Due to its geographical location, Bangladesh and all its cities are vulnerable to both geological and climatic disasters. These risks and vulnerabilities have been aggravated by poverty and rapid and unplanned urbanization. According to Bashar and Rashid (2012), the urbanization rate in Bangladesh is one of the highest in the world, and this fast pace is expected to continue for the foreseeable future. The urban population of Bangladesh now exceeds 40 million and is growing at 3.4%

Box 2.3 Tokyo Metropolitan Plan for Disaster Preparation

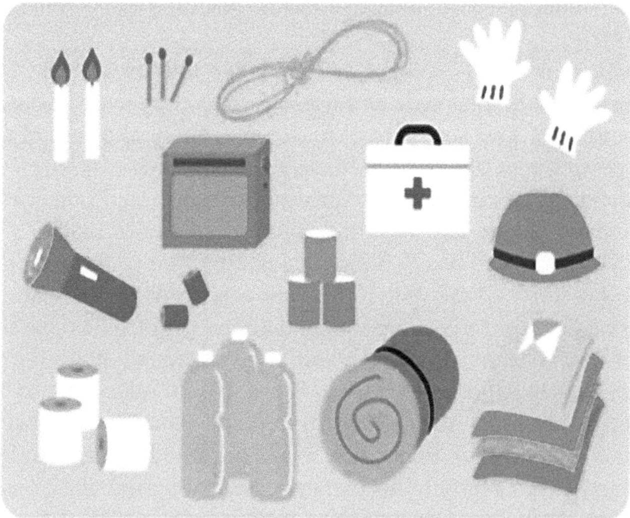

When a disaster happens, you may not be able to use electricity, gas, or water. Put aside vital supplies for life and keep them in place. Prepare water for 3 days and foods for 1 week for whole family. Check those freshness dates and make sure the battery works. Also prepare for

- *First-aid kit, medicine, etc.*
- *Portable radio, flashlight, batteries*
- *Cash and valuables*
- *Clothes*

http://www.bousai.metro.tokyo.jp/foreign/english/index.html.

per annum (Rahman, 2011). It is estimated that about 40% of Bangladesh's people will be living in urban areas by 2030 (Khan, 2008). Due to poverty, low income, and natural disasters, rural migration to cities has been rapid and continuous in the nation. This migration is responsible for 40% of the increase in urban populations, and for Dhaka and some coastal cities, this figure is as high as 70% (Islam, 2006, cited in Khan, 2008). It is estimated that Dhaka alone contains nearly 40% of the country's urban population, and in recent decades, it has been identified as one of the fastest-growing cities, with an urbanization rate of over 2.5% (Hossain, 2008; Roy, 2009).

After migrating to Dhaka, many people (especially the poor, who are the majority of these new residents) find shelter in marginal settlements with substandard housing and limited infrastructure and services. Therefore, Dhaka is growing into the flood-prone, low-lying areas. Since urbanization has marked effects on global and local environmental conditions, including climate change and related hazards, Dhaka is increasingly experiencing environmental degradation and disaster risks due to its uncontrolled and unplanned rapid urbanization (Grimm, 2000, cited in Dewan & Yamaguchi, 2009)

Research papers and news records both indicate that every year, Dhaka faces a number of disasters (both manmade and natural). This number of disasters is increasing in an alarming rate. This section will focus on the natural disasters and the role of the municipal government of Dhaka to manage disaster risk.

Floods, which are associated with river water overflow and rain-water stagnation, earthquakes, and tornadoes are the most common natural disasters experienced in Dhaka. The city experienced heavy flooding at least 10 times from 1954 to 2007, and the floods in 1988, 1998, 2004, and 2007 were catastrophic (Rabbani, 2009). These catastrophic floods affected the entire city and almost all of the population. The slum areas, where almost 30% of Dhaka's people live, are found along the water's edge, so they are particularly exposed to floods (Shaw, 2013). According to Shaw, (2013), after a flood occurs, different parts of the city become waterlogged for several days. In so-called normal levels of flooding, many roads become inaccessible for as much as 8 h, while during heavier rainfall, this can increase to a period of 12 h or even more.

In recent years, Dhaka has seen extensive waterlogging during the monsoon season (May–October). The main cause of these floods was the rise in water levels of the rivers bordering the city (Faisal et al., 2003; Shaw, 2013). Rivers and canals, including the tributaries of the Brahmaputra and Meghna rivers, surround Dhaka. In addition to the rise of river water, internal drainage congestion and uncoordinated operation of flow regulation structures contributed to the flooding. Rapid and unplanned urban growth and an uncontrolled real estate boom in the city causes serious encroachment of natural drainage and retention areas, hindering the natural flow of water and causing substantial waterlogging and flooding in almost every year of the past decade (Parvin et al., 2013; also see Figure 2.1).

Apart from flood risks, Dhaka is considered as one of the most vulnerable cities to earthquakes. The Earthquake Risk Index (EDRI) for Dhaka put it at the top of its list of the 20 highest-risk cities in the world (Khan & Phibbs, 2005). According to the UN IDNDR-RADIUS initiative, Dhaka and Tehran, Iran, are the cities with the highest relative earthquake risks in the world (Jahan et al., 2011). Several faults have been found in and around the city.

More than 75 earthquakes of high magnitude occurred within a 200-km radius of Dhaka between 1885 and 2015. These frequent earthquakes indicate the possibility of future major earthquakes in the vicinity. Based on the calculation of recurrence year of 1885 Bengal earthquake, it is predicted that in the year around 2017 Dhaka would experience an earthquake with a magnitude about 7.3 (Khan et al., 2005). The Earthquake Contingency Plan for Dhaka City (ADPC Center, 2009) estimated that if a 7.5-magnitude earthquake originating in the Madhupu fault occurred, 51% of Dhaka's buildings would be damaged. The city's inherent vulnerability to building infrastructure, dense population, haphazard and unplanned development, poor emergency response, and low recovery capacity make it highly vulnerable after such a high-magnitude earthquake (Al-Hussini, 2003, cited in Shah & Murao, 2011; Jahan

Figure 2.1 Dhaka's urbanization and disaster risks: (a) dense population; (b) rapid urbanization; (c) unplanned growth; (d) slums located in hazardous areas.

et al., 2011). In addition to these direct impacts, earthquakes can cause liquefaction of soil, rendering built-up areas even more in danger of structural collapse.

2.4 Role of City Government in Enhancing Dhaka's Resilience

In a number of Bangladesh municipalities, city corporations and urban development authorities [e.g., Rajdhani Unnayan Katripokkho-RAJUK in Dhaka City, Chittagong Development Authority (CDA) in Chittagong City, Khulna Development Authority (KDA) in Khulna City, and Rajshahi Development Authority (RDA)] are introducing as well as enforcing DRR tools in their urban areas. The Ministry of Food and Disaster Management (MoFDM) in Bangladesh has designed a local government framework to define specific organizations and their responsibilities as functional tools in response to any natural or manmade disaster. According to the MoFDM (cited in Parvin et al., 2013), these disaster preparedness actions are:

- Enforce planning and development law and regulation to protect land use in vulnerable areas
- Adopt and enforce building codes
- Develop hazard map based one sites
- Monitor and revise the land-use plan and development regularly

- Create and launch a public awareness campaign
- Control epidemiological outbreaks during disasters
- Provide medical facilities to emergency camps and relief centers

The National Plan for Disaster 2010–2015 proposed that there will be a plan for each Pourashava/City Corporation, to be prepared by the Pourashava/City Corporation Disaster Management Committee, with links to the National Plan. These plans should highlight overall risk mapping, preparedness and risk reduction approaches, and disaster response and recovery at all stages of an event.

2.4.1 Disaster Management by the Dhaka City Corporation (DCC)

Although the National Plan for Disaster 2010–2015 has advocated for the preparation of City Corporation Disaster Plans, neither of the Dhaka city corporations has been able to develop a disaster plan of its own. The Dhaka City Corporation (DCC) has no typical master plan, development plan, or disaster management plan. Every fiscal year, ward commissioners submit their development plans indicating the sectorial development budget for each ward. Similarly, each zone prepares a maintenance budget and plan.

The Disaster Management Bureau of the Bangladesh government published and circulated a Standing Order on Disasters (SOD) in 1999, which was revised in April 2010. The SOD includes a provision for establishing a Disaster Management Committee at the City Corporation level and outlined detailed roles and actions for it to take to prepare and respond to disasters. In accordance with the SOD, the DCC formed a Disaster Management Committee (City Profiles Series, 2006) chaired by the DCC mayor and composed of engineers, zonal officers, ward commissioners, conservancy officers and staff, and volunteers. In addition to the internal members of DCC, the Fire Service and Ambulance Department, Water and Sewerage Supply Authority, Electric Supply Authority, Red Crescent, Metropolitan Police, and National Cadet Core are part of the Disaster Management Committee. This committee is vested with duties related to predisaster, disaster, and postdisaster initiatives and management (Parvin & Shaw, 2011). Disaster management activities are mostly focused on flood-related disasters and are described next

The Local Government (City Corporation) Act of 2009 (Amendment-2011) divides the DCC into the Dhaka South City Corporation (DSCC) and Dhaka North City Corporation (DNCC). The DSCC consists of 56 wards covering the thanas of Dhaka Kotwali, Motijheel, Sutrapur, Ramna, Bangsal, Wari, Gendaria, Chwokbazar, Lalbagh, Hazaribagh, Dhanmondi, Shahbagh, New Market, Khilgaon, Kamrangirchar, and some others. The DNCC consists of 36 wards covering the thanas of Mirpur, Mohammadpur, Sher-E-Bangla Nagar, Pallabi, Adabor, Kafrul, Dhaka Cantonment, Tejgaon, Gulshan, Banani,kallanpur, Badda, Uttara, and some others. DCC was divided 4 years earlier in 2011, and since then there has been no elected mayor within it. Therefore, disaster management and risk reduction related tasks conducted or initiated by any of these specific city corporations (DSCC or DNCC) have not been well documented. In the following section, the overall role of the DCC for flood and earthquake risk reduction is highlighted.

2.4.2 Flood Risk Reduction Activities by DCC

Dhaka City faces heavy rainfall, tornadoes, waterlogging, and flooding during the rainy season from July to October. In preparation, during July and August DCC arranges campaigns, meetings, and seminars at the ward level to increase awareness. People are generally advised about how to increase their preparedness, including knowing how and when they will go to the nearest shelter, how and where they will register (with DCC, Ward Office) their families, what types of emergency food, materials, and help they may need, and whom they should contact during emergencies. Various departments of DCC also make preparations and stock appropriate emergency items to help citizens respond to flooding and waterlogging. Speedy dissemination of flood and cyclone forecasts to relevant officials and organizations is ensured by the Disaster Management Committee of the DCC. The DCC also identifies safe centers and emergency shelters for evacuation, assigns responsibilities, and arranges disaster drills.

In order to minimize loss, damage, and health risks during flooding, the DCC operates 500 wireless phone sets in the field and three control rooms at different parts of the city, where officers and staff work around the clock during actual events. A display board is hung in the control room to tally the damage and needs in different parts of the DCC. Relief and rehabilitation works are monitored, and emergency instructions are provided to other organizations and service departments, such as the Dhaka Electric Supply Authority (DESA), Water Supply and Sewerage Authority (WASA), Fire Service, and Red Crescent. (Parvin & Shaw, 2011).

Relief materials like dry food, candles, purified drinking water, clothes, and saline water are distributed among the affected people. In order to limit potential health crises, conservancy staffs of DCC provide clean temporary toilets, dispense bleaching powder, and spray insecticides to control mosquitoes and other pests. Additionally, DCC deploys 25 medical teams that move from camp to camp and are sufficiently equipped with oral saline, fever medicine, diarrhea medicine, and other primary health care supplies.

Immediately after a disaster strikes, DCC sends a report outlining the estimated work required to the relevant departments, along with commentary. It also organizes a meeting to exchange views, experiences, problems, and difficulties during the disaster. After a flood occurs, DCC conducts surveys and estimates the number of canals and drains (both on the surface and underground) that require clearing. It also prepares eviction/demolition lists of shops, houses, and all types of unauthorized structures, as well as lists of roads that are to be raised above the floodwater level (Parvin & Shaw, 2011).

2.4.3 DCC Role in Earthquake Contingency Plan

With the Comprehensive Disaster Management Programme (CDMP), the Ministry of Food and Disaster Management in the Bangladesh government has developed an Earthquake Contingency Plan for the DCC in 2009. This contingency plan includes 9 clusters of different tasks and service groups for earthquake preparedness, response, and recovery. Among these, the DCC is responsible for taking the lead role in seven clusters, which are responsible for command and coordination, search, rescue and evacuation, relief service, shelter service, water supply, sanitation and hygiene, restoration and urban service, and transportation.

In addition to these regular and routine activities, over the last few years, DCC has organized workshops, exercises, drills, awareness campaigns, and rallies as part of its disaster management and risk reduction activities. Since 2004, disaster preparedness and response training has been provided to 120 ward commissioners of DCC and some schoolteachers and religious leaders (imams of mosques). A Disaster Management Volunteer Group in every ward is being prepared, and the National Plan for Disaster Management 2010–2015 has proposed the preparation of a Disaster Management Plan by the Disaster Management Committee.

Disaster management activities of the DCC are mostly related to preparedness (especially for floods), distribution of relief, and some postdisaster reconstruction. There are few attempts at the DCC or community level to reduce disaster risk and enhance the disaster resilience of the DCC. In order to reduce the disaster risk faced by the DCC and enhance the city's resilience, the government needs to learn the lessons imparted from various disaster-prone cities like Tokyo, Kobe, Jakarta, and Manila.

2.5 Conclusion

Scholars (e.g., Basu et al., 2013) argue that the disaster risk of a city comes from the failure of its government. Further, disaster loss can be differentiated at the city level. A city's vulnerability to disaster is determined by its level of economic development and disaster preparedness (UNHABITAT, 2007). On the other hand, the level of economic development and disaster preparedness primarily depends on the role and efficiency of the city government. Therefore, the Safer Cities program of UN agency (UN-Habitat) supports to strengthen cities' efficiency to deal with urban safety issues and reduce any associated risks. Manyena, (2006) also supports the idea that the development of disaster resilience is largely dependent on the capacity and role of local authorities, and this emphasizes the need for city governments to develop their capacities in order to implement proper DRR initiatives.

Local governments are considered as the key stakeholders in DRR and disaster preparedness. A city government typically is in charge of critical development functions to reduce disaster risk, such as land use planning, urban development planning, public works, construction safety and licensing, social services, responding to the needs of the poor and underprivileged, and the implementation and strengthening of the decentralization process (UNISDR, 2010; UNHABITAT, 2007).

Although city governments are primarily responsible for taking the initiative for DRR, they cannot do all the necessary jobs effectively by themselves. Since cities are hubs of diversified socioeconomic activities and the residence of many different types of people, the risk reduction approaches of city government depend on and need to incorporate the input of diversified stakeholders. The World Economic Forum (2014) stated that as urban populations grow, multi-stakeholder processes of inclusive governance will increasingly be needed to make cities resilient against these complex and

Figure 2.2 Multistakeholder support for city governments to help build resilient cities.

interconnected risks. SFDRR 2015–2030 also stated one of the key goals as follows: "Incorporate disaster risk reduction measures into multilateral and bilateral development assistance programmes within and across all sectors, as appropriate, related to poverty reduction, sustainable development, natural resource management, environment, urban development, and adaptation to climate change" (UNISDR, 2015, p. 23).

As advised by SFDRR, for a large city like Dhaka, it is impossible for the city government to do everything by itself. Therefore, the city's governance system should accommodate outsourcing and assigning some of this work to others (Figure 2.2). This approach of partnership and sharing should be attempted by pursuing the following:

- Development of a public-private partnership scheme, where some of the activities can be outsourced to private agencies
- Development of a community governance scheme, community volunteer network, or community-based group registration
- Development of a city-to-city sharing and learning approach
- Development of a networking partnership with city government, civil society, and nongovernmental organizations (NGOs)
- Development of mechanisms for some of its crucial services to be linked more strongly to DRR (like solid waste management and sanitation, especially with flood issues; and land use and building codes better suited for earthquakes)

With the support and partnership of multiple stakeholders, Dhaka needs to work on approaches for multihazard risk reduction. Investment, policy planning and implementation, law enforcement, capacity building, decentralization, public-private partnerships, and gender equity all should be the key focuses of the city's DRR efforts.

As many scholars have advised, the DCC Disaster Management Committee is trying to incorporate multiple stakeholders to handle the city's many hazards. However, a Disaster Management Plan has not yet been prepared. Therefore, adequate and proper guidance to deal with multiple hazards and to approaches for DRR through involving multiple stakeholders is lacking. According to the chief town planner of DCC, who is one of the primary members of the Disaster Management Committee, the existing personnel and logistics are not sufficient to prepare and implement a disaster management plan and hazard map for each zone of the DCC.

In spite of the absence of a Disaster Management Plan for Dhaka, there are many different risk reduction tools in place, like land-use planning, zoning, development control, and planning regulation. In many Bangladeshi municipalities, city corporations and urban development authorities are introducing and enforcing DRR tools in their respective urban areas. These responsible authorities in Dhaka's local government need to understand the hazards in a particular place, and the characteristics, frequency, and magnitude of those hazards so they can design appropriate tools to reduce the risks. World Economic Forum (2014) recommended for substantial improvement in Dhaka`s urban governance capacities in order to address disaster risks. Dhaka is the hub of emerging economies, where future urbanization will mostly take place and which is especially vulnerable to disaster risk. Therefore, urban governance of Dhaka must need extensive capacity for risk reduction.

The UNHSP (2004) advocates that good governance, which includes participatory decision making, transparency and accountability and integrates all these elements into development initiatives, would have a substantially positive impact in managing urban risks and vulnerabilities (Tran et al., 2009). The "Making Cities Resilient" campaign also advocated increasing knowledge, investing wisely and building more safely by including DRR in participatory urban development planning processes and protecting critical infrastructure. Dhaka's government must follow these suggestions to reduce its disaster risks and make the city resilient.

Acknowledgments

The first and fourth author acknowledges support from Sumitomo Foundation and JICA JST SATREP project support for this research.

References

Asian Disaster Preparedness Center (ADPC). (2009). Published by Comprehensive Disaster Management Programme (CDMP). Department of Disaster Management, Ministry of Disaster Management and Relief, Bangladesh Government. Available online at http://kmp.dmic.org.bd/handle/123456789/130.

Asian Development Bank (ADB). (2008). *Managing Asian Cities: Sustainable and Inclusive Urban Solutions*. Metro Manila, Philippinnes: ADB. http://www2.adb.org/documents/studies/managing-asian- cities/mac-report.pdf. Accessed 07.07.15.

Baker, J. L. (Ed.). (2012). *Climate change, disaster risk, and the urban poor: Cities building resilience for a changing world.* World Bank Publications. Washington DC, USA.

Basu, M., et al. (2013). Making cities and local governments ready for disasters: A critical overview of recent approaches. *Risk, Hazards, & Crisis in Public Policy, 4*(4), 250–273.

Bashar, T., & Rashid, S. (2012). Urban microfinance and urban poverty in Bangladesh. *Journal of the Asia Pacific Economy, 17*(1), 151–170.

City Profiles Series. (2006). *Dhaka Bangladesh, Disaster Risk Management Profile.* http://emi.pdc.org/cities/cp_Dhaka-July 2006.pdf. Accessed 06.06.10.

Dewan, A. M., & Yamaguchi, Y. (2009). Land use and land cover change in Greater Dhaka, Bangladesh: Using remote sensing to promote sustainable urbanization. *Applied Geography, 29*, 390–401.

Faisal, I. M., Kabir, M. R., & Nishat, A. (2003). The disastrous flood of 1998 and long-term mitigation strategies for Dhaka City. *Natural Hazards, 28*, 85–99.

Fernandez, G., Takeuchi, Y., & Shaw, R. (2011). From resilience mapping to action planning. Climate and disaster resilience in cities. *Community, Environment, and Disaster Risk Management, 6*, 149–161.

Hossain, S. (2008). Rapid Urban Growth and Poverty in Dhaka City. *Bangladesh e-Journal of Sociology, 5*(No.1), 1–24.

International Strategy for Disaster Reduction (ISDR). (2004). *Living with risk: A review of global disaster reduction initiatives.* www.unisdr.org/eng/about_isdr/bd-lwr-2004-eng.htm.

Jahan, I., et al. (2011). Assessing social vulnerability to earthquake hazard in Old Dhaka, Bangladesh. *Asian Journal of Environment and Disaster Management, 3*(3), 285–300.

Khan, A., & Phibbs, P. (2005). Housing and education: An example of a non-shelter outcome. In *The refereed 46th annual conference of American Collegiate Schools of Planning (ACSP).* Kansas City: MO. 27–30 October 2005.

Khan., et al. (2005). *Oriental Geographer, 49*(2), 205–216.

Khan, H. (2008). *Challenges for Sustainable Development: Rapid Urbanization, Poverty and Capabilities in Bangladesh.* Available online at http://mpra.ub.uni-muenchen.de/9290/.

Manyena, S. B. (2006). The concept of resilience revisited. *Disasters, 30*(4), 434–450.

Matsuoka, Y., et al. (2012). Partnership between city government and community-based disaster prevention organizations in Kobe, Japan. *Community-Based Disaster Risk Reduction, 10*, 151–184.

Parvin, G. A., Ahsan, S. R., & Shaw, R. (2013). Urban risk reduction approaches in Bangladesh. In R. Shaw et al. (Eds.), *Disaster risk reduction approaches in Bangladesh* (pp. 235–257). Tokyo: Springer.

Parvin, G. A., & Shaw, R. (2011). Climate disaster resilience of Dhaka City Corporation: An empirical assessment at zone level. *Risk, Hazards, & Crisis in Public Policy, 2*(2) Article 6 (pp. 1–30).

Rabbani, M. D., & Golam (2009). Environmental Risks in Dhaka: Present Initiatives and the Future Improvements. In R. Shaw, H. Srinivas, & A. Sharma (Eds.), *Urban Risk Reduction: An Asian Perspective* (pp. 319–338). United Kingdom: Emerald.

Rahman, M. M. (2011). The problem of poor pressure. *Daily Star Forum, 5*(1), 38–41.

Roy, M. (2009). Planning for sustainable urbanization in fast growing cities: Migration and adaptation issues addressing in Dhaka, Bangladesh. *Habitat International, 33*(3), 276–286 Elsevier Ltd.

Shah, F. M., & Murao, O. (2011). Understanding seismic risk recognition and intention for safety measures of residents in Dhaka, Bangladesh. *Asian Journal of Environment and Disaster Management, 3*(3), 357–372.

Sharma, A., Srujan, A., & Shaw, R. (2011). Chapter 1: Overview of urban development and associated risks. In R. Shaw, et al. (Eds.), *Climate and Disaster Resilience in Cities (Community, Environment, and Disaster Risk Reduction, Vol. 6)* (pp. 1–16). Bingley, UK; Emerald Group Publishing Limited.

Shaw, R. (2013). Urban Disaster Risk Reduction Framework: Assessing Urban Resilience in World Vision Project Sites in Bangladesh, China, and Indonesia—Final Report, 1 April 2013. *World Vision International*. Singapore.

Shaw, R., Srinivas, H., & Sharma, A. (2009). Preface. In *Urban risk reduction: An Asian perspective* (p. xvii). Bingley, UK: Emerald Group Publishing.

Surjan, A., Takeuchi, Y., & Shaw, R. (2011a). Chapter 1: Introduction. In *From disaster and climate risk to urban resilience: Approaching through community-based environmental improvement* (pp. 1–18). Singapore: Research Publishing Services.

Surjan, A., Takeuchi, Y., & Shaw, R. (2011b). Chapter 3: Urban disaster risk reduction. In *From disaster and climate risk to urban resilience: Approaching through community based environmental improvement* (pp. 49–62). Singapore: Research Publishing Services.

Tran, P., et al. (2009). Urban disaster risk analysis, action planning, and implementation management. In R. Shaw et al. (eds.), *Urban risk reduction: An Asian perspective* (pp. 13–36). Bingley, UK: Emerald Group Publishing.

United Nations Human Settlements Programme (UNHABITAT). (2007). *Global Report on Human Settlements 2007: Enhancing Urban Safety and Security*. London: Earthscan. http://www.unhabitat.org/downloads/docs/GRHS2007.pdf.

United Nations International Strategy for Disaster Reduction (UNISDR). (2015). *Sendai Framework for Disaster Risk Reduction 2015–2030*. Geneva, Switzerland: UNISDR.

United Nations International Strategy for Disaster Reduction (UNISDR). (2010). *Local Governments and Disaster Risk Reduction: Good Practices and Lessons Learned. A Contribution to "Making Cities Resilient" Campaign*. UNISDR. Accessed April 28, 2010, at http://www.UNISDR.org/preventionweb/files/13627_LocalGovernmentsandDisasterRiskRedu.pdf.

United Nations Human Settlements Program (UNHSP). (2004). Reducing urban risk and vulnerability: A thematic paper submitted for discussion at the UN-HABITAT/UN-ISDR. In *Working meeting on vulnerability assessment and reducing urban risk, Madrid, 7–9 September 2004*.

World Economic Forum. (2014). *Global risks 2014* (9th ed.). Geneva, Switzerland: World Economic Forum.

Cities, Vulnerability, and Climate Change

3

Atta-ur-Rahman[1], Gulsan Ara Parvin[2], Rajib Shaw[3], Akhilesh Surjan[4]
[1]Associate Professor, Institute of Geography, Urban and Regional Planning, University of Peshawar, Peshawar, Pakistan; [2]Researcher, Graduate School of Global Environmental Studies, Kyoto University, Kyoto, Japan; [3]Professor, Graduate School of Global Environmental Studies, Kyoto University, Kyoto, Japan; [4]Associate Professor, Faculty of Engineering, Health, Science and the Environment, Charles Darwin University, Darwin, Australia

3.1 Introduction

Globally, the intensity and frequency of disasters are on the rise. In urban areas, more than 50% of the world's population is living on just 2% of Earth's land mass (Rahman & Shaw, 2015). Asia has 60% of the world's population, and these people are residing on just 30% of the Earth's total land mass (UNHABITAT, 2010a). Despite the low level of urbanization, over 53% of the world's urban population is reported to be living in Asia alone, a continent made up predominantly of developing nations. However, this area also has the largest diversity in terms of socioeconomic and physical characteristics. Of the top 20 megacities, 13 are in Asia, most of them in developing nations (UN, 2015). It is estimated that by 2020, half of the Asian population will be living in cities.

According to the United Nations Department of Economic and Social Affairs (UNDESA, 2010; UN, 2010; UN, 2015), Asia's urban population will increase from 1.9 billion in 2012 to 3.3 billion by the year 2050. Furthermore, it has been predicted that the highest rate of urbanization is expected in India, China, Indonesia, Pakistan, and Bangladesh. It has been estimated that by 2050, India will add another 497 million, China 341 million and Indonesia 92 million people to their urban populations (UNDESA, 2010). Similarly, in Asian cities, more than 30% of the people live in slums and squatter settlements (Douglass, 2013). However, the situation is worst in Cambodia, Bangladesh, Nepal, and Mongolia, where more than half of the urban population lives in slums and blighted areas. In Pakistan, nearly half of its total urban population resided in slum areas in 1990, and they have shown only a little improvement since then, with 47% living in slums in 2007 (UNHABITAT, 2010a). However, in terms of numbers rather than percentages, slum dwellers have actually increased, from 17.6 million in 1990 to 27.5 million in 2007 (UNHABITAT, 2010b). In the wake of achieving Millennium Development Goals (MDGs) related to slums, city governments have extended their services in an attempt to reduce slum vulnerability and miserable housing conditions, but significant efforts are still needed.

Worldwide, the impact of climate change has increased the recurrence and intensity of both meteorological and hydrometeorological phenomena (Shaw & Sharma

Urban Disasters and Resilience in Asia.

2011; Carmin et al., 2013). It has been noted that the changing climate has further enhanced the unpredictability and intensity of sessional rainfall, heat waves, flood hazards, storm surges, drought, and cyclonic storms, and it poses major threats to the food, water, and energy security of cities, where half of the world population is living (Douglass, 2013). In addition, the coastal and marine environment, agriculture and livestock sector, forests, biodiversity, and health are other areas that will be seriously affected as the climate-induced melting of glaciers, heat waves, cyclonic storm surges, diseases/epidemics, floods, intense precipitation, droughts, and variable monsoons turn into an inevitable harsh future reality for Asian cities.

Existing scientific research reveals that unforeseen events hit both developing and developed nations alike, but developing nations experience much higher vulnerability and intensification. In the past, numerous cities have been affected by natural and human-induced disasters, with thousands of inhabitants either buried under debris or washed away (Shaw et al., 2009; Rahman et al., 2014; Rahman & Shaw, 2015). Eventually, urban disasters have put unprecedented pressures on city budgets. In the world of today, cities function to empower societies because urban centers are the hub of industrial and commercial activities. Whenever any disaster hits an urban area, the entire surroundings suffer immense and widespread human losses and adverse consequences. This in turn shifts the city budget from urban development to emergency response and early recovery.

Disaster risk reduction (DRR) methodologies are fully established within the scientific community (Mercer, 2010) but we still have a long way to go to expand the DRR approaches in reducing vulnerabilities through prevention, mitigation, preparedness, emergency response, and early recovery. The DRR practitioners and the scientific community are struggling to discover and introduce innovative methodologies to respond effectively to extreme natural events; however, today stresses remain in the urban areas (Campanella, 2006). In many instances, the disaster manager has learned from past experience and recovered by creating better strategies for next time. Therefore, DRR approaches for building resilience have been considered a priority by the Sendai Framework for Disaster Risk Reduction (SFDRR) 2015–2030. This framework also highlighted two other priorities: (i) strengthening disaster risk governance to manage disaster risk; and (ii) investing in DRR for resilience.

In Asia, with few exceptions, nations are on the right path now, and a paradigm shift in DRR has been noted from being reactive to being proactive (Rahman & Shaw, 2015). These countries are in the process of preparing urban policies, programs, strategies, and plans, and they are still struggling to endorse DRR in their urban planning. Regional DRR authorities are working on vulnerability and risk assessment, establishment of emergency response forces and mechanisms, undertaking multihazard early warning systems, raising awareness, and disaster and climate change education to proactively respond to increasing disasters and climate change adaptations (Shaw & Sharma 2011; Douglass, 2013). Parallel to this, city governments are working on urban disasters and building city resilience. However, the attention of policy makers in building institutional and structural resilience may help in managing urban risk effectively despite a country's limited financial resources (Rahman & Shaw, 2015).

3.2 Urban Linkages

Cities are the hubs of economic activy and core areas of innovation and creativity (Rahman & Shaw, 2015). These metropolises are both regionally and globally connected, so they have comparatively more vulnerability to multiple hazards, the anticipated impact of climate change, and negative food, energy, and financial impacts than do rural areas. In this regard, Asian cities are more exposed and vulnerable to the impacts of disastrous events. The frequency of multifarious hazards are on the rise due to weak city governance, rapid urban growth, lack of disaster and climate change education, poor economic structure, and increasing environmental pollution. Keeping the stated situation in mind, both frequency and intensity are escalating as a result of climate change, reducing the economic capacity of city governments to respond effectively to ever-increasing hazards. In an urban setup, cities largely contribute to the economic prosperity of a country, but they also face numerous interrelated challenges, which need to be effectively mainstreamed to achieve sustainability (Rahman & Shaw, 2015).

For any urban system, efficient and swift means of transportation are among the key factors determining economic growth. Roads, railways, waterways, and airways are vital transport corridors that link a city to other regions. However, in Asian cities, there is lack of swift (or even fully operational) transportation and traffic systems that can contribute to the effective exchange of goods and services. Research literature further confirms that the existing link between regional transport systems and intracity transport networks is inadequate (Chaolin & Sunsheng, 2010). The existing highways and high-speed railway stations are isolated from each other, lacking effective linkages. In urban planning, intracity swift means of transportation is required for a city to experience healthy economic growth, effective emergency response, and early recovery from crisis events. In Asian cities, the shortage of urban transport corridors is a major hurdle in achieving desired levels of emergency response and early recovery. It is important that city administrations need to work toward strengthening both intracity and intercity transportation networks to achieve resilience (Campanella, 2006).

In Asia, urban poverty is the root cause of a number of underlying disaster risk factors. Some countries have achieved a remarkable decrease in urban poverty, whereas others have registered an increase. For instance, Pakistan has substantially reduced urban poverty due to its sustained economic growth, which has created employment opportunities. In Pakistan, the poverty level dropped from 65% in 1990 to 23% in 2004, and this achievement is remarkable relative to other countries in the Economic and Social Commission for Asia and the Pacific (ESCAP) region (ESCAP, 2008). In this perspective, some of the key factors in reducing the number of poor people include increases in foreign remittance and capital investment in business and the construction sector, which created abundant employment opportunities (UNHABITAT, 2010b).

In Asia, almost every city is exposed to single or multiple hazards and frequently suffers in one way or the other from the impact of disasters. Historically, numerous cities on the continent have been hit hard in terms of physical, economic, and social damage. However, these cities effectively recovered and have been rebuilt with much more resiliency. The megacities of Tokyo, Delhi, Karachi, Dhaka, Bombay, Chittagong, Jakarta, Manila, Bangkok, Shanghai, Yangon, Calcutta, and Ho Chi Minh City have come across serious episodes of either single or multiple disasters of floods, tsunami, earthquake, heat wave, storm surges, intense rainfall during the past two decades. It is noteworthy that in the face of changing climate, these cities suffered more frequently from the impacts of hydrometeorological disasters (Shaw & Sharma 2011). In order to effectively respond to these multifarious hazards, it is time that the city governments in consultation with the regional disaster management authorities need to prepare disaster management plans at the city level, which should clearly specify the role of institutions and communities.

3.3 Impact of Urban Disasters

In Asian cities, both urbanization and industrialization processes provide a base for socioeconomic gain (Shaw et al., 2009; Douglass, 2013). However, such processes considerably increase the potential risk of damage from environmental pollution, ecological disruptions, hydrometeorological disasters, and climate change exacerbations (Carmin et al., 2012). It is important to note that in Asia, the majority of its urban centers are located on the coastal belt of the Indian and West Pacific Oceans, where tsunamis and climate change have substantially increased the risk of cyclones, coastal floods, urban floods, sea level rises, etc. (Shaw & Sharma 2011). The World Economic Forum (WEF) has estimated that by the year 2050, the exposure of city dwellers to various hazards, including earthquakes, tsunamis, urban floods, cyclones, and storm surges will double (WEF, 2014).

Like other continents, Asia is exposed to a variety of hydrometeorological extreme events such as floods, storm surges, intense rainfall, heat waves, and drought. During the past three decades, the frequency and intensity of such disasters are increasing, and in turn, more and more citizens are suffering from their effects. The toll, in terms of both human losses and property damages, is also escalating. Coastal cities are exposed to tsunamis, storm, sea surges, and urban floods. Some cities, such as Dhaka, Bangladesh, are vulnerable to frequent flood disasters. The unprecedented and haphazard increase in urban populations has further accelerated the frequency and intensity of climate-related disasters (Rahman & Shaw, 2015). With particular reference to floods, in megacities, little attention has been given to enforcing land-use regulations and restricting construction in high-risk zones, regulating the smooth flow of drainage systems, and uprooting encroachments onto the river channel can help regulate the laminar flow of floodwaters and minimize the impacts of urban floods.

It has been reported that in Asia, almost every city is vulnerable to a hazard or a list of hazards (Shaw et al., 2009). In the past, some cities were fully ruined, while others were partially collapsed and some are still vulnerable to shocks and disturbances of unforeseen events. The mega-cities of Shanghai and Guangzhou in the deltaic belt of eastern China; the deltaic city of Ho Chi Minh, Vietnam; Bangkok, Thailand; Dhaka; Bombay, India; and Karachi, Pakistan (among others) have already experienced a series of flooding episodes during the past two decades. In these megacities, the frequency and intensity of flooding are gradually increasing. These coastal cities are also vulnerable to cyclones and storm surges, of which the recent 2013 Super Typhoon Haiyan (costing $13 billion) and Typhoon Fitow in China and Japan (costing $10 billion) are the most notable examples (Caulderwood, 2014) It needs the proper attention of policy makers to address DRR and climate change issues and to build city resilience as a priority because cities are the hub of commercial and industrial activities.

3.4 Cities' Vulnerabilities and the Underlying Risk Factors

Worldwide, both the frequency and intensity of urban disasters are on the rise. Over the past three decades, the average impact of disastrous events each year has increased from US$50 billion in 1980 to over US$200 billion in the 1990s. In cities, the increasing rate of urbanization, economic development, and industrial establishment has led to new problems and risks. It has been estimated that the costs of these urban risks are strongly linked to the increasing urban populations and industrial development. Such risks have swelled particularly in those cities where the existing urban services and infrastructure is under capacity or altogether lacking. As a result, these cities are facing parallel challenges of limited urban infrastructure and environmental risks, mainly because of rapid population growth, industrial development, excessive waste generation, and intensive utilization of natural resources. Water, land, and air contamination is other key environmental risk in urban areas. This problem requires prompt actions by both policy makers and city authorities to formulate smart city plans and mainstream urban risk reduction approaches.

The Hyogo Framework for Action (HFA) 2005–2015 defined vulnerability as the conditions determined by physical, social, economic, and environmental factors or processes, which increases the susceptibility of a community to the impact of hazards. In Asia, almost every city has a series of human-induced or natural incidents that repeatedly caused damage to its residents. Several cities were ruined and then rebuilt with much more resilient structures and fabrics (Shaw et al., 2009). In addition, the number of partially collapsed cities was more than double that of fully damaged ones. China, India, Indonesia, Pakistan, Japan, Bangladesh, Iran, Philippine, Turkey, Vietnam, Thailand are the nations that are exposed to multiple hazards and experienced serious disasters of various intensities and magnitudes. The frequency of hydrometeorological disasters (notably floods, intense rainfall, storm surges, drought, extreme temperatures, and heat waves) exceeded all other types of disaster. Surjan et al. (2011) stated that cities are as powerful and important as they are vulnerable; therefore, their governments need to think rationally in order to devise effective disaster risk management strategies for increasing urban resilience through logically developing institutional frameworks, urban risk reduction strategies with a particular focus on climate change adaptation, and mitigation strategies.

In Asia, several urban hazards pose numerous challenges to sustainable urban development. The extent of vulnerability varies from city to city, depending upon the coping capacity and efficiency of a city's government (Shaw et al., 2009). Similarly, increasing urbanization is consistently linked to increasing urban problems, which indicates that urbanization and urban risks have a positive corelation (Parvin et al., 2013). Every year, hundreds even thousands of urban dwellers are severely affected by the impacts of extreme hydrometeorological and geophysical events. These disasters place tremendous burdens on citizen access to urban amenities, including water supplies, education, health, and sanitation. It is the prime responsibility of city authorities to withstand and resist unforeseen events to the greatest extent possible. The city government should incorporate urban risk reduction approaches and ensure that economic development continues steadily, with no setbacks. The HFA priority 1 and 4 plans specifically focus on including urban risk reduction strategies.

3.4.1 Cities and Substandard Housing

In most Asian cities, the majority of citizens are residing in dilapidated, substandard housing. Such houses have low resilience even against the lowest-scale hazards. It has been reported that almost all these cities have developed building codes that would help alleviate this situation, but they are rarely enforced by their governments. Usually, rural migrants to cities want low-cost housing, which almost inevitably are in poor condition. That is why millions of urban citizens are at high risk during disasters, and poor sections of society are hit the hardest. Similarly, unplanned development has been carried out in areas with high tsunami risk, proximity to active flood channels, potentially active landslides and earthquakes (Rahman et al., 2014). Such neighborhoods lack even basic urban services (Rahman & Shaw, 2015). It is the prime responsibility of city governments to stop illegal development and formulate and enforce safe building bylaws.

3.4.2 Cities and Flood Vulnerability

In Asia, urban flooding is a frequent phenomenon. Almost every city is vulnerable to urban flooding in one way or another, and urban dwellers are at high risk. In urban environments, farmland, vegetation cover, and bare soil have been converted into built-up areas. As a result, water runs off of the concrete structures, sometime known as *pluvial flooding* or *urban flooding*. In urban areas, with rapid increase in impermeable surfaces and urban development, the likelihood of flooding has increased. Similarly, the intensity of urban floods becomes higher with prolonged rainfall. The probability of urban flooding is expected to further increase with changing climate (Carmin et al., 2013). In Asia, numerous coastal cities are exposed to both urban and coastal floods, whereas inland cities are vulnerable to flash floods, river floods, or urban floods. In developing countries, human encroachments onto the active flood channel, poor flood management strategies, lack of flood early warning systems, and disposal of solid waste in drainage lines are the major causes of urban flooding.

In Asian cities, populations grow gradually, and these new residents need shelter to accommodate their needs. As urban populations increase, the existing resources and infrastructure are more and more pressured. At the same time, land values increase, and higher buildings are constructed to fulfill the housing demand. Finally, the most pronounced aspect of cities in the developing world is haphazard urban expansion over valuable natural resources. In all these cases, poor people are the ones who have no choice but to build shelter on illegal, unsafe land in high-risk zones. This situation, where people consistently encroaching onto the flood prone areas narrow the channel and in turn reduce the channel carrying capacity, is very common in almost every megacity. These are some of the major factors contributing to enormous flood damage in urban areas.

3.4.3 Urbanization and Growing Vulnerability

In Asia, the explosive growth and pace of urbanization is one of the most important underlying risk factors. The population of the existing cities grows through natural increase, rural-urban migration, and annexation of land, with natural increase as the most dominant factor. High population growth poses numerous challenges for urban authorities to provide amenities to their residents. The situation gets worse during emergencies due to the presence of heavy vehicles on roads that do not have the capacity to handle them, traffic jams along the main corridors, and the fact that residents of these slum areas are already suffering from poverty. Similarly, the number of dwellers in high-risk places is gradually increasing, both in urban and peri-urban areas. Most of these citizens live in the poorer sections of society, and they have no other choice but to live in such dangerous areas (Rahman et al., 2014).

Poor people initially occupy low-cost areas that were not formally designed for residential purposes (Rahman & Shaw, 2015). In 1990, for example, 51% of the total urban population of Pakistan resided in slum areas, and these regions have shown very little improvement over the years, as already noted in this chapter (UNHABITAT, 2010a).

In Asian countries, the amount and nature of solid waste collection and disposal largely varies from city to city. With few exceptions, in the majority of mega-cities, mixture of organic and inorganic solid waste is collected, which creates problems in sorting, recycling, and reusing (Rahman & Shaw, 2015). In all cities in the developing world, in spite of committed efforts by their governments to improve services, the proper way to handle and dispose of solid waste collection is still a challenging issue. Meanwhile, toxic waste generated by hospitals and industries is another problem that must be addressed. In these cities, solid waste is either openly burned or dumped in ecologically important areas, which generates bad smells, environmental pollution, and other problems. City planners have recognized such issues and determined that the fact that financial resources are limited is a major determination in urban waste collection and disposal (Samiullah, 2012). It has been pointed out that environmental protection policy and plans are intact, but weak enforcement and insufficient environmental education are specific areas of neglect. There are model cities as well, which, apart from other sectors, also have focused on proper collection and disposal of waste.

Climate change has strongly influenced hydrometeorological phenomena. Some intensifying factors include deforestation, overgrazing, rapid urban development at the cost of degrading farmland, and heavy carbon emissions (Rahman & Shaw, 2015). As a consequence, urban peripheral land, which once acted as the "breadbasket" for the nearby city dwellers, has been rapidly consumed by construction (Samiullah, 2012). With the growing population, commercial activities expanded; as a result, the underlying risk factors multiplied daily.

3.5 Cities and Climate Change

Globally, urban areas are among of the major contributors to greenhouse gas emissions. In 2006, it was estimated that globally these urban hot spots account for over 66% of primary energy demands and have produced 71% of energy-based greenhouse gas emissions (IEA, 2008). Compared to the other major continents, it was found that Asia was the leading producer of greenhouse gas emissions (IEA, 2008). Recently, it has been estimated that on average, the ESCAP region produces 75 billion tons of untreated solid waste and generates huge amounts of carbon dioxide to the atmosphere every year, a prominent contributor to greenhouse gas emissions (UNESCAP, 2007). Similarly, in China, over 40% of energy related carbon dioxide emissions are contributed by 35 major urban centers (Dhakal, 2010; Chaolin & Sunsheng, 2010). Furthermore, it has been estimated that in Asia, the per capita footprint is much lower in developing countries than in developed nations. However, among developing nations, the growing rate of greenhouse gas emission is an alarming trend that requires prompt action and policy decision making. It has been calculated that the per capita carbon foot print at the national level in Asian countries is far less than the per capita carbon footprint in their urban areas. This statistic clearly indicates that urban areas are the hot spots of greenhouse gas emission and the hub of industrial and commercial activities.

Climate change is a major challenge for scientists to find the root causes and devise innovative policy responses (Chaolin & Sunsheng, 2010). The Intergovernmental Panel on Climate Change (IPCC, 2007) reported in great detail that human intervention in the natural ecosystem plays a major role in global warming and contributing to greenhouse gas emissions. In Asia, major urban centers are the center of economic development and a dominant source of greenhouse gas emissions. These megacities are playing a key role in enhancing national economic development and growth, but they also need innovative planning to halt greenhouse gas emissions.

During the past decade, increasing scientific research on building urban disaster resilience has remained a major area of debate related to climate change and DRR. Joerin et al. (2012) adopted the Climate Disaster Resilience Index (CDRI), with 5 dimensions (natural, physical, economic, social, and institutional), 25 parameters, and 125 variables, which focuses on institutional and human capacity and response to potential climate-related disasters in Chennai, India. By applying the CDRI, one can appraise a city's regional resilience against climate-related disasters from the community perspective. Similarly, Prashar et al. (2013) also applied CDRI to the concept of city resilience, community action planning, and building climate-related disaster resilience in eastern Delhi. It specifically attempts to determine the policy and root gaps in individual sectors as well as overall, and provide direction on how to bridge loopholes and build city resilience (Rahman & Shaw, 2015). Scores on a 1 to 5 scale (with 1 being low resilience and 5 being high) were calculated for all 25 parameters and 125 variables.

With reference to growing cities and growing vulnerabilities, industrialization and urbanization are the two interlinked processes that largely contribute to climate change exacerbations (Carmin et al., 2013). Even today, in the process of industrialization, the use of fossil fuels and emission of carbon dioxide occurs on a large scale. And industrialization has resulted in excessive urbanization. There are frequent changes in land use in cities, especially in the periphery, where natural resources have been encroached upon by urbanization and construction. Of the Asian countries, China, India, and Japan are the major contributors of greenhouse gas emissions (Chaolin & Sunsheng, 2010).

Natural disasters hit poor and rich alike, but the poor suffer more, for many reasons. Poor urban dwellers are at the greatest risk of experiencing the impacts of climate change and hydrometeorological phenomena (UNHABITAT, 2010b). These urban dwellers are either exposed to storm surges or other coastal hazards. In Pakistan, a study reveals that during the past two decades, the increasing trend of flash floods in *nallah Lai* and the resultant human losses and other damage in Islamabad and Rawalpindi are examples of the growing number of climate change disasters (Rahman & Khan, 2013; Rahman & Shaw, 2015). Recent scientific studies reveal that the occurrences and intensity of climate-related extreme events have increased and are expected to further escalate in the future (Rahman & Khan, 2013). Such climate change–related hazards need attention from city governments in the form of disaster management plans and strategies to address climate change issues in sustainable urban planning and building city resilience.

In Asia, almost half of the urban populations are living in low-lying coastal zones that are prone to coastal hazards (UNHABITAT, 2008; Shaw et al., 2009). Some of the major Asian deltaic coastal cities (namely, Dhaka, Karachi, Bangkok, Ho Chi Minh City, Shanghai, Jakarta, and Manila) are highly vulnerable to sea-level rises, cyclones, stormwater surges, acid rain, heat islands, and coastal floods (Rahman & Shaw, 2015). Similarly, parallel to the direct negative impacts of climate change, these urban areas are also affected by climate-induced crises in the urban periphery, like droughts, floods, soil erosion, and desertification. This in turn poses a threat to increasing food insecurity in cities and putting stress on the urban dwellers. It is the poorest sections of society that contribute the least to greenhouse gas emissions, but have the highest exposure to and will suffer the most from the consequences of climate change.

3.6 Why Should Cities Become More Resilient?

It is estimated that in Asia, half of the population will be settled in urban areas over the next decade (Samiullah, 2012). Of this total, two-thirds will be living in megacities. Urban areas (especially the major cities) have more attractions in terms of financial and physical resources. However, the urban authorities in these metropolises are hard-pressed to provide services to their growing populations; and as a result, the risks are increasing by the day (Shaw & Sharma 2011). In addition, if disasters and climate change exacerbations are not effectively addressed in these cities' programs and strategies, the risks will increase even more and have more serious repercussions for residents, particularly the poorest ones, who often lack the benefit of safety nets (WEF, 2014). It is the primary responsibility of city governments to enhance the resilience of their communities through risk assessment and reduction, preparedness, emergency response systems, and early recovery (Rahman & Shaw, 2015). To address this growing concern, in 2010, the United Nations Office for Disaster Reduction (UNISDR) launched a campaign called "My City Is Getting Ready," with the goal of "making cities resilient." The intention of this initative was to prepare disaster-resilient cities that could serve as role models, and as a result of these successful efforts, the number of model cities in Asia is increasing.

It was recently estimated that at the global level, 98% of the internal displacement of populations was attributed to the impact of climate change, mainly hydrological and meteorological phenomena (IDMC, 2013). In Asia, several devastating disasters, including a tsunami in India in 2004; an earthquake in Pakistan in 2005; Cyclone Nargis in Myanmar in 2008; Cyclones Sidr and Aila in Bangladesh in 2007 and 2009, respectively; Typhoon Haiyan in Pakistan in 2010; Typhoon Fitow in 2013 in China and other parts of Asia; and recent floods in China, India, Bangladesh, and Thailand displaced millions of victims (IDMC, 2013). Urban areas in close proximity to disaster-hit locations were where the majority of the victims were found (UNHCR, 2006). Therefore, with the climate variability and resultant increasing frequency and intensity of hydrometeorological disasters, the number of displaced populations is projected to increase further in the future (IDMC, 2013; Rahman & Shaw, 2015).

In urban economic systems, cities serve as a main hub of revenue. With the impacts of disasters, cities' economic situations badly suffer (Rahman & Shaw, 2015). Similarly, supply chains are also directly affected, which in turn increase the prices of daily essentials. As a result, the revenue generation system is negatively affected, and loss of economic revenue system occurs. Parallel to other impacts of urban disasters, damage to ecocultural systems also happen. An area's cultural heritage and natural scenic beauty are both affected by recurrent floods, earthquakes, landslides, intense rainfall, and depletion of groundwater aquifers, cyclones, and storm surges. In order to protect these vital areas, there is a need for cities' DRR plans to reduce vulnerabilities and enhance urban resilience (Campanella, 2006) and give due weight to archaeological monuments.

3.7 Approaches to Urban Resilience

The scientific literature defines urban resilience as the physical, institutional, social, and economic abilities of cities to recover from the impact of disaster (Campanella, 2006). Many recent studies have considered resilience the policy and planning that a development applies in order to bounce back from a disaster, or the potential to mitigate the effects of an extreme event (Coaffee & Hare 2008). It is actually function of multiple factors including the key role of citizen coping capacity and resilience itself (Shaw and Sharma 2011). Increasing urban resilience is part of urban planning process that intends to mitigate the potential threats. Similarly, the urban resilience is referred to the potential capacity against the hazards (Campanella, 2006). Such kind of planning activities have been recently incorporated in urban development activities to enhance urban resilience and implement strategies for a wide range of potential events to cities including floods, earthquake, storm surges, heat waves and predicted impacts of climate changes are part and parcel of urban resilience (Coaffee & Hare 2008).

In the process of developing urban resilience, various design alterations and reinforcements are made, such as architectural and structural improvements, land-use regulation and enforcement, new building regulations, and management strategies that intend to minimize the impacts of extreme events. Some cities have deep historical roots and cannot be easily changed, however, which complicates this process and can lead to bad results. For example, a city with a strong and diverse economic setup will be able to rebound more quickly than a city with a weak economic, social, and political setup. Worldwide, Tokyo is the most populous metropolitan city, and it is exposed to multiple hazards. However, effective disaster prevention strategies make the city the best in terms of emergency response and early recovery. The Tokyo metropolitan area has developed disaster prevention plans and urged citizens to keep in place food and water for at least three days, in addition to batteries, a first aid kit, medicine, flashlights, and other important supplies. The plan also includes guidelines for disaster prevention, with tips on how to react to events such as earthquakes, tsunamis, and floods. Tokyo can be considered a role model for city resilience.

In discussions of urban risk reduction, the stress remains on building city resilience by enhancing the capacity of human resources, organizations, infrastructures, and communities to respond promptly to both human-induced and natural extreme events. Recent studies have recommended that in the face of climate change, cities should build resilience at all levels to the increasing urban risk (Carmin et al., 2013). Urban authorities need to quantify their cities' physical resilience to potential events, assess infrastructural and environmental resilience, and evaluate the extent of their systems in terms of early recovery and the status of their community resilience (Leichenko, 2011).

3.8 Conclusions

A community's resilience can be measured using standard parameters at the household level (Campanella, 2006). In Asian cities, there are different zones where poor, lower-class, middle-class and high-class families reside; every section of the society bears a different level of hazard resilience, but all need to be protected from disaster. It has been noted already in this chapter that poor and lower-class families reside in areas vulnerable to numerous hazards. The latest studies have focused on exploring various strategies and mechanisms to increase the hazard resilience of poor sections of urban communities in the developing world, particularly in Asia. After a disaster, it is important to rebuild devastated cities with a commitment to bringing back their physical and socioeconomic resources. The involvement of urban dwellers at the community level and in every stage of rebuilding should be the prime aim of developing early recovery plans, reducing vulnerabilities, and building city resilience.

References

Campanella, T. J. (2006). Urban resilience and the recovery of New Orleans. *Journal of American Planning Association*, *72*(2), 141–146.

Carmin, J., Dodman, D., & Chu, E. (2013). Urban Climate Adaptation and Leadership: From Conceptual Understanding to Practical Action. *OECD Regional Development Working Papers 2013/26* OECD Publishing, http://dx.doi.org/10.1787/5k3ttg88w8hh-en. Accessed 16.06.15.

Carmin, J., Nadkarni, N., & Rhie, C. (2012). *Progress and challenges in urban climate adaptation planning: Results of a global survey*. Cambridge, MA: MIT Press.

Caulderwood K. (2014). Report: The Ten Most Expensive Natural Disasters in 2013. *International Business Times*. January 15. Accessed on March 20, 2014, from http://www.ibtimes.com/report-ten-most-expensive-natural-disasters-2013-1540058.

Chaolin, G. U., & Sunsheng, H. A. N. (2010). Climate change and China's mega urban regions. *Frontiers of Architecture and Civil Engineering in China*, *4*(4), 418–430.

Coaffee, J., & Hare, P. O. (2008). Urban resilience and national security: The role for planning. *Urban Design and Planning*, *161*(4), 173–182.

Dhakal, S. (2010). GHG emissions from urbanization and opportunities for urban carbon mitigation. *Current Opinion in Environmental Sustainability*, *2*(4), 277–283. Accessible from www.sciencedirect.com.

Douglass, M. (2013). *The Urban Transition of Environmental Disaster Governance in Asia*. Working Paper Series No. 210. Singapore: Asia Research Institute, National University of Singapore. Available at http://www.ari.nus.edu.sg/publication_details.asp?pubtypeid=WP&pubid=2334.

International Displacement Monitoring Centre (IDMC). (2013). *Global Estimates 2012: People Displaced by Disasters*. Geneva, Switzerland: A report of the Norwegian Refugee Council.

International Energy Agency (IEA). (2008). *World energy outlook*. Paris: International Energy Agency. Available at www.worldenergyoutlook.org/2008.asp.

Intergovernmental Panel on Climate Change (IPCC). (2011). Summary for Policymakers. In C. Field, et al. (Ed.), *Intergovernmental Panel on Climate Change special report on managing the risks of extreme events and disasters to advance climate change adaptation* (pp. 1–20). Cambridge, UK: Cambridge University Press.

Joerin, J., Shaw, R., Takeuchi, Y., & Krishnamurthy, R. (2012). Assessing community resilience to climate-related disasters in Chennai, India. *International Journal of Disaster Risk Reduction*, *1*, 44–54.

Leichenko, R. (2011). Climate change and urban resilience. *Current Opinion in Environmental Sustainability*, *3*, 164–168.

Mercer, J. (2010). Disaster risk reduction or climate change adaptation: are we reinventing the wheel? *Journal of International Development* *22*, 247–264.

Parvin, G. A., Ahsan, S. M. R., & Shaw, R. (2013). Urban risk reduction approaches in Bangladesh. In R. Shaw, et al. (Ed.), *Disaster risk reduction approaches in Bangladesh, 2013* (pp. 235–257). Tokyo: Springer.

Prashar, S., Shaw, R., & Takeuchi, Y. (2013). Community action planning in East Delhi: A participatory approach to build urban disaster resilience. *Mitigation and Adaptation Strategies for Global Change*, *18*, 429–448.

Rahman, A., & Khan, A. N. (2013). Analysis of 2010 flood causes, nature, and magnitude in the Khyber Pakhtunkhwa, Pakistan. *Natural Hazards*, *66*(2), 887–904.

Rahman, A., Khan, A. N., & Collins, A. E. (2014). Analysis of landslide causes and associated damages in the Kashmir Himalayas of Pakistan. *Natural Hazards*, *71*(1), 803–821.

Rahman, A., & Shaw, R. (2015). Urban risk and reduction approaches in Pakistan. In A. Rahman, et al. (Ed.), *Disaster risk reduction approaches in Pakistan* (pp. 295–314). Tokyo: Springer.

Samiullah. (2012). *Expansion of built-up area and its impact on urban agriculture: A case study of Peshawar, Pakistan. PhD thesis, Institute of Geography, Urban and Regional Planning.* Pakistan: University of Peshawar.

Shaw, R., & Sharma, A. (2011). *Climate and disaster resilience in cities: Community, environment, and disaster risk reduction*. Bingley, UK: Emerald Group Publishing Ltd.

Shaw, R., Srinivas, H., & Sharma, A. (2009). *Urban risk reduction: An Asian perspective*. Bingley, UK: Emerald Group Publishing Ltd.

Surjan, A., Takeuchi, Y., & Shaw, R. (2011). *From disaster and climate risk to urban resilience: Approaching through community-based environmental improvement*. Singpore: RPS.

United Nations (UN). (2010). *World urbanization prospects: The 2009 revision*. CD-ROM Edition—Data in Digital Form. New York: Population Division, Department of Economic and Social Affairs, United Nations.

United Nations (UN). (2015). Department of Economics and Social Affairs Division, Population Ddivision. *World Urbanization Prospects, 2014 Revisions*. Accessed on July 8, 2015, from http://esa.un.org/unpd/wup/CD-ROM/.

United Nations Department of Economic and Social Affairs (UNDESA). (2010). *World urbanization prospects: The 2009 revision* United Nations Department of Economic and Social Affairs, Population Division ESA/P/WP/215. New York.

United Nations Economic and Social Commission for Asia and the Pacific (ESCAP). (2007). *Report: Regional Seminar and Study Visit on Community-Based Solid Waste Management*. Quy Nhon City, Vietnam: United Nations ESCAP and others. 15–16 December 2007. Accessible at www.housing-the-urban-poor.net/Docs/SWMreport.pdf.

United Nations Human Settlements Programme (UNHABITAT). (2008). *State of the world's cities 2008/2009: Harmonious Cities*. Nairobi, Kenya: UNHABITAT. HS/1031/08E.

United Nations Human Settlements Programme (UNHABITAT). (2010a). *The state of Asian cities 2010/11*. Japan: United Nations Human Settlements Programme (UNHABITAT), Regional Office for Asia and the Pacific ACROS Fukuoka Building.

United Nations Human Settlements Programme (UNHABITAT). (2010b). UNHABITAT launches new tools to strengthen urban resilience. *UNHABITAT for a Better Urban Future, Barcelona*. Accessed on February 9, 2014, from http://www.unhabitat.org/content.asp?cid=12813&catid=5&typeid=6&subMenuld=0&goback=%2Egde_1988319_member_5829979954695127042.

United Nations High Commissioner for Refugees (UNHCR). (2006). Current dynamics of displacement. In N. Merheb, & S. Loughna (Eds.), *The state of the world's refugees 2006—Human displacement in the new millennium*. Geneva, Switzerland: UNHCR. (pp. 9–29).

World Economic Forum (WEF). (2014). *Global Risks 2014*. Geneva: World Economic Forum. Available at www.weforum.org/risks.

Resilient Homes Make Cities Resilient

4

Kenji Okazaki[1], *Krishna S. Pribadi*[2]
[1]Professor, Kyoto University, Japan; [2]Professor, Bandung Institute of Technology, Indonesia

4.1 Introduction

Due to the rapid urbanization and rapid spread of modern life, it appears that earthquakes are becoming more complicated and deadly. As cities are growing and becoming more crowded than ever and infrastructure is often insufficient, they are becoming more vulnerable to natural disasters. The urban poor, who have no choice but to live in hazardous areas, are particularly in danger. Substantial rapid urbanization has taken place worldwide over the last 50 years, while the return period of a large earthquake would be usually more than 50 years, meaning that most large cities and megacities have not experienced a major earthquake in recent years. A medium-size earthquake could even devastate a city if it takes place beneath or near it, such as in the case of an earthquake that struck Haiti in 2010.

The earthquake, including tsunamis that it can trigger, is the most deadly natural disaster in modern society. Of the 10 deadliest disasters in the past 30 years, 7 were induced by earthquakes: Haiti in 2010; Sichuan in 2008; Pakistan in 2005; Sumatra, Indonesia, in 2004; Iran in 1990 and 2003; and Armenia in 1988. It seems that earthquake-related disasters have been more destructive recently, as more than half of the deadliest earthquakes have taken place since 2000. While huge progress in seismology and earthquake engineering has been made and international groups have reiterated the importance of disaster reduction in recent years (such as at the World UN Conference on Disaster Reduction in 1994 and 2005, and many other conferences), it does not seem that such progress has reduced the damage caused by earthquakes.

Earthquake risk is on the rise in urban areas, where people and homes are increasingly densely concentrated. In earthquakes, most deaths and injuries are caused by total or partial collapse of buildings, particularly homes. If a large earthquake occurs at night, when people are at home, almost all deaths are caused by collapse of the houses. Even during the daytime, more than half of all earthquake deaths are caused by collapse of homes. For example, about 5000 direct deaths (approximately 90% of the total) from the Kobe earthquake in 1995, which occurred at 5:46 a.m., were caused by the collapse of deteriorated or otherwise vulnerable homes. Most of the remaining deaths were caused by fires that trapped people in their collapsed homes.

It is ironic that the shelters that are supposed to protect people from various hazards often end up killing them. Collapsed homes block streets, hampering evacuation, relief, and firefighting activities. Further, people are distressed when they lose their home after a disaster. Most of the tragedies and difficulties that disaster management must contend with are attributed to the loss of lives and homes due to the destruction

Urban Disasters and Resilience in Asia.

of homes. In particular, the loss of a home usually means loss of family life and memories and prolongs the evacuation process, which is very harsh and stressful. Once a home is lost, it would be very costly and time consuming to reconstruct it. Governments are also forced to allocate tremendous resources to constructing emergency shelters, temporary homes, and permanent homes to replace any destroyed ones.

Because earthquakes cannot be predicted precisely, even by applying the most advanced science and technology, it is essential to make vulnerable homes safer in order to reduce the number of victims and the amount of severe damage caused in such incidents. The more resilient the existing homes are, the lower the death rate in the event of a quake, and the less drastic will be the disruptions to economic and social activities in the affected areas. No matter how effectively emergency management and relief activities are conducted, those lives lost can never be regained.

Therefore, the "resilient homes make cities resilient" concept can be summarized as follows:

- If homes can be made physically resilient, people would not be killed by the collapse of their homes in earthquakes. If most homes in a city are resilient, the communities would be resilient in turn, and the death toll becomes minimal.
- If homes are not destroyed, they will not block roads, which would make evacuation, relief, and firefighting activities easier.
- If people do not lose their homes, they will maintain their family life.
- If people survive (along with their resilient homes), they can help facilitate the relief, recovery, and reconstruction activities in their communities.
- If homes are resilient, governments will not need to allocate a huge amount of resources on constructing temporary houses and permanent houses, and so they can focus on other urgent issues.

In developing countries, a large number of people live in traditional or nonengineered homes, which are very vulnerable to earthquakes. This is the most critical reason why so many people are killed by earthquakes in developing countries (Coburn, 2002, Grundy, 2007, Narafu et al., 2009). Hence, nonengineered homes make cities vulnerable. In other words, making nonengineered homes safer would make cities resilient against earthquake. These vulnerable homes may be characterized as follows:

- Homes made of locally available and cheaper materials
- Relatively low quality of construction work
- No or little engineering intervention
- No building regulation or no conformity to the regulations, and thereby, no check from the formal sector

As construction techniques and practices, skill of masons, and construction materials differ from country to country, the strength of nonengineered construction differs as well. However, there is little knowledge of the comparative strengths of various developing countries.

In this chapter, the authors first introduce the current conditions of the design and characteristics of nonengineered construction in several countries. A comparative survey was conducted in seven selected developing countries: Egypt, India, Indonesia, Nepal, Pakistan, Peru, and Turkey.

Second, the authors introduce efforts in Indonesia (representing developing countries) and Japan (representing developed countries) to reduce the damage caused by earthquakes by securing home safety. In Indonesia, like many other developing countries, it is crucial to secure the safety of newly constructed houses through the appropriate implementation of building codes and dissemination of technical guidebooks. In Japan, like many other developed countries, it is crucial to improve the safety of existing homes by retrofitting. In Japan, there are still a lot of old (i.e., deteriorated) and vulnerable wooden houses, while it is believed that newly constructed homes meet the building codes and therefore are strong against earthquakes.

4.2 Nonengineered Construction in Developing Countries

4.2.1 Research Objectives

A joint research on nonengineered construction was conducted in 2009–2010 in order to better understand and compare the current practices of nonengineered buildings in selected developing countries, aiming to improve policies and develop appropriate technology for reducing vulnerability against earthquakes. Based on the information obtained, general problems were identified and recommendations for improving nonengineered construction in developing countries were formulated (Okazaki et al., 2012).

This research was jointly conducted with the following institutions in the selected seven developing countries:

- National Research Institute of Astronomy and Geophysics (NRIAG), Egypt
- SEEDS Technical Services (STS), India
- Research Center for Disaster Mitigation-Institut Teknologi Bandung (RCDM ITB), Indonesia
- Designmen Consulting Engineers Ltd/ETSSR Center, Pakistan
- Japan-Peru Center for Earthquake Engineering Research and Disaster Mitigation of National University of Engineering (CISMID), Peru
- National Society for Earthquake Technology (NSET), Nepal
- Istanbul Technical University, Turkey

4.2.2 Selected Sites

The partner institutions in the seven developing countries in this study were assigned to identify several types of nonengineered homes that would represent the current practice of nonengineered construction in their respective countries, and to select several sites for the field survey. Some countries conducted the survey on the existing building due to the rarity of nonengineered construction projects, supported by secondary data from literature studies. In Turkey, the survey targeted the existing old nonengineered buildings because nonengineered construction is not common in urban areas recently. The selected sites are shown in Table 4.1.

4.2.3 Typical Nonengineered Building Construction

Nonengineered masonry structures can be divided into three categories: unconfined masonry, confined masonry, and reinforced concrete with infill masonry walls. Unconfined masonry building relies on the wall as the only load-bearing structural element.

Table 4.1 **Location of the Study Sites**

Country	City/Region
Egypt	15th May City, Helwan City, Giza Square, 6th October City/Haram City and El-Marg City
India	Balasore, Dehradun, Barmer, Portblair, and Shimla
Indonesia	Bandung City
Pakistan	Potohar Plateau and Plains of Punjab
Peru	Puente Piedra, Carabayllo, Independencia, Huachipa and San Juan de Miraflores
Nepal	Balkot, Bhaktapur, Nankhel, Kirtipur, Kathmandu, Imadole, Lalitpur, Hattiban, Lalitpur of Kathmandu Valley
Turkey	Yenikapi, Sirkeci, Uskudar

Confined masonry building relies on the masonry wall (both vertical and lateral) as the main load bearing structural element. The confinement also contributes to maintaining the integrity of the wall when loads are applied to the structures. Reinforced concrete with infill masonry wall building relies on reinforced concrete columns and beams as the main load-bearing (both lateral and gravity) structural elements. Typical nonengineered buildings in the selected countries are shown in Figure 4.1.

4.2.3.1 Egypt

The most common types of nonengineered buildings in Egypt are reinforced concrete skeleton type buildings, wall-bearing limestone buildings, and combined reinforced concrete and limestone wall buildings.

Figure 4.1 Typical nonengineered buildings in the survey, labeled by country.

4.2.3.2 India

The most common nonengineered building in India is unconfined masonry building of various types of bricks with 1–2 stories. Most of these brick masonry buildings use mud brick (adobe), burned brick, or quarry stone as the materials.

4.2.3.3 Indonesia

The three most popular nonengineered constructions found in Indonesia are unconfined brick or concrete block masonry, confined masonry, and reinforced concrete frame with infill masonry. The structure used most commonly in Bandung is confined masonry confined by reinforced concrete practical columns and beams.

4.2.3.4 Nepal

In Nepal, there are two types of nonengineered brick masonry buildings: unconfined brick masonry buildings and confined masonry buildings.

4.2.3.5 Pakistan

Three types of nonengineered techniques (confined masonry, unconfined masonry, and reinforced concrete with infill masonry) are mostly used to construct nonengineered buildings in Pakistan.

4.2.3.6 Peru

In Peru, there are three types of nonengineered buildings: confined masonry building with horizontal and vertical confinements that support the brick walls, unconfined masonry walls building without reinforced beams or columns and reinforced confined elements, and reinforced concrete frame with concrete shear walls or infill masonry.

4.2.3.7 Turkey

There are three types of nonengineered buildings in Turkey: reinforced concrete frames with clay hollow brick infill wall, unconfined brick masonry, and wooden structures.

4.2.4 Building Regulations and Codes

Most of the selected countries, such as India, Indonesia, Pakistan, Peru, and Nepal, have building regulations, codes, and guidelines on nonengineered construction at the national level. Unfortunately, the building regulations, codes, and guidelines on nonengineered structures are mostly not implemented by these countries, except in a few big cities. It was also found that some countries have problems with disseminating these regulations to construction workers. In Turkey and Egypt, nonengineered building codes are not available at the national level. However, both countries have local offices in charge of building administration in the surveyed cities. In Turkey, the national building code is only used to build engineered structures.

4.2.5 Project and Project Site Facts

Most of the surveyed sites are located on a flat/gentle slope. Most of the nonengineered buildings found in Nepal, India, Turkey, and Peru are constructed by artisans assigned by the building owner, using simple tools. In Egypt and Pakistan, nonengineered buildings are constructed by hired contractors. In general, there is no problem or limitation on the availability of construction materials for the nonengineered buildings. One-third of buildings are owned by the private sector, and any design intervention mostly comes from owners. More than 80% of nonengineered buildings are found to have design intervention and be supervised by their owners. Typically, a nonengineered building is constructed within 6 to 17 weeks.

4.2.6 General Building Information

Most of the nonengineered edifices are used as residential buildings and small commercial sites. Some of them are also used for schools. Strip foundations as well as isolated pads are mostly found in the nonengineered buildings surveyed, which were constructed from various materials, such as stone and reinforced concrete. For confined masonry, most of the buildings use reinforced concrete confinement. Wood is the most common material used for door- and window-framing systems.

The study found that most of the nonengineered buildings in the selected countries are made of unconfined masonry, except in Indonesia and Peru, where confined masonry is more popular. The building area depends on its main function, with average values varying from 80 m^2 to 187 m^2. The typical number of rooms ranged from 3 to 7.

4.2.7 Technical Characteristics

Foundation dimensions varied among the countries. The widest strip foundation was found in Turkey (0.8 m), while the narrowest were found in India and Peru (0.4 m). In Turkey, foundations as deep as 1 to 2 m are common, while in Egypt, foundations are shallower (0.7 m).

Most of the buildings (i.e., 70%) utilized fired clay bricks as wall material, with a one-brick thickness with exceptions of Indonesia (one-half brick) and Egypt (two bricks). In terms of the wall height–to–thickness ratio, the highest ratio is found in Indonesia (19.8), while the smallest is found in Egypt (9.0). The higher value of the ratio indicates the lower wall stiffness (see Figure 4.2). The ratio of the total length of the wall to the floor area indicates the capacity of the building to absorb lateral forces, and the highest value was found in Nepal. The highest wall opening ratio is found in Turkey.

Most nonengineered constructions provide beams, and few of them have columns. This depends on the structural system adopted in the respective countries. In Indonesia, most of the surveyed sites exhibit confined masonry, so both columns and beams are available; while in Pakistan, Egypt, and India, where most of the selected sites are unconfined masonry, the buildings are only provided with beams or lintels. In all the selected countries, most nonengineered construction had poor detailing on the connection of the structural elements.

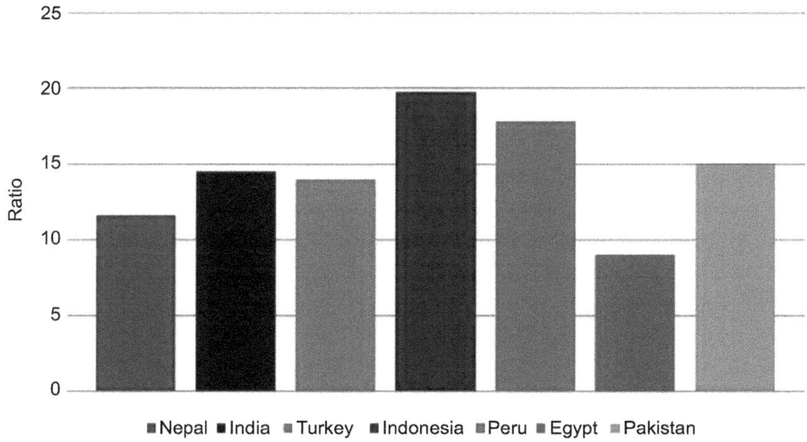

Figure 4.2 Wall height–to–thickness ratio.

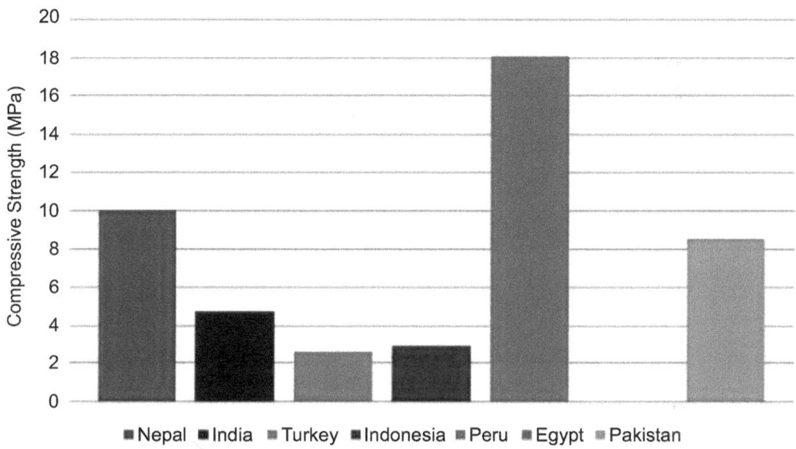

Figure 4.3 Brick's compression strength.

4.2.8 *Masonry Materials*

Most nonengineered construction in the selected countries use fired clay brick or stone masonry for the wall material. Brick sizes in Turkey, Nepal, Indonesia, Peru, and Pakistan are relatively similar, meanwhile in India and Egypt bricks have different sizes compared to the others. Peru has the highest brick compressive strength, while Turkey has the smallest brick compressive strength. Test results from sites in each country showed that some did not have adequate strength for the brick (see Figure 4.3).

Most of these countries used ordinary Portland cement as the plaster and mortar cementing agent. Pakistan has the highest mortar strength, even though the mix is similar to other countries. The mortar thickness varied from 10 to 25 mm. The mortar thickness in Egypt was found to be the thickest (25 mm), while Turkey and Pakistan had the thinnest mortar layer (10–20 mm and 11.5 mm, respectively). The common plaster mix is either 1:6 or 1:4 (cement:sand), except in Peru, where the mix is 1:1. Turkey has the thickest plaster (20–30 mm), while Nepal has the thinnest plaster (10 mm).

4.2.9 Concrete Materials

Some of the nonengineered constructions use very poor concrete strength for the structural elements. The highest concrete compressive strength was found in Indonesia (22 MPa), while the smallest was found in Turkey (8–10 MPa). All of these countries used ordinary Portland cement as construction materials. Most of the aggregates were taken from rivers and mountain quarries. A concrete mix of 1:2:4 (cement:sand:-aggregate) was used in Nepal, India, and Pakistan, while in Indonesia, a mix of 1:2:3 was common, and a mix of 1:1.2:aggregate was found in Peru. Both concrete mixes in Indonesia and Peru produced relatively high compressive strength. In some countries, workers on some sites did not use any measurement to mix the concrete, which negatively affected the accuracy and quality.

4.2.10 Contractors/Builders

In Egypt, most of the buildings are built by a contractor in conjunction with forepersons and workers. Most workers in the selected countries did not receive any training in construction; they gained their skills only from experience. The number of workers involved in the project varied from 4 to 10. While Egypt has the most workers on a project (10), the work experience of these people was the lowest (5 years) in Egypt. In Indonesia and Nepal, workers in nonengineered buildings had the most working experience (i.e., more than 20 years).

4.2.11 General Issues

Many building owners and artisans in these countries had limited knowledge about proper construction methods, and they did not consider earthquakes as a potential hazard. Most of these owners paid more attention to the construction costs rather than building safety. Some of the artisans and masons had relatively little formal education or training on proper construction methods. They gained their skills only from guidance from the foreperson and their own experience. This may be one of the reasons why improper detailing on the building's structural elements occurred.

All of the selected countries had codes, or at least guidance, governing earthquake-safe construction. However, it is deplorable that the dissemination of the code or guidelines does not seem to be followed properly. Most of the workers did not know that their country had such a code or guideline. Hence, the quality of the work cannot be assured. Lack of awareness by local communities about earthquake-safe construction and earthquake hazard adds to the problem.

Lack of integrity in the building's structural elements, improper detailing on structural elements, and low quality of construction materials are typical problems found on many sites, probably because workers do not know how to build earthquake-safe buildings or, even if they have the knowledge, they cut back on quality due to intervention from owners to reduce construction costs.

In many cases, there is no particular construction quality control system from local or national authorities. Therefore, even though some countries have a particular system governing quality control, workers at the construction sites were never or seldom investigated or supervised by authorities.

Some quantitative figures for these parameters can be summarized as follows:

- Average area of nonengineered construction: Less than 200 m^2
- Average number of rooms: 5
- Average project duration: 11 weeks
- Average wall height–to–thickness ratio: 14.5
- Average mortar thickness: 19 mm
- Average plaster thickness: 19 mm
- Average concrete strength: 17 MPa
- Average number of workers: 6

4.2.12 Limitations

Some selected sites might not probably represent the actual conditions of nonengineered construction of their specific country, even though the sites were judged to be representative models from under-construction sites. In some countries like Turkey, however, existing buildings were selected as samples as there were no nonengineered buildings under construction in the survey sites. This might not represent the current condition of nonengineered practice in these countries, considering that the selected sites were built long time ago.

4.3 Efforts to Secure Home Safety in Indonesia

4.3.1 Housing and Building Laws and Standards

In Indonesia, houses are prone to various hazards, in particular (urban) fires, floods, earthquakes, landslides, and strong winds. Table 4.2 shows a comparison of the number of earthquake-damaged houses after several specific disasters.

To ensure the safety of houses, there are laws on housing and building, as well as on land division. Besides laws, there are guidelines and standards related to house and building safety. The hierarchy of laws in Indonesia is as follows (from the highest level to the lowest):

- Constitution
- Decree of People Council
- National law
- Government regulation
- Presidential regulation
- Ministerial regulation
- Provincial regulation
- City/district/municipality regulation

Ministerial regulation is used by ministers to exercise their mandate set by higher level regulations. Standards and guidelines by themselves are not laws, but when they are stipulated in these regulations, they become enforceable. To be operational, national laws need to be followed by implementing regulations in the form of government regulations. Building standards and guidelines are not part of laws; they become mandatory only when stipulated in laws and regulations. Government regulations include national-level and local-level regulations, as well as ministerial regulations.

The local governance system of Indonesia is divided into provincial governments and regency or city governments (shown in Figure 4.4 as of 2015). Apart from these, there are also subdistricts (Kecamatans) and administrative villages (Kelurahans), which are internal agencies of the regency and city governments. There are also natural villages, which perform traditional self-governance. Each local government has administrative organizations such as secretariat offices, administration offices, and technical organizations and agencies under the local chief executive and the vice local chief executive, and each body performs various assigned tasks. For example, BAPPEDA (Badan Perencanaan Pembangunan Daerah - Regional Development Planning Agency) is in charge of city planning and infrastructure development.

Table 4.2 Impact of Some Major Earthquakes in Indonesia

No	Event	Magnitude	Loss of Life (Persons)	Displaced Persons	Partially Damaged Houses (Housing Units)	Totally Damaged Houses (Housing Units)
1	Aceh earthquake (and tsunami)—December 26, 2004	M 9.4	110,000	700,000	57,137	69,932
2	Nias earthquake—March 28, 2005	M 8.6	850	40,000	71,891	12,010
3	Yogyakarta earthquake—May 27, 2006	M 6.8	5,700	100,000	260,000	154,000
4	Bengkulu earthquake—September 12, 2007	M 8.5	35	N/A	390,825	19,375
5	West Java earthquake—September 2, 2009	M 7.4	81	178,490	216,424	46,697
6	West Sumatra–Padang earthquake—September 30, 2009	M 7.6	1,117	N/A	249,833	114,797

Source: Pribadi et al. (2014)

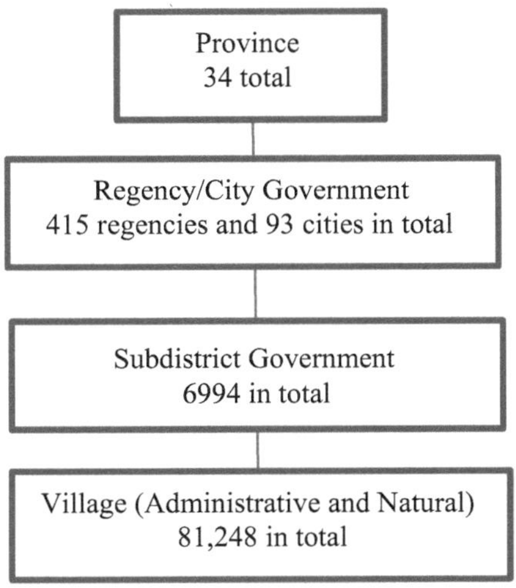

Figure 4.4 Local governance in Indonesia.

Related to buildings and houses, the pertinent ministry is the Ministry of Public Works and Housing, while at the local level (provincial , district, or city), there are local agencies with the mandate to oversee the implementation of laws on buildings and houses at the local level, such as the local Department of Public Works, Building, and Housing.

Local governments are responsible for implementing building and housing laws through local regulations. These governments (district/city/municipality) are responsible for establishing local government regulations, set up institutional support, and implement these regulations. Laws and regulations that stipulate building and house safety have already been systematically developed and put in place since 1998. Table 4.3 shows a list of most recent laws related to building and house safety in Indonesia.

Regarding building standards that are relevant to building and house safety, there are currently national standards (SNI - Standar Nasional Indonesia) that are published by the National Standardization Agency (BSN Bandan Standarisasi Nasional), technical guidelines and technical manuals that are issued by the Ministry of Public Works and Housing and endorsed by ministerial regulations, as discussed previously. Some examples of the most prominent standards and guidelines related to building and house safety are shown in Table 4.4.

The laws, regulations, and standards on building safety are implemented through local government authorities, which issue specific local regulations in accordance to Law 23 on local governance. Examples of local government regulation of buildings are the Bandung City Government Regulation No 5 (passed in 2010) and the Jakarta Province Government Law No. 7 (passed in 2010). Both of these regulations regulate buildings in each jurisdiction, including building permit processes, technical requirements, and standards. However, the establishment of local laws is still far from complete. Currently, out of 440 jurisdictions in Indonesia (district and city governments), there are only 246 (less than 49%) that have local government building regulations in place. In addition, many of the existing local building regulations are not yet in line with what is mandated by national regulations, particularly on technical requirements and local

Table 4.3 **Most Recent Indonesian Laws Related to Building Safety**

Year	Law No.	Title	Type	Content Related to Building Safety
2014	23	Local governance	National law	Role of local government in building administration, building permits, and occupancy permits
2011	1	Housing and residential areas	National law	Technical and administrative requirements related to planning, design, construction, and operation of housing and residential areas
2011	20	Multistory housing	National law	Requirement on the technical, administrative, and environment aspects of multistory buildings
2002	28	Buildings	National law	Building functions, technical and administrative requirements, process, and role of community and government
2005	36	Implementing regulation of Law 28 on buildings	National government regulation	Technical requirements related to building safety, standards and guidelines to be followed, and role of local government in ensuring control of building process and occupation (not detailed)
2007	26	Guidelines on building advisory team	Ministry of Public Work regulation	Establishment and role of advisory teams at the local government level for building implementation
2008	26	Technical requirements for fire protection for buildings	Ministry of Public Work regulation	Access and water for firefighting, fire escapes, passive and active protection, utilities, inspection, and control
2007	25	Guidelines on certification for building occupation permit	Ministry of Public Work regulation	Requirements and procedures for certifying building for occupation and role of local government
2007	24	Guidelines on building permit process	Ministry of Public Work regulation	Procedures and requirements for building permit issuance and role of local government for building control
2007	6	Guidelines on arrangement of building and its environment	Ministry of Public Work regulation	Detailed planning of urban development zoning , which includes subdivision plans, siting, green buffer zones, access, building mass forms, infrastructure, and lifelines
2007	5	Guidelines on the construction of multistory, low-cost residential buildings	Ministry of Public Work regulation	Related to building safety, technical requirements for structural design, fire, and electrical installation safety.
2006	30	Technical guidelines on building facility and accessibility	Ministry of Public Work regulation	Detailed technical requirements for building facilities, in particular for accessibility to different type of users (including the disabled), including assignment of specific technical standards
2006	29	Guidelines on building technical requirements	Ministry of Public Work regulation	Detailed technical requirements for building design, construction, and operations, including assignment of specific technical standards (design, loadings, materials etc.)

Table 4.4 **Examples of Building Standards in Indonesia**

Code Number	Year	Title	Remarks
SNI-1726	2012	Design Procedure for Earthquake Resistance Structure of Buildings	National standard, design code
SNI 2847	2013	Requirements for Structural Concrete for Buildings	National standard, design code
SNI 1729	2013	Design Procedure for Steel Structure for Buildings	Draft of national standard, design code
SNI 7833	2012	Design Procedure for Precast and Prestressed Concrete for Buildings	National standard, design code
SNI 2847	2013	Requirements for Structural Concrete for Buildings	National standard for material
SNI 2052	2014	Steel Reinforcement Bars	National standard for material
SNI 7973	2013	Design Procedure for Wooden Structure	National standard, design code
SNI 04- 0225	2000	Requirements for Electrical Installation	National standard
SNI 03-1736	2000	Design Procedure and Passive Protection for Fire Prevention in Buildings	National standard, design code

building advisory teams (MPW, 2014). On the issue of local building advisory teams, due to their limited capacity in terms of resources as well as comprehension of the system, most local government authorities are not yet able to implement this mechanism, excepting the Special Capital Region of Jakarta and the city of Surabaya.

4.3.2 Efforts to Improve Housing Safety

The implementation of building safety regulations and standards is still a major issue in Indonesia. The lack of capacity of local government authorities, in terms of both human and financial resources, has hampered the effective implementation of building administration. Compliance to building codes varies for nonengineered structures, while for formal and engineered structures, compliance ranges from relatively fair to good. This is particularly the case with the housing sector, where most housing stocks rely on informal or nonengineered structure approaches, which are less regulated in terms of technical requirements. In this regard, it is awareness and the motivation of homeowners, builders, and workers dealing directly with homeowners that dictate compliance to building codes and standards.

In many cases, inspection systems are more oriented toward administrative compliance rather than technical, so it is understood that homeowners are mostly left to their own understanding of good practices for building safer houses. In this regard, a massive campaign on the dissemination of building and house regulation, building codes and standards, and good practices is of paramount importance and should be prioritized. On the other hand, local government authorities on building and housing safety can lack the motivation and resources for this kind of effort; hence, a national program for supporting and motivating local governments to promote local building and house regulations, building codes, and standards needs to be a priority of the affected ministries.

Past experience shows that confined masonry residential housings in Indonesian rural and urban areas are more earthquake resistant than unconfined masonry structures, although many are still vulnerable due to lack of proper reinforcement detailing and mediocre building materials. There seems to be a large gap between the good practices provided in many technical guidelines produced by various organizations and the actual practices in the field. There have been efforts to produce more user-friendly documentation, including popular media such as videos and TV or radio broadcasts in order to fill the gap. Examples of this include an Australian Indonesian Facility for Disaster Reduction (AIFDR) and Earthquake Safer House (Rumah Aman Gempa), an initiative sponsored by the BNPB Indonesian National Disaster Management Agency in West Sumatra after an earthquake in 2009 (Romo-Murphy, 2013, information can be accessed from https://sites.google.com/site/rumahamangempa/), and a Japan International Cooperation Agency (JICA) sponsored initiative to support the Directorate of Building Administration, MPWH from 2007 to 2010, information can be accessed from http://www.jica.go.jp/indonesia/english/office/topics/safe_house.html)

User-friendly unofficial guidelines and manuals related to house-building safety were produced after major earthquakes by government organizations, the United Nations (UN) and nongovernmental organizations (NGOs), and prominent individuals. These publications include user-friendly documentation such as pocket books, flyers, and brochures (MPW, 2013, 2011, 2009, 2006; Boen, 2009). These are available widely and can be downloaded from the Internet.

Training about safer house construction has always been a popular type of intervention after disasters since the Indian Ocean earthquake and tsunami in 2004. Government agencies, UN organizations, international and national donors, and NGOs such as Save the Children, CARE, AusAid, Oxfam, RedCross, and BuildChange, as well as academics from various international and national universities, have been very active in providing safer house construction practices to local masons and artisans, as well as homeowners. The focus is mostly on good practices and skills for proper craftsmanship, including knowledge of appropriate building materials and techniques for repairing damaged houses. However, a serious question remains, about whether the knowledge gained from such training is retained and implemented by the recipients in their daily practices. A more important point is whether the knowledge and skills related to building safer houses are transferred and disseminated locally to other fellow workers and masons.

Retrofitting practices are rarely conducted in Indonesia, except for a few demonstration projects (Ando et al., 2009; Kusumastuti et al, 2008), and postdisaster reconstruction interventions. There are currently no regulations or building codes available that deal with retrofitting buildings, except for some guidelines and practical manuals developed by some NGOs and individual consultants, such as a guideline about the retrofitting of simple buildings and a related interesting video demonstration, which was uploaded to YouTube in 2013 (Boen, 2010, 2013)

4.4 Efforts to Secure Home Safety in Japan

4.4.1 Building Standard Law

Historically, the highest concern in Japan with regard to building safety is fire safety rather than earthquake safety. While urban fires have taken place almost every year, large earthquakes take place relatively infrequently—only once every dozen years, or sometimes even hundreds of years. Building regulations to reduce damage caused by fires was first established in Edo (now Tokyo), more than 300 years ago because buildings in Japan were made mainly of wood and paper (for partition), and dried grass (for roof and mat), which are quite flammable. Urban Building Law was enacted for the first time in 1919 to provide minimum requirements for structural safety. However, it did not consider seismic loads. It was amended after the Great Kanto Earthquake in 1923, which claimed the lives of more than 100,000, so that the buildings could withstand earthquakes. The amended law prohibited brick masonry, which had been introduced to Japan from Europe, because most of the brick buildings collapsed during the 1923 earthquake. Most cities in Japan were devastated by bombing during World War II. Then, there was a great need to reconstruct buildings orderly and efficiently. Thus, the Building Standard Law (BSL) was enacted in 1950 with the goal of safeguarding the life, health, and property of people by providing minimum standards concerning the site, structure, equipment, and use of buildings.

With regard to the structural requirements, the BSL has been revised several times following large earthquake disasters, reflecting the lessons and findings gleaned from these disasters. For example, the maximum spaces between tie hoops were reduced in the revision of 1971 to improve the ductility of reinforced concrete columns after the Offshore Tokachi Earthquake took place in 1968. *Shin-taishin,* a method that contains the current design principle against earthquakes, was introduced in 1981 after the Offshore Miyagi Earthquake in 1978. The principle of the new design method is as follows:

1. The building should not be damaged by snowfall, storms, or earthquakes of a medium scale, which can infrequently happen.
2. The building should not collapse or fall by snowfall, storms, or earthquakes of a large scale, which can very rarely happen.

In Japan, wooden houses are traditional and still common even today. In the case of the Great Hanshin-Awaji Earthquake in 1995, 5500 lives were lost immediately, but 88% of the total death toll was caused by people being crushed or suffocated, mainly due to collapse of the wooden homes. The remaining deaths, most of which were caused by fire, were attributed to the collapse of the homes, where people were trapped and then burned.

In the central area of Kobe City, where a detailed research on damage was conducted, 95% of the collapsed homes (mainly wooden ones) were built before 1981, when the seismic requirements of BSL were revised. Approximately 30% of these homes built before 1981 collapsed or were heavily damaged, while less than 10% of the homes built since 1981 did. It can be summarized that the older the homes, the more vulnerable they would be to earthquakes because they may not meet the current structural requirement and some of them may have deteriorated due to inappropriate maintenance.

It is generally believed that newly constructed buildings would meet the requirements of BSL, so buildings constructed after 1981 were probably safe enough against earthquakes. It should be pointed out, however, that a number of homes constructed after 1981 may not meet the current seismic safety requirements either, due to poor consideration on seismic safety and poor construction work. Aging of homes also imposes serious problems in seismic safety, as the wood that forms the usual structure of homes is susceptible to degradation due to inadequate maintenance. It should be also pointed out that technical requirements for ordinary wooden homes are rather simplified (i.e., with no need for detailed structural calculations) and conformity to the technical requirements would not be checked in building confirmations (i.e., permits) for practical reasons. Needless to say, though, the plan must be drawn only by qualified architects.

It is estimated that thousands of people will be killed by the collapse of vulnerable homes and buildings due to large earthquakes that will probably strike in coming years, apart from deaths by fires and others. Therefore, the Japanese government has decided to promote the retrofitting of vulnerable buildings, particularly homes. The current goal of retrofitting is as follows. In 2003, the total number of housing units in Japan was 47 million (excluding vacant homes), of which it was estimated 25% (or 11.5 million homes) would not meet the current structural requirements, making them vulnerable against earthquakes, according to the Ministry of Land, Infrastructure, Transport, and Tourism (MLIT). The Japanese government, therefore, set a goal to reduce the ratio from 25% to 10% by 2013 so the casualties caused by earthquakes would be drastically reduced. However, the target year has been extended to 2015, and it still seems difficult to achieve this goal by then. The next section introduces various policies and strategies to promote the improvement of building safety, particularly home safety, in Japan.

4.4.2 Policies and Strategies to Improve Building Safety

4.4.2.1 Act for Promotion of Retrofitting

The Japanese government enforced the Act for Promotion of Retrofitting immediately after the 1995 Great Hanshin-Awaji Earthquake Disaster in Kobe, aiming to promote the retrofitting of important facilities and existing vulnerable homes, based on the lessons learned from the disaster. According to the act, the owners of buildings that are used by a large number of people are urged to make efforts to have their buildings undergo seismic performance evaluation and seismic retrofitting. Concerning certified seismic retrofitting work under the act, provisions of the current BSL (other than those related to earthquake resistance) may not be retroactively applied. Without such a certificate, renovation works must meet the current BSL retroactively. Local governments must provide guidance, advice, and instruction concerning the retrofitting of buildings.

4.4.2.2 Housing Quality Assurance Act

The Housing Quality Assurance Act (HQAA) was enacted in 1999 to ensure housing quality, including structural safety; to protect the interests of home purchasers; and to resolute promptly disputes related to homes. Building suppliers are now responsible for the quality of homes for 10 years after delivery. The Housing Performance Indication System was introduced for nine quality components such as structural safety, fire safety, thermal and acoustic environments, and consideration for the aged and disabled. The HQAA was amended in 2006 such that building suppliers or sellers of new homes must have sufficient financial resources to execute defect warranty liability, either by deposit or insurance.

4.4.2.3 Amendment of BSL

In 2005, it was discovered that some newly constructed buildings did not meet the BSL requirements, and thus were very unsafe in case of earthquake, due to illegal fabrications of structural calculations by architects with the intent to reduce construction costs. The BSL was amended in 2006 to strengthen the building confirmation (permit) and inspection system after this scandal. For specific high-rise buildings, peer review process for structural calculation by the third party was introduced. A guideline for building confirmation, interim inspection, and final inspection was developed for stricter quality control. In order to ensure professional design capacity for specific buildings, the Kenchiku-shi (Architect) Law was also amended to create two new classifications: "1st-class structural design Kenchiku-shi" and "1st-class equipment design Kenchiku-shi." They must have more than 5 years of experience and complete a specified training course.

4.4.2.4 Financial Assistance for Those Who Retrofit Their Buildings and Homes

Immediately after enactment of the 1995 Act for Promotion of Retrofitting, the national government initiated a program to subsidize those who retrofit their buildings, which are either collective homes or specific buildings that are used by a large number of people. The detached houses were included in the subsidy program in 2002, and since then, it has been revised several times so that the subsidy ratio was increased and the number of targeted homes increased.

The current subsidy program can be summarized as follows. The national government subsidizes the local government, which subsidizes those who conduct the seismic performance assessment of homes or buildings constructed before 1981. In this case, a third of the cost for assessment must be borne by the building owner, while a third must be borne by the local and national governments apiece. The national government also subsidizes the local government, which subsidizes those who retrofit their homes or buildings. In this case, the owner must bear 77% of the cost for retrofitting, while the local and national government would bear 11.5% apiece. If the home or building in question is located along emergency or evacuation roads, the subsidy distribution would change to one-third by the owner, one-third by the local government, and one-third by the national government. If the local government provides more subsidy, then the owner's share decreases accordingly, but the subsidy of the national government remains the same.

In addition to the subsidy program, the national government has revised the taxation program so that the income tax would be deducted by 10% for those who retrofit their homes, with a maximum of 200,000 yen. Low-interest loans have been made available from the Housing Finance Corporation for retrofitting. The Japanese government established a unique insurance policy against earthquakes. It is a kind of reinsurance system by the Japan Insurance Corporation, which was established only for earthquake insurance by private insurance corporations and the Japanese government. It is heavily subsidized by the national government, which may cover up to 95% of the premiums, when the scale of the disaster is highest. Homes that were constructed after 1981 or retrofitted may enjoy premiums discounted by 10–20%.

4.4.2.5 Development of Technologies to Promote Retrofitting

The national and local governments of Japan are currently encouraging technology development to secure housing safety against earthquakes, particularly for retrofitting. They include affordable and practical retrofitting methods, securing safe space in a home, and anchorage of furniture. The Japan Building Disaster Prevention Association published *The Standard for Seismic Evaluation* and *Guidelines for Seismic Retrofitting*. It also developed a "Simplified Seismic Diagnosis" for wooden houses. By answering a series of simple questions in the questionnaire, people can easily understand whether their home is safe or not. Once it turns out that their home is not safe enough, they would be encouraged to consult with experts for a more detailed seismic assessment.

4.4.2.6 Effectiveness and a Way Forward

Have these acts and related policies really promoted the retrofitting of vulnerable homes and buildings? MLIT has reported that as of March 2009, about 38,300 housing units and 4800 specified buildings had received subsidies from both the national and local governments for retrofitting since 1996. These numbers are very low, compared with the numbers of homes and buildings that do not have sufficient structural safety. As mentioned earlier, 11.5 million homes were estimated to be unsafe in 2003, and the number was not reduced drastically. In addition, a number of homes constructed after 1981 would not meet the BSL requirements, due to deterioration after 20–30 years since the construction. Hence, the total number of unsafe homes might be increasing (or at least remain unchanged), rather than decreasing.

4.5 Conclusion

According to joint research on nonengineered construction in developing countries, all of the selected countries have certain codes or guidelines on building construction in place, but these codes seem to have only minimal impact on practices in the field. The major types of nonengineered construction are unconfined masonry, confined masonry, and concrete frame with masonry infill. Some similarities were found, such as in design intervention, building owners, availability of materials and workers, supervision, construction tools, foundation type, wall thickness, type of plaster/mortar of cementing agent, type of cement for concreting, and training obtained by workers.

Most of the nonengineered buildings in developing countries, technically, are not properly constructed. The survey results reveal that most nonengineered buildings do not pay attention to the detailing, quality of materials, configuration, and quality of construction. Recommendations from the joint research on nonengineered construction in developing countries are therefore summarized as follows:

- Guidelines and codes on earthquake-safe construction should be simpler and disseminated widely and properly to homeowners, builders, and artisans.
- Awareness-raising programs need to be conducted to improve the community awareness in the community.
- Strengthening is needed for poor-quality existing structures. Some elements of the buildings, such as walls, connections, columns, and beams, need to be strengthened in order for buildings to react well when subjected to future earthquakes.
- Quality control or inspection is needed from local authorities in order to control the implementation of building guidelines and codes and good construction practices.
- A certification program to masons, artisans, and forepersons will ensure the quality of workers.
- Experts should communicate more often with workers and contractors on topics related to safe construction. This will reduce the gap in knowledge between experts and field practitioners.

The current Indonesian regulations and standards for buildings seem to be already quite developed, although improvements are still needed, particularly to cater to the needs of the informal and small-scale construction industry, which is the main supplier to the housing sector, involving mostly semiskilled builders and artisans. However, there is a big gap between developing information and translating it into easily understood practical materials to local government officials and communities, including local builders all over the country. It is a great challenge to try to increase the understanding of the need for safer houses with the huge population of this country, dispersed in a vast geography of more than 13,000 islands.

Action plans to reduce this gap should include important efforts to develop the local government capacity in building proper building administration schemes in accordance with the requirements of the laws, and also in developing the local capacity to conduct massive dissemination efforts to the local communities, which include continuous training schemes for local masons and artisans. More efforts to produce and distribute dissemination materials need to be considered and implemented. Finally, a monitoring and evaluation scheme for effective tracking of the efforts at the national and local levels should be put in place if improvements are to be made with proper accountability.

In Japan, it is generally believed that all newly constructed homes meet the technical requirements of the BSL, and therefore are safe enough against earthquakes. However, the vulnerable old wooden houses that still remain keep cities vulnerable; it is estimated that the majority of deaths from earthquake will be caused by the collapse of such houses. Therefore, the Japanese government has been making every effort to promote the retrofitting of these houses via various policies, such as technical development, financial assistance, insurance, and education. It is found, however, that even these strong supporting policies of the Japanese governments are failing in motivating people for home safety.

References

Ando, S., Subedi, J., Fujieda, A., & Nakamura, H. (Eds.). (2009). *Reducing Vulnerability of School Children to Earthquakes: A project of School Earthquake Safety Initiative (SESI).* Kobe, Japan: United Nations Centre for Regional Development (UNCRD).

Boen, T. (2009). *Constructing seismic-resistant masonry houses in Indonesia.* UNCRD: Kobe, Japan.

Boen, T. (2010). *Retrofitting simple buildings damaged by earthquakes.* World Seismic Safety Initiative (WSSI) & UNCRD: Kobe, Japan.

Boen, T. (2013). A video film: *Guideline on retrofitting simple masonry house.* JICA-MPW-WSSI, www.youtube.com/watch?v=f6TNCHR0H34.

Coburn, A., & Spence, R. (2002). *Earthquake protection* (2nd ed.). John Wiley and Sons. West Sussex, England.

Grundy, P. (2007). *Report on International Symposium on Disaster Mitigation and Community Based Reconstruction, Yogyakarta, August 8–9, DRC.* Monash University.

International Association for Earthquake Engineering (IAEE). *Guidelines for Earthquake Resistant Non-Engineered Construction,* NICEE-IIT, Kampur, India.

Kusumastuti, D., Pribadi, K. S., & Rildova (2008). *Reducing Earthquake Vulnerability of Non-Engineered Buildings: Case Study of Retrofitting of School Building in Indonesia* 14th World Conference on Earthquake Engineering (14 WCEE), October, Beijing, China.

Ministry of Public Works (MPW). (2006). *Technical guidelines for earthquake-resistant house.* (http://ciptakarya.pu.go.id/dok/hukum/pedoman/pedoman_teknis_bangunan_tahan_gempa.pdfhttp://ciptakarya.pu.go.id/dok/hukum/pedoman/pedoman_teknis_bangunan_tahan_gempa.pdf).

Ministry of Public Works (MPW). (2013). *Practical guidelines to repair damaged house by earthquake* Research Institute on Human Settlement.

Ministry of Public Works (MPW). (2014). *Monitoring and evaluation of local building regulations in Indonesia* Directorate of Building Administration, Prospera Consulting Engineers.

Ministry of Public Works and Housing (MPWH). (2014). Monitoring and evaluation of local building regulations in Indonesia. Presentation Material from Prospera Consulting Engineers to the Directorate of Building and Environmental Administration, MPWH, https://www.academia.edu/9771340/Presentasi_Monitoring_dan_Evaluasi_Peraturan_Daerah_Bangunan_Gedung_di_Indonesia.

Ministry of Public Works (MPW)-JICA. (2009). *Minimum requirements for safer houses: confined masonry, timber structure* Pocket book and poster, Jakarta, http://www.jica.go.jp/indonesia/english/office/topics/safe_house.html.

Narafu, T., Ishiyama, Y., Okazaki, K., Ando, S., Imai, H., Pribadi, K. S., Dixit, A. M., Ahmad, N., Ali, Q., & Turer, A. (2009). A proposal for a comprehensive approach to safer non-engineered homes. *Journal of Asian Architecture and Building Engineering*, 315–322, 2010.

Okazaki, K. (2011). *Incentives for Safer Buildings – Lessons from Japan* Global Assessment Report on Disaster Risk Reduction 2011. May, United Nations Geneva, Switzerland.

Okazaki, K., & Saito, T. (2012). *Seismic Risk Perception Concerning Non-engineered Houses* 15th World Conference on Earthquake Engineering (15 WCEE), Lisbon, Portugal, September.

Okazaki, K., Pribadi, K. S., Kusumastuti, D., & Saito, T. (2012). *Comparison of Current Construction Practices of Non-Engineered Buildings in Developing Countries* 15th World Conference on Earthquake Engineering (15 WCEE), Lisbon, Portugal, September.

Romo-Murphy, E. (2013). Evaluating Disaster Preparedness in West Sumatra through Media, GRF Davos Planet@Risk, Volume 1, Number 1, October 2013.

Pribadi, K. S., Kusumastuti, D., Sagala, S., & Wimbardana, R. (2014). *Post-disaster housing reconstruction in Indonesia: Review and lessons from Aceh, Yogyakarta, West Java, and West Sumatra earthquakes.* Disaster Recovery: Used or Misused Opportunity (pp. 197–224), Springer, Tokyo, Japan.

Urban Regulation and Enforcement: A Challenge

5

Akhilesh Surjan[1], Atta-ur-Rahman[2], Gulsan Ara Parvin[3], Rajib Shaw[4]
[1]Associate Professor, Faculty of Engineering, Health, Science and the Environment, Charles Darwin University, Darwin, Australia; [2]Associate Professor, Institute of Geography, Urban and Regional Planning, University of Peshawar, Peshawar, Pakistan; [3]Researcher, Graduate School of Global Environmental Studies, Kyoto University, Kyoto, Japan; [4]Professor, Graduate School of Global Environmental Studies, Kyoto University, Kyoto, Japan

5.1 Introduction

Asia is urbanizing fast, especially in recent years; however, the quality of life available to most urban residents in the continent's developing nations remains a matter of concern. The pattern, form, and scale of urban development in Asia often receive criticism, and not without justification. Issues such as traffic congestion, severe air pollution and low visibility, and lack of open spaces are regularly featured in newspaper articles and other forums. Urban areas can be looked at as bundles of properties of different design types, sizes, construction materials and technologies, and levels of quality. Be it a house or a mansion, a shopping mall or an office complex, an informal crowded bazaar or an art gallery, each of these edifices must be built and operated subject to certain rules and regulations, although the level of detail may vary from place to place.

From time immemorial, humans have demonstrated certain principles upon which they founded and expanded their settlements. Respecting the natural environment and climatic conditions of the area in question was one of the first such tenets that guided locational choices for settlement. Even the oldest civilizations were paying attention to hill slopes, natural drains, floodplains, low-lying areas, and fertile land and were basing settlements in a manner that confronted nature as little as possible. The zoning of activities, housing construction material, and design in coastal areas, hilly areas, floodplains, and other geographical areas reflected a careful consideration of natural topography and ecosystems. As the populations of these settlements grew, compromises were made, but still with the goal of not confronting or altering nature.

The advancement of city planning and management was exhibited in civilizations such as those in the Indus Valley, ancient Egypt, and Mesopotamia. Urban forms, local governance, civic infrastructure and services, community facilities, road network, and drainage systems were key features unearthed from archeological findings from cities of this era. Ancient Greek and Roman cities also incorporated public amenities and defense into their city planning. Walls surrounded these cities to protect them from invaders and also to define city boundaries. Aqueducts demonstrated engineering efforts to bring water to the city from faraway areas and provide sanitation. Social, economic, and political might was reflected throughout these cities.

Urban Disasters and Resilience in Asia.

5.2 Influence of Disasters on Settlement-Planning Norms

Historically, the affluent and/or ruling class has been the main influence on city planning. The safety and protection of territories and assets from possible invasion always remained a key concern. The construction of high walls all around settlements is one of the common features seen across continents. It is difficult to ascertain if disasters and their prevention were major elements of city planning in ancient times, but wide roads, sanitation, and aesthetic features suggest that provision of healthy environments and prevention of disease were considered, at least to some extent.

Fires have had a disastrous impact on cities for centuries. Major incidents of fire, often referred to as "great fires," devastated many cities, including London, Tokyo, Pittsburgh, Montreal, Chicago, Boston, Vancouver, and Kobe. Indeed, the great fire of Rome in 64 A.D., which destroyed 70% of the city, became infamous in history due to the weak leadership response of Nero, the ruler of the Roman Empire at that time, giving birth to the popular expression, "fiddling while Rome burned." This was probably the first documented account of an urban fire that caused widespread devastation; however, whether this resulted in change in urban planning norms or building construction techniques in that era cannot be confirmed.

Tokyo, which is the largest megacity of the world today, has been shaped and reshaped through disasters through the years. In 1657, the great fire of Meireki destroyed two-thirds of the city (then known as Edo), with an estimated loss of 100,000 lives. The Japanese military government of that time, the shogunate, made many changes to its city planning system during the reconstruction period of the two years following the blaze. Street widening, relocating temples and shrines to river banks because they used fire during prayers, reorganizing some districts, and leaving large spaces between castles and other areas to act as firebreaks were some of the features incorporated after the great fire. Tokyo's trade and commerce district was restored with careful consideration of fire protection and is considered to have lifted the overall national economy as a result.

However, in 1923, the Great Kanto Earthquake once again devastated Tokyo and its adjoining regions, killing more than 100,000 people. This earthquake of 7.9 magnitude occurred around lunchtime on September 1, when many people were using fire for cooking their meals. Fire quickly spread to several locations and engulfed major portions of the city in flames. The impact of this disaster was huge, but the nation took it as an opportunity to rebuild the city and make it even better. The reconstruction of Tokyo incorporated the construction of modern road networks, public services, and trains. One noteworthy aspect of the city planning that took place after the earthquake was the provision of parks as refuge areas. Construction standards also were revisited, and in particular, public buildings were built to stricter standards than private buildings, with the idea that they could be used as temporary accommodation for refugees. "Public utilities, including running water, and waste removal services were slow to recover. As a result, sanitary conditions deteriorated rapidly, particularly affecting those living in shelters and barracks. Water-borne infectious diseases (especially typhoid and dysentery) spread rapidly" (Nagashima, 2004). Urban engineering projects after the earthquake paid particular attention to public health and hygiene.

In 1960, Japan designated September 1 as Disaster Prevention Day. "On Disaster Prevention Day emergency drills organized by local governments are held throughout the country. Some of these drills consist of ducking under desks to escape falling objects and evacuating from buildings. At many elementary and middle schools, September 1 is the first day of school after summer vacation. So a lot of schools carry out an evacuation drill as part of the back-to-school ceremony" (Web Japan, n.d.).

Another massive earthquake, striking the Japanese city of Kobe more than seven decades later, further influenced city planning and construction practices in the country. The damage from this earthquake cost 2.5% of Japan's gross domestic product (GDP) at that time. Massive damage to buildings and key infrastructure due to the earthquake was followed by fire, which affected large portions of the city. Structural damage patterns were visible in high-rise buildings built before the modern building code of 1981, and those edifices suffered seriously, while high-rise buildings constructed following this code had very little damage. Traditional houses in Kobe used heavy roof tiles as a response to the typhoons that frequently hit the area, but this feature caused those houses to suffer the most damage via pancake collapsing that took place during the earthquake. This taught policy makers to consider a multihazard approach when planning and designing human habitats from then on. The Kobe earthquake also highlighted the need for retrofitting of key infrastructure to the latest building and construction codes.

Many countries in the Asia-Pacific region have experienced disasters of unprecedented scale in recent decades. Disasters have not only become more frequent and intense, but they also affect larger populations living in the urban areas of this region. For instance, every major disaster in Japan has led to revisiting urban planning, building safety and construction codes, policies of key lifeline infrastructure, and use of recreational spaces as places for people to congregate during emergencies. To reduce disaster response time and enhance disaster resistance, the Japanese government designated special disaster response routes; further increased the space between buildings, further reinforced public building and infrastructure construction practices, maintained necessary supplies in a decentralized manner, redistributed and reinforced fire and response agencies, and other tactics. Volunteerism as a major form of public engagement, recognizing soft measures such as - raising community awareness, promoting self-help and mutual help, etc. as important as technological measures, community- and school-based disaster recovery, and building a culture of prevention were just some of the vital outcomes of disasters in Japan.

In other countries, such as Indonesia, Vietnam, India, Bangladesh, Nepal, Pakistan, Thailand, the Philippines, Iran, and Myanmar (Burma), national and provincial governments have created special departments and ministries focusing on disaster risk reduction (DRR). One of the major duties of these agencies is to review building construction practices and introduce planning rules, regulations, codes, laws, and bylaws to reduce the loss of lives and assets after disasters strike. One of the greatest problems faced by developing countries, even today, is diluted implementation of these legislative tools in real life. Gradually, efforts are being put in place in many countries to address the enforcement of rules more strictly, but there is a long way to go due to severe structural problems within the system of governance that are there even in calm times. The following section of this chapter will explore some examples from the Asia-Pacific region to discuss the challenges and possible solutions.

5.3 Disaster Management in India

India is large in population, area, and geographical diversity, and is exposed to a wide variety of hazards, including drought, floods, cyclones, earthquakes, heat waves, and landslides. Distribution of relief, both in cash and kind, following a major disaster has been the prevalent form of practice by governments from the pre-independence era (i.e., before the late 1940s) until recently. Disasters were considered to be primarily affecting rural areas (including agricultural land), and thus, until 2001, the Agriculture Ministry remained a nodal agency for disaster-related affairs in India.

The Orissa Super Cyclone of 1999 and Bhuj Earthquake of 2001 are two relatively recent major events that catapulted a paradigm shift in disaster management in India. Multihazard-prone zones have been identified, and a multidisciplinary, multisectoral, and multistakeholder approach has been instituted for DRR. The Indian government enacted the Disaster Management Act in 2005, which led to the establishment of the National Disaster Management Authority (NDMA), headed by the prime minister; State Disaster Management Authorities (SDMAs) headed by the states' chief ministers; and District Disaster Management Authorities to spearhead and implement a holistic and integrated approach to disaster management in India (NDMA, 2013). As an apex organization, NDMA spearheads the formulation of policies, plans, and guidelines on disaster management, as well as dealing with capacity building. The National Policy on Disaster Management was also approved by the government in 2009 (NDMA, 2009). This policy includes a section entitled "Techno-Legal Regime," which elaborates specific aspects of planning and implementation, as detailed in Table 5.1.

In India, building regulations and bylaws fall under the purview of state legislation; national laws are only advisory in nature. State governments, with the help of local bodies, formulate the rules and regulations, and it is the responsibility of local bodies to enforce them. Under various planning and development related acts, state governments must notify the residents of areas for planned growth in advance. The model town and country planning act was formulated in 1960 and revised in 1985 to provide guidelines to states. The *Vulnerability Atlas of India,* in which earthquake, cyclone, and flood hazard maps were prepared for every province, was published in 1997. In every district, the risk to various types of buildings from varying intensities of these three hazards was clearly presented in this atlas (MHA/BMTPC, n.d.). Despite impressive progress made in recent decades, enforcement of DRR related spatial planning and construction legislation in India remains a challenge.

5.4 Disaster Management in Japan

Japan has been the site of disaster incidents for centuries. The cases of earthquakes and large fires in Tokyo and Kobe were discussed previously; this section will specifically delve into the Great East Japan Earthquake and Tsunami (GEJET) of 2011 and its influence on Japan's DRR efforts. On March 11, 2011, a massive earthquake of magnitude 9.0—on the Richter scale—occurred in East Japan. "A half-hour later, a tsunami of unprecedented force broke over 650 kilometres of coastline, toppling

Table 5.1 **Salient Features of the Techno-Legal Regime for Disaster Response in India**

Aspects Covered	Features	Stakeholders Involved
Municipal regulations	Revisiting development control regulations; building bylaws and structural safety features Periodical review of regulations to identify safety gaps and suggest suitable modifications to align them to building codes Undesirable practices compromising safety from disasters being addressed in the regulations Appropriate compliance mechanisms to prevent construction in unsuitable areas without necessary safeguards Introduction of suitable regulations for rural areas Suitable financial incentives to local bodies for the preparation of appropriate regulations Sensitization of governmental organizations at all levels, including local authorities and the community at large	Bureau of Indian Standards National, provincial, and local government bodies Communities
Land-use planning	Carrying out analysis of environmental and hazard data to help with the formulation of alternative land-use plans for different geographic and administrative areas with a holistic approach Applying such analysis specifically to megacities, metros, and high-density urban settlements for safer location of habitats and other critical facilities Reviews of master plans and their compliance being the priority responsibility of the states At a macro level, land-use planning based on an inventory database of various activities to take place in these locations Future land-use in urban areas to be assessed based on estimated intensity of development	Central ministries and departments; provincial governments; scientific institutions

Continued

Table 5.1 Salient Features of the Techno-Legal Regime for Disaster Response in India—cont'd

Aspects Covered	Features	Stakeholders Involved
Safe construction practices	Ensuring safe construction of new buildings and retrofitting of selected older lifeline buildings as per earthquake guidelines Reexamine the design and specifications of houses that government provides via development and welfare to beneficiaries, to ensure safety Building codes to be updated every five years as a requirement and also put in the public domain Observance of national building codes made mandatory in all state/ municipal building bylaws Training of engineers, architects, small builders, construction managers, and artisans to be intensified at both the state and district levels Safe schools and large-capacity hospitals, national monuments, and critical lifeline buildings to be regarded as a national priority Government-funded school buildings and hostels shall be designed with earthquake-resilient features and fire safety measures	National, state, district governments; municipal agencies; Bureau of Indian Standards; education departments; public infrastructure agencies; health departments
Compliance regime	Putting in place a sound compliance regime, with binding consequences, to ensure effectiveness of techno-legal and techno-financial provisions Monitoring, verification, and compliance arrangements put in place at the national and state levels Adoption of best management practices such as self-certification, social auditing, and external compliance regimes, including audits conducted by professional agencies to be promoted through information technology These practices done in consultation with various stakeholders and knowledge institutions for adoption after due trial and validation	National and state government agencies; local stakeholders
Enforcement	State governments ensuring the enforcement of techno-legal and compliance systems by establishing an effective mechanisms under the provisions of the act	State governments

Source: National Policy on Disaster Management 2009 (NDMA, 2009).

sea walls and other defences, flooding more than 500 square kilometres of land, and washing away entire towns and villages" (Ranghieri & Ishiwatari, 2014).

Suppasri et al. (2013) provides a detailed account of the lessons that Japan learned from GEJET. According to the authors, this mega earthquake was the greatest in Japanese history—it actually destroyed most of the structural countermeasures that the nation had put in place to prevent tsunamis, such as breakwaters, seawalls, water gates and control forests. Tsunami breakwaters had successfully protected the Tohoku region of Japan in the past, but the 2011 quake produced a tsunami that exceeded the design capacity, causing major damage to breakwaters and severely affecting the settlements of Kamaishi and Ofunato city, although "it helped to reduce the impact of the tsunami (both tsunami height and arrival time) on the cities" (Suppasri et al., 2013). Similarly, seawalls that were put in place to protect settlements from high tides and typhoons were completely destroyed, as tsunami overflows were greater than 3–4 m. The tsunami gate at Fudai village, which was constructed to close the river mouth in case of tsunami, could successfully protect the village. But in Minami-Sanriku town, the seawalls and tsunami gates were no help, and 95% of the town was destroyed. Many firefighters lost their lives while battling to close the tsunami and seawall gates. As a result of this tragedy, the Japanese government is planning to create a new system that could be operated remotely.

In addition, control forests were wiped out by 20-m-high tsunamis, and this caused Rikuzentakata city to be devastated. However, the control forests survived in Ishinomaki city, and together with breakwaters, that reduced the negative consequences to that city to some extent. Suppasri et al. (2013) also stated that although the earthquake-resistant building code was revised in 1981 and 2000, tsunami loading was not taken into account, and hence several high-rise reinforced concrete and steel-reinforced concrete buildings were overturned. Many public buildings that were designated as evacuation shelters also did not remain intact due to unexpectedly high tsunami waves and runup. These researchers concluded that modern technology can offer structural protection from the type of huge tsunamis that occurs every 500–1000 years, but doing so is highly impractical due to the budget and time involved in erecting such structures. Instead, land-use planning offers a much more viable solution. It is also important to understand that the structural strength of tsunami barriers can (and should) be further improved, and they can reduce damage, but they should not be considered a guarantee of complete protection. Therefore, awareness of the imminent arrival of catastrophic events and evacuation to high grounds after receiving early warnings should not be foregone. The paper recommends adapting a good balance of soft and hard measures to save lives. Japan has demonstrated that its learning from disasters and gradually incorporating countermeasures into policies, plans, and actions at all levels has helped to build the country's resilience. "The Meiji-Sanriku Tsunami of 1896 killed 40% of the population in the affected zone, whereas GEJET claimed 4%"; this offers great lessons for the world community living in hazard-prone zones (Ranghieri & Ishiwatari, 2014).

5.5 Garment Factory Collapse in Bangladesh

In April 2013, Rana Plaza, an eight-story garment factory in Dhaka, the capital city of Bangladesh, collapsed, claiming 1100 lives. It is not difficult to conclude that this

Table 5.2 Recommendations for Safe Factories

Short-term actions	Rapid assessments of structural, electrical, fire, and worker safety Identification and prioritization of risks Preparation of approaches to address identified risks Management sensitization about companywide safety and lean principles
Medium-term actions	Fixing problems with structural, electrical, fire, and worker safety immediately Comprehensive disaster management planning for ongoing assessments, surveillance, maintenance, preparedness, and response Training of managers and orientation of key personnel to disaster prevention and safety measures Initiating a factory improvement program that helps identify and eliminate "7 waste"
Long-term actions	Detailed assessments (including nondestructive testing for structural safety) Retrofitting as needed Training of task forces based on the comprehensive disaster management plan Building the awareness of all involved personnel based on the comprehensive disaster management plan Consolidating gains and starting companywide improvement efforts

Source: STS/RBC (n.d.).

building was neither constructed not maintained well, and structural safety norms must have been flouted. "High death toll from fire incidents reflects not only substandard buildings but poor emergency procedures, inadequate and blocked fire exits, and overcrowded workplaces" (Maher, 2012). The ready-made garment industry, one of the largest export industries in the country, employs mostly women, mainly from low-income families, as its workforce. A good number of ready-made garment industrialists also have a strong influence in government, so they have the power to suppress the rising tide of voices advocating for minimum wages and health and safety standards for workers. SEEDS Technical Services and Rajesh Bheda Consulting conducted a joint study and proposed a "Safer Factories Initiative" that would integrate workplace safety with DRR (STS/RBC, n.d.). The study's recommendations are presented in Table 5.2.

5.6 Urban Planning and Disasters

The root of modern urban planning can be traced to the human suffering that has been prevalent in the rapidly industrializing towns of the Western world. The poor socioeconomic conditions of workers, together with their substandard living conditions, became a serious public concern. Health, convenience, and aesthetics were the three pillars of urban planning, and they remain so, but there are added challenges facing cities in developing countries. Most, if not all, Asian cities of the 21st century are enjoying enormous opportunities, but they also face numerous challenges that are complex and have no simple solution. Cities are experiencing disasters more often than previously seen, and climate change has added another twist to this already-grave situation.

Urban local bodies, development authorities, town and country planning organizations, and municipal agencies are some of the key stakeholders responsible for planning and implementing cities' path for development. Undoubtedly, the process of planning happens in a highly contested space where economic, political, and historical factors play specific roles in shaping the future of a city and its adjoining region. Traditionally, urban and regional planners envisage how a city and area will look like after, say, 10–20 years, or even longer. However, the vibrant and extremely dynamic nature of growth, coupled with haphazard development in cities, jeopardize development and create a space full of risky locations.

The informal sector plays an important role in the economy of developing countries and is considered one of the largest contributors to unplanned development in cities. However, one needs to look at the problem from a different perspective, where rural areas are deprived of livelihood opportunities, which forces large populations to migrate to urban areas in an attempt to improve their lives. If spatial planning does not consider this painful reality, it will continue to proliferate slums, ghettos, and informal sector.

5.7 The Enforcement Challenge

The enforcement of rules, regulations, and laws is always a matter of public discourse. No plan that affects a city can be completed without winning the full confidence of at least most of its residents. Planning is a process that involves contesting and competing challenges, which often leaves many dissatisfied. Strictly adhering to rules and regulations is not always possible. Therefore, city councils considers each and every case within the larger framework of planning, and thus the approval process can be long and very complicated. In order to avoid undue delays and create a seemingly comfortable habitat, people tend to circumvent enforcement procedures—frequently to their eventual detriment.

Local government bodies are often underemployed and have a rather limited capacity to pay even one visit to all the construction sites in their jurisdictions. And the public at large, although aware of local environmental risks, can miss the big picture of creating a safer environment for themselves and future generations. Engineers, architects, artisans, and construction managers do not often communicate with each other well, perhaps because they lack a common language. Most of them also lack formal training in hazard-resistant construction. Thus, the amalgamation of these circumstances lead to having to rebuild unsafe structures over and over again; often, it is only after disaster strikes that people think to correct this process—but by then, they have paid a very heavy cost.

The private business sector, as well as other nongovernmental entities, play a major role in driving the economy and social change, but so far, they are not considered participants in urban enforcement. Incentive-based mechanisms also have not been sufficiently explored as a means of building a culture of risk reduction. Housing finance companies and infrastructure financing agencies in developing countries often do not employ effective risk reduction measures.

5.8 If There Is a Will, There Is a Way

Urban law enforcement agencies often receive great criticism for being inefficient and easily swayed by the powerful players. In many ways, this belief cannot be dismissed.

In the context of developing countries, there exists a large gap in even the basic level of awareness of appropriate information among the general public. Schools and higher-education institutions play an important role in imparting such knowledge, but their reach is too limited to make a substantial enough change. National and provincial governments often advocate for better enforcement, but more often than not, they do little to improve enforcement mechanisms at the local level.

Local agencies seriously need encouragement, empowerment, and support in order to be effective in enforcement. Training of masons and construction workers could be a major factor in creating a safer-built habitat. Nongovernmental organizations (NGOs), community-based organizations (CBOs), and the private sector should work hand in hand with local governments to devise innovative strategies and incentive mechanism to make the change possible. If "Rome was not built in a day," then building a culture of prevention may also take time. But with escalating risks due to the connections between disaster, environment, and climate, rapid upscaling of efforts from multiple actors, across multiple disciplines, and at a wide variety of scales is required to build cities that can be temples of the modern world. Otherwise, the goal for creating a sustainable and safe human society will remain a distant dream.

References

Maher, S. (2012, November undated). *Clean Clothes Campaign—Hazardous Workplaces: Making the Bangladesh Garment Industry Safe*. Retrieved June 28, 2015, from http://www.cleanclothes.org/resources/publications/2012-11-hazardousworkplaces.pdf.

Ministry of Home Affairs and Building Materials and Technology Promotion Council (MHA and BMTPC). (n.d.). Building a New Techno-Legal Regime for A Safer India. New Delhi: Ministry of Home Affairs and Building Materials and Technology Promotion Council, Government of India.

Nagashima, T. (2004). Sewage disposal and typhoid fever: The case of Tokyo 1912–1940. *Annales de démographie historique*, 2(108), 105–117.

National Disaster Management Authority (NDMA). (2009). *National Disaster Management Policy, 2009*. New Delhi: NDMA.

National Disaster Management Authority (NDMA). (2013). *Evolution of NDMA;* September 26. Retrieved July 3, 2015, from http://www.ndma.gov.in/en/about-ndma/evolution-of-ndma.html.

Ranghieri, F., & Ishiwatari, M. (2014). *Learning from megadisasters: Lessons from the Great East Japan Earthquake*. Washington, DC: World Bank.

SEEDS Technical Services (STS)/Rajesh Bheda Consulting (RBC). (n.d.) Risk *reduction and emergency preparedness in the apparel industry*. New Delhi: SEEDS Technical Services and Rajesh Bheda Consulting.

Suppasri, A., Shuto, N., Imamura, F., Koshimura, S., Mas, E., & Yalciner, A. C. (2013). Lessons learned from the 2011 Great East Japan Tsunami: Performance of tsunami countermeasures, coastal buildings, and tsunami evacuation in Japan. *Pure and Applied Geophysics*, *170*, 993–1018.

Web Japan. (n.d.). Disaster Prevention Day. Retrieved June 12, 2015, from http://web-japan.org/kidsweb/explore/calendar/september/bousai.html.

Expanding Coastal Cities: An Increasing Risk

6

Akhilesh Surjan[1], Gulsan Ara Parvin[2], Atta-ur-Rahman[3], Rajib Shaw[4]
[1]Associate Professor, Faculty of Engineering, Health, Science and the Environment, Charles Darwin University, Darwin, Australia; [2]Researcher, Graduate School of Global Environmental Studies, Kyoto University, Kyoto, Japan; [3]Associate Professor, Institute of Geography, Urban and Regional Planning, University of Peshawar, Peshawar, Pakistan; [4]Professor, Graduate School of Global Environmental Studies, Kyoto University, Kyoto, Japan

6.1 Introduction

Water is not just an everyday need for humans, but also one of the most essential components of their survival. Throughout history, humans have found bodies of water, including rivers and oceans, as attractive sites to establish settlements. Although living conditions and lifestyles have changed over time, the fascination that people have for water has remained the same. Apartments, condominiums, villas, houses, and other establishments at the waterfront, riverfront, lakeside, and seaside are considered prime properties and fetch maximum real-estate prices. This trend can be observed all over the world, from New York City to Dhaka to Lagos to London. Despite growing concerns about the consequences of climate change, such as sea level rise, coastal erosion, coastal inundation, intense rainfall, and frequent flooding, proximity to water still remain the single largest denominator of spatial primacy.

Coastlines add another important dimension to human settlement (namely, money). For many centuries, the exploration of sea routes and new maritime routes for trade led to the development of sea transport, ports, and harbors, as well as the whole range of activities that has laid the foundation of today's coastal megacities. Trade between historical civilizations across continents, which is often referred to as *spice trade,* has also furthered cultural exchange. Maritime trade routes and sea transport fueled commercial activities and offered tremendous opportunities of growth to a new class of human settlement, which depended on opportunities of import and export.

Oceans cover about 70% of the Earth's surface and offer major food support to human populations. A sizeable share of global population is engaged in livelihoods that have direct or indirect dependence on oceans. Human settlements, both rural and urban, have been experiencing increase in population and the supporting infrastructure. The majority of new economic zones, especially in developing countries, are established near the coast and are experiencing massive amounts of investment. Cities along the coastline are swelling in size and population and expanding in both the horizontal and the vertical directions.

Urban Disasters and Resilience in Asia.

6.2 The Population Challenge

According to the United Nations Population Fund (UNFPA) "It took hundreds of thousands of years for the world population to grow to 1 billion, and in just another 200 years or so, it grew sevenfold. In 2011, the global population reached the 7 billion mark, and it is projected to climb to over 9 billion by 2050. This dramatic growth has been driven largely by increasing numbers of people surviving to reproductive age, and has been accompanied by major changes in fertility rates, increasing urbanization and accelerating migration. These trends will have far-reaching implications for generations to come" (UNFPA, 2015).

This statement clearly highlights two key dimensions of population growth—increasing urbanization and accelerating migration. Coastlines and coastal areas have registered phenomenal growth recently. This has helped create new cities, grown the population in existing cities, and turned some already-large cities into megacities. Both push and pull factors are responsible for the migration of sizable populations to urban areas. While permanent migration is a prominent form of migration, seasonal migration also causes a significant impact on cities, especially in the developing countries in Asia. Population growth rates are highest in coastal cities due to their locational advantages, which influence social change, create diverse employment opportunities and offer unique cultural amalgamation.

During the first decade of the 21st century, world population witnessed a landmark shift in its population distribution. In 2008, the world urban population crossed the 50% mark, and for the first time in human history, more people were living in urban areas than rural areas. The UN Department of Economic and Social Affairs stated, "Globally, more people live in urban areas than in rural areas, with 54 per cent of the world's population residing in urban areas in 2014. In 1950, 30 per cent of the world's population was urban, and by 2050, 66 per cent of the world's population is projected to be urban" (UN Department of Economic and Social Affairs, Population Division, 2015). Although the percentage of people living in rural areas of Asia and Africa is higher than in urban areas, both continents are urbanizing fast compared to other regions.

Urbanization and the challenges associated with it, especially among developing economies, have been receiving increased attention from policy makers as well as development practitioners. Due to the increasing concentration of population in cities, together with a complex web of infrastructure and services, urban risk is also emerging as an area of special interest among disaster managers. Despite growing concerns over the impact of human activity on the coastal systems, there seems to be an absence of updated and reliable data on people living in the coastal zone. Even within UN agencies, various data are cited with regard to coastal population. For example, the United Nations Atlas of the Oceans, which is a collaborative initiative of UN agencies and global scientific partners, created a web-portal containing maps, facts, data, and statistics related to oceans. As per this atlas, "About two-thirds of the world's population live within 60 kilometres of the coast, and almost half of the world's cities with more than 1 million people are sited in and around the tide-washed river mouths known as estuaries" (United Nations Atlas of the Oceans, 2011). Another interagency UN paper, on the other hand, states, "More than 40% of the world's population (more than 2.8

billion people) live within 100 kilometres of the coast" (IOC/UNESCO, IMO, FAO, UNDP, 2011). And yet another UN leaflet (published by UNEP and UNHABITAT) says, "Half of the world's population lives within 100 km of the sea, and three-quarters of all large cities are located on the coast" (UNEP and UNHABITAT, 2005). It is clear from these statistics that different sources have different data. There is an urgent need to direct international efforts to observe coastal population and related data on a consistent basis.

6.3 The Young and the Old

Two recent reports highlight a different kind of population dynamics that the world is experiencing now. A report entitled "The State of World Population 2014" states that "our world is home to 1.8 billion young people between the ages of 10 and 24, and the youth population is growing fastest in the poorest nations" (UNFPA, 2014). It reveals that of the total 7.3 billion people in the world, about a quarter of the world population is young. Almost 9 out of 10 young people live in less-developed countries. In the Asia Pacific region, 25% population is aged 10–24 and will be the "main agents of tomorrow's resilience" (UNFPA, 2014). Although there is no specific data available about young population in coastal regions, it is safe to assume that the availability of dynamic livelihood opportunities might be resulting in higher proportion of young people near the coastline. Targeted investment toward young people's education, health, and protection of rights, as well as more employment opportunity, will have long-term benefits in making social and economic transformation possible.

In 2012, a report stated that "with one in nine persons in the world aged 60 years or over, projected to increase to one in five by 2050, population ageing is one of the most significant trends of the 21st century" (UNFPA and HelpAge International, 2012). In 2012, 11% population in Asia was 60 years and over, which is expected to rise to 24% by 2050. However there are marked differences between countries. For example, over 30% population of Japan's population is already aged 60 years or older. The report further notes, "Older persons are the world's fastest-growing population group" and population aging is "happening fastest in developing countries, including in those that also have a large youth population" (UNFPA and HelpAge International, 2012). In the absence of data, it is difficult to make assumptions about the share of aging people in the population of coastal regions. However, the provision of essential health care and social services, income security, and inclusive age-sensitive developmental planning approaches, especially in coastal human settlements will ripe benefits today, tomorrow, and over the longer term.

6.4 Aspirations and Missing Links

Human settlements historically have mirrored the aspirations of their societies at a given period of time. However, today's rate of dynamic change coupled with the growing reach of the digital revolution paint a completely different picture. Cities in Asia's developing countries present many contrasts in the spatial, economic, social, and cultural spheres. Recently, over 7 million people (representing 1 out of every 1,000 people in the world) took the My World Survey, a questionnaire conducted by the United Nations (UN) to find out development goals from individual perspectives. Participants in this survey picked 6 out of 16 priorities for better future, which in turn informed post-2015 processes, presented in Table 6.1.

This was one of the largest surveys conducted in the history of the United Nations. Its results reveal that education, health, employment, good government, security, and food are the highest-ranked aspirations of the community at large, both in Asia and the world at large (United Nations et al. (undated)). A notable outcome is that protection of the environment and prevention of climate change are least important. This is rather surprising for two reasons. First, the populations of Asian nations are by far the major victims of the negative consequences of climate change and natural disasters. Second, from communities to subnational, national, and regional levels, significant efforts toward disaster risk reduction (DRR) and climate change adaptation have already been made and are continuing. It is evident that there is a long way to go before society becomes fully aware of the need to become resilient and safe. However, in particular, these are urgent priorities for settlements in coastal regions before they face intensified dire consequences from natural disasters.

Table 6.1 My World Survey Results

Sixteen Options to Prioritize	Ranking of Priorities		Author's Interpretation
	Global	Asia	
A good education	1	1	Top priority
Better healthcare	2	2	Top priority
An honest and responsive government	4	3	Top priority
Better job opportunities	3	4	Top priority
Protection against crime and violence	6	5	High priority
Affordable and nutritious food	5	6	High priority
Equality between men and women	10	7	Medium priority
Access to clean water and sanitation	7	8	Medium priority
Support for people who can't work	8	9	Medium priority
Reliable energy at home	11	10	Medium priority
Better transport and roads	9	11	Medium priority
Freedom from discrimination and persecution	12	12	Low priority
Phone and Internet access	15	13	Low priority
Political freedom	13	14	Low priority
Protecting forests, rivers, and oceans	14	15	Low priority
Action taken on climate change	16	16	Low priority

Source: United Nations and partners (n.d.). Results updated as on June 3, 2015.

6.5 Growth of Coastal Cities

The growth of cities adjoining the coast has been a matter of academic discourse for a long time. How humans and oceans interact, offering both opportunities and threats to each other, has been a subject of inquiry across various scientific disciplines. Urbanization offers unique challenges, and so do opportunities to improve human well-being and improve the overall quality of life. Timmerman and White (1997) used a unique term, "megahydropoli," to describe the rise of large coastal cities. This paper highlighted that coastal zone management and urban planning lack focus on coastal cities, and an ecosystem approach can provide a good way to fill this gap. A follow-up study by Yeung (2001) made further inquiry in this area and concluded that "present patterns of coastal development and utilization are not sustainable for coastal megacities in Asia." This study recommended education and awareness raising of community and policy makers as a first step toward improvement.

UNHABITAT (2012) said that Asia added 0.88 million new urban dwellers every week during the first decade of this century. This report added that Asia is home to half of the world's urban population and 7 out of the 10 most populous cities. Interestingly, 14 out of 19 largest cities in the world are port cities. According to the World Resources Institute (n.d.), "By 2030, the world's cities are expected to add 1.5 billion people, up to 1 billion cars, and build more urban area than all of humanity has built throughout history. The majority of this growth will occur in the developing world." China adapted a "city cluster" approach, which encouraged the growth of second-tier cities around large cities, to attract investors; this "resulted in dramatic differences in productivity between coastal cities and their counterparts in the hinterland, with the growth differential surpassing 6%" (UNHABITAT, 2012). Undoubtedly, population centers and their supporting infrastructure, including economic assets, are disproportionately concentrated in Asia's coastal cities (World Bank, 2010).

Within the overarching discussion about coastal areas, low-elevation coastal zones (LECZs) have received special attention due to their location-specific vulnerability to climate-related hazards. The term *LECZ* is defined in the literature as a contiguous area along the coast that is less than 10 m above sea level. McGranahan et al. (2007) published one of the most cited studies, which stated that LECZs "cover 2 per cent of the world's land area but contain 10 per cent of the world's population and 13 per cent of the world's urban population."

Sea-level rise (SLR) is one of the major concerns caused by global warming. It is expected that coastal zones will face a heightened threat due to SLR, and the consequences (although not completely estimated) will be significant. A study by Nicholls and Cazenave (2010) recommends enhancing the "scientific understanding of observed and future climate-induced sea-level rise and develop pragmatic impact and adaptation scenarios that capture the uncertainties of future sea-level rise." Figure 6.1 clearly illustrates that almost the entire coastal zone of the Asia-Pacific region is highly vulnerable to sea-level rise.

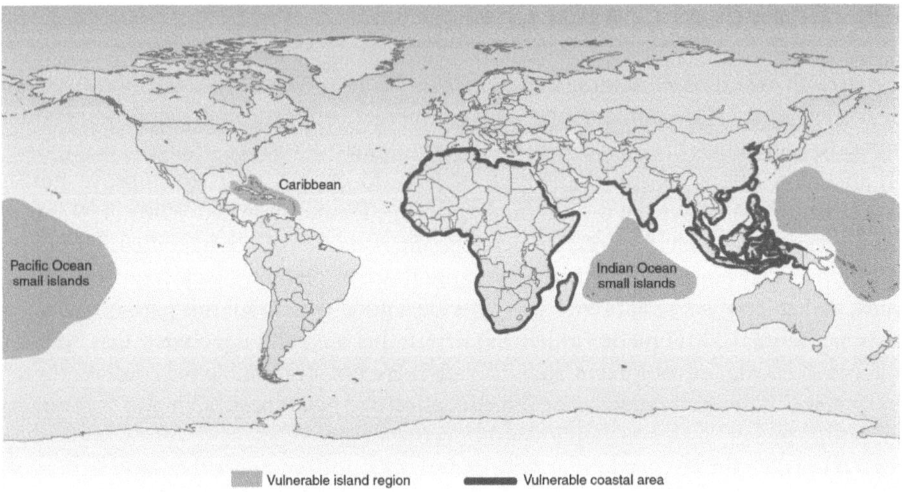

Figure 6.1 Coastal areas of the world vulnerable to SLR. (*Source:* Nicholls & Cazenave, 2010).

6.6 Research on the Resilience of Coastal Areas

A sizable number of studies have been done by academic, scientific, and international researchers investigating numerous aspects of risk and resilience in coastal areas. Some of these studies attempt to enhance the conceptual understanding of the complex interaction of human and environmental system at the coast, while others provide detailed accounts of real-world scenarios and situations for the purpose of recommending possible mitigation and adaptation options.

Bryant et al. (1995) evaluated development-related pressures on coastal ecosystems and biodiversity and found that compared to other regions, European and Asian coastlines were at the greatest risk. They estimated that "over half of the world's coastal ecosystems are at high or medium risk from development-related activities." An index developed as part of this study evaluated human pressure on coastlines through five globally available georeferenced indicators, presented in Table 6.2.

Sekovski et al. (2012) used a Drivers-Pressures-States-Impacts-Responses (DPSIR) framework to explore the role of coastal megacities in environmental degradation. One of the unique aspects recognized by this study is the financial capacity of coastal megacities, which has the potential to bring about environmental improvement. In this study's analysis, the drivers are urbanization, energy consumption, transportation, industry, agriculture, water consumption, and tourism. The pressures, on the other hand, are identified as waste generation, gas emission, alteration of hydrological and sediment flux, pressure on groundwater recharge and supply, habitat loss, and pressure on fish stocks. The negative impacts include coastal erosion, a decline in biodiversity, altered ecosystem function, human health impacts, and socioeconomic impacts.

Another program, entitled "Climate Adaptation in Coastal Megacities" was undertaken by the World Bank in collaboration with the Asian Development Bank (ADB)

Table 6.2 Indicators of Human Pressure on Coastal Ecosystems

Indicator	Potential Threat Measured
Cities with populations over 100,000	Potential threats from coastal development, sewage, and industrial pollution
Major ports	Potential threats resulting from species introductions, oil spills, and industrial pollution.
Population density	Potential threats from coastal development, degradation through overharvest (e.g., destruction of mangroves), and pollution
Road density	Indirect measure of access to coastal resources and coastal development
Pipeline density	Potential threat of oil pollution and spills of other industrial wastes

Source: Bryant et al. (1995).

Table 6.3 Objectives and Methodology of the World Bank (2010) Study

Objectives	Methodology	Local Information Used
To highlight the scale of climate-related impacts and vulnerabilities at the city level	Determining climate variables at the level of the city/watershed through downscaling techniques	Climate variables such as changes in temperature, precipitation, SLR, and storm surge
Provide estimates of associated damage costs	Estimating impacts and vulnerability through hydrometeorological modeling, scenario analysis, and geographic information system (GIS) mapping	Socioeconomic and developmental factors such as land subsidence, land use, and population increases
Examine potential adaptation options	Preparing a damage/loss assessment and identification/prioritization of adaptation options	Local topographical and hydrological information

Source: World Bank (2010).

and the Japan International Cooperation Agency (JICA). This program was implemented in three cities, Manila, Ho Chi Minh City, and Bangkok, and the results were published in 2010 (World Bank, 2010). These urban areas were described as "low-lying coastal cities situated in the deltas of major river systems in the East Asia region" and are "highly vulnerable to climate-related risks" (World Bank, 2010). The objectives and methodologies of this study are summarized in Table 6.3.

This city-level study estimated climate risks up to 2050 by using a high- and a low-emission scenario and two Intergovernmental Panel on Climate Change (IPCC) scenarios. However, the study also cautioned about a number of uncertainties involved in risk forecasting, such as pathways of greenhouse gas (GHG) emissions and increases in extreme and seasonal precipitation under different scenarios. One specific observation of the study is the absence of modeling economic growth and linking it with urban development. Key findings and recommendations of the study are summarized in Table 6.4.

One of the interdisciplinary studies currently in progress, entitled "Coastal Cities at Risk: Building Adaptive Capacity for Managing Climate Change in Coastal Megacities," is studying ways "to protect people, communities, and vital economic sectors

Table 6.4 Key Findings and Recommendations of the World Bank (2010) Study

	Key Findings	Recommendations		
		1	2	3
1	Frequency of extreme events likely to increase	Better management of urban environment and infrastructure will help manage potential climate-related impacts.	Climate-related risks should be considered an integral part of city and regional planning.	Targeted, city-specific solutions combining infrastructure investments, zoning, and ecosystem-based strategies are required.
2	Increase in flood-prone areas due to climate change			
3	Increase in population exposed to flooding			
4	Substantial damage cost [may range from 2%–6% of regional gross domestic product (GDP)]			
5	Damage to buildings alone accounts for 70% of the total estimated flood-related costs			
6	Poor and vulnerable will be disproportionately impacted, but everyone will be affected			
7	Land subsidence (non-climate-related factor) may account for a greater share of the damage costs from flooding compared to climate-related factors			

Source: World Bank (2010).

that are most at risk from the effects of climate change" (About CCaR, n.d.). This collaborative study being conducted during 2011–2016 by START (an international nongovernmental organization), Integrated Research on Disaster Risk (IRDR), and partners focuses on four cities, Bangkok, Lagos, Manila, and Vancouver. The research is expected to contribute toward building capacities for increased risk reduction and adaptation, improved planning, and reduced negative impacts of climate change. Studies like this are indicative of the heightened awareness of the risk built over a period of time in coastal areas, especially in highly populated urban areas.

Researchers from the International Institute for Applied Systems Analysis (IIASA) investigated recent examples of pooling of risks in Latin America and the Caribbean and its application to Asian megacities. The study recommends "focusing on stimulating a broad process-based dialogue," similar to the strategy initiated in the Caribbean, to make a case for risk pooling (Hochrainer & Mechler, 2011). City-level studies have specific relevance to local governments. A study by Dasgupta et.al. (2013) assessed the vulnerability of the city of Kolkata, India. After thorough investigation and in-depth modeling, the study concludes that simple measures, such as desilting of the main sewers of the city, have the potential to reduce vulnerable population estimates by at least 5%.

IPCC's fifth assessment report predicted the unfavorable impacts of relative SLR on coastal systems and low-lying areas that will be experienced in the form of submergence, coastal flooding, and coastal erosion. It reconfirmed that Asian coastal cities are at a very high risk, and socioeconomic development itself increases the exposure of populations and assets to coastal hazards. Cities in South, Southeast, and East Asia are particularly vulnerable to SLR, but the economic growth of these countries also boosts their "monetary capacity to adapt" (IPCC, 2014).

6.7 Conclusion and Way Forward

Asian coasts offer an incredible concentration of opportunities to transform the quality of life of their residents, but they also are becoming more vulnerable to both climate- and non-climate-related shocks and stresses. While there is no "one size fits all" solution available for coastal cities, some general recommendations can be drawn from experiences accumulated over a period of time, as seen in Table 6.5.

The year 2015 is a decisive milestone of this century. A courageous, acceptable, equitable and binding climate agreement has been due for a long time, and this year, such a deal should be struck by global leaders. The world community adopted the successor to Millennium Development Goals (MDGs). Concerted efforts are required from every citizen of the earth, albeit at various levels, to instigate collective action. From changes in everyday behavior of individuals to responsible strategies and sensible policy making at the global level, an all-encompassing solution space need to be carved for a better future. The cost of inaction could be unprecedented and losses could be irreversible, especially for coastal settlements that are going to be at greater risk until a deal for their sustainable future is sealed and implemented.

Table 6.5 Selected Recommendations for Resilient Urban Development in Asia's Coastal Areas

Problems/Concerns	Key Recommendation	Scale of Intervention	Stakeholder Engagement
Lack of reliable data on human settlement growth patterns on the coast in general	Organizations responsible for creating repository of data for urban development should carve out a separate data set for settlements at the coast	Efforts required from district level to global level	Local/provincial/national governments, science and technology agencies, and regional and international organizations
Absence of city-region approach, hence peri-urban and rural areas were ignored	Footprints of coastal cities should be assessed considering the intensity of linkages with adjoining areas on both land and sea	Regional, city, district, and zone levels	Private sector should be playing a greater role, working hand in hand with local and regional authorities
Nonexistence of mechanisms to track coastward migration	Combination of mobile, closed-circuit TV cameras, and digital and manual techniques for precise tracking of in- and out-migration	Provincial and national governments must strategically invest in tracking human movement	Census agencies, water-land-air transport authorities, private sector, and science and technology agencies
Underinvestment in robust risk management strategies	Risk management should be given high priority and monetary provision should enhance significantly	Efforts required from local level to international level	Local government, investment agencies, provincial and national governments, and donor agencies
Poor implementation of coastal zone management provisions	Coastal zones should be subject to strict management as per agreed regulations	National, subnational, city, district, and zone levels	Spatial development monitoring agencies, financial institutions to devise mix of incentive- and penalty-based approaches
Lack of sensitization to coastal sensibilities among people in general	Awareness programs to sensitize communities and policy makers alike about coastal and environmental aspects	From the school level to ward, city, subnational, and national levels	Print and social media, community-based organizations (CBOs), grassroots and local organizations, schools, nongovernmental organizations (NGOs), and mass-media organizations
Lack of peer-to-peer learning opportunities	Network of coastal settlements should be established to enhance colearning	National and provincial levels	National and provincial governments and the corporate sector

References

About CCaR. (n.d.). Retrieved July 8, 2015, from Coastal Cities at Risk: http://coastalcitiesatrisk.org/wordpress/about-ccar/.

Bryant, D., Rodenburg, E., Cox, T., & Nielsen, D. (1995). *Coastlines at Risk: An Index of Potential Development-Related Threats to Coastal Ecosystems.* Washington, DC: The World Resources Institute.

Dasgupta, S., Gosain, A. K., Rao, S., Roy, S., & Sarraf, M. (2013). A megacity in a changing climate: The case of Kolkata. *Climate Change, 116*(3–4), 747–766.

Hochrainer, S., & Mechler, R. (2011). Natural disaster risk in Asian megacities: A case for risk pooling? *Cities, 28*(1), 53–61.

Intergovernmental Panel on Climate Change (IPCC). (2014). In C. Field (Ed.), *Climate Change 2014: Impacts, Adaptation, and Vulnerability. Part A: Global and Sectoral Aspects. Contribution of Working Group II to the Fifth Assessment Report of the Intergovernmental Panel on Climate Change.* Cambridge, United Kingdon and New York, NY: Cambridge University Press.

IOC/UNESCO, IMO, FAO, UNDP. (2011). *A blueprint for ocean and coastal sustainability.* Paris: IOC/UNESCO.

McGranahan, G., Balk, D., & Anderson, B. (2007). The rising tide: Assessing the risks of climate change and human settlements in low elevation coastal zones. *Environment and Urbanization, 19*(1), 17–37.

Nicholls, R. J., & Cazenave, A. (2010). Sea-level rise and its impact on coastal zones. *Science, 328*, 1517–1520.

Sekovski, I., Newton, A., & Dennison, W. C. (2012). Megacities in the coastal zone: Using a driver-pressure-state-impact-response framework to address complex environmental problems. *Estuarine, Coastal, and Shelf Science, 96*, 48–59.

Timmerman, P., & White, R. (1997). Megahydropolis: Coastal cities in the context of global environmental change. *Global Environmental Change, 7*(3), 205–234.

United Nations et al. (n.d.). *Methodology.* Retrieved June 3, 2015, from http://www.myworld2015.org/?page=methodology.

United Nations et al. (n.d.). *My Analyticsy.* Retrieved June 3, 2015, from http://data.myworld2015.org.

United Nations Atlas of the Oceans. (2011). *About.* Retrieved July 6, 2015, from http://www.oceansatlas.org/servlet/CDSServlet?status=ND1lY29zeXN0ZW1zJjY9ZW4mMzM9Ki-YzNz1rb3M~.

United Nations, Department of Economic and Social Affairs, Population Division. (2015). *World urbanization prospects: The 2014 revision (ST/ESA/SER.A/366).* New York, NY: United Nations.

United Nations Environment Programme (UNEP) and United Nations Human Settlements Programme (UNHABITAT). (2005, September undated). Coastal Area Pollution. *The Role of Cities* Retrieved July 7, 2015, from http://www.unep.org/urban_environment/PDFs/Coastal_Pollution_Role_of_Cities.pdf.

United Nations Human Settlements Programme (UNHABITAT). (2012). *State of the world's cities 2012/2013: Prosperity of cities.* Nairobi, Kenya: UNHABITAT.

United Nations Population Fund (UNFPA). (2015, April 8). *Overview.* Retrieved July 6, 2015, from World population Trends http://www.unfpa.org/world-population-trends.

United Nations Population Fund (UNFPA). (2014). *The state of world population 2014.* New York, NY: UNFPA.

United Nations Population Fund (UNFPA) & HelpAge International. (2012). *Ageing in the twenty-first century: A celebration and a challenge*. New York, NY: UNFPA and HelpAge International.

World Bank. (2010). *Climate risks and adaptation in Asian coastal megacities: A synthesis report*. Washington, DC: International Bank for Reconstruction and Development/World Bank.

World Resources Institute. (n.d.). *WRI Ross Center for Sustainable Cities*. Retrieved July 7, 2015, from http://www.wri.org/wri-ross-center-sustainable-cities.

Yeung, Y.-M. (2001). Coastal mega-cities in Asia: Transformation, sustainability, and management. *Ocean and Coastal Management*, *44*(5–6), 319–333.

Impact of Urban Expansion on Farmlands: A Silent Disaster

7

Atta-ur-Rahman[1], Akhilesh Surjan[2], Gulsan Ara Parvin[3], Rajib Shaw[4]
[1]Associate Professor, Institute of Geography, Urban and Regional Planning, University of Peshawar, Peshawar, Pakistan; [2]Associate Professor, Faculty of Engineering, Health, Science and the Environment, Charles Darwin University, Darwin, Australia; [3]Researcher, Graduate School of Global Environmental Studies, Kyoto University, Kyoto, Japan; [4]Professor, Graduate School of Global Environmental Studies, Kyoto University, Kyoto, Japan

7.1 Introduction

This chapter analyzes the rapid urban growth in south Asia that has resulted from urban expansion, chaotic development, and encroachment over prime agricultural land. It has posed a serious threat to food security and carbon-sinking hotspots. The loss of farmland to urbanization has been especially rapid in the developing countries of South Asia (Rahman & Shaw, 2015). The serious implications of this development have been recognized as a silent disaster, and attempts are being made to address this situation. The situation is particularly serious in developing countries like Pakistan, India, Thailand, and Bangladesh. This is mainly due to the lack of urban planning policies and investments, which has led to unplanned city expansion and undesirable growth of land utilization and its ramifications. In this context, a case study on Peshawar, Pakistan, is analyzed in the chapter to explore the factors responsible for haphazard urban expansion over the farmland in the urban periphery and the resultant negative implications in the form of threats to food security and loss of vegetation cover.

In 1900, around 14% of the world's population resided in cities, and it was only in 2008 when the urban population number actually surpassed the rural (Rahman & Shaw, 2015). Presently, over half of the world's citizens live in cities, and it is projected that by 2025, the ratio of rural to city dwellers will be 2:3 (Satterthwaite et al., 2010). This rapid pace of urbanization has resulted in a growing number of built-up areas (Bhatt et al., 2005). Globally, urban areas will triple and urban populations will double by 2030, and as a result, urban expansion will consume a large part of the available prime agricultural land (SCBD, 2012). With an expected increase in economic growth, these urban centers will expand even further (Lata et al., 2001) and this expansion is projected to reduce the amount of productive agricultural land around cities (Bhatta 2010). Sudhira et al. (2004) stressed that there is an urgent, immediate need to monitor these changes and promote city growth in a sustainable fashion. In 1990, the active labor force employed in the manufacturing sector exceeded the labor force engaged in primary economic activities (Khan, 2003).

Urban Disasters and Resilience in Asia.

Land-use pattern in a region is the outcome of natural and socioeconomic factors (Khan, 2003). Nowadays, due to massive agricultural and demographic pressure, land is gradually becoming a scarce resource (Opeyemi, 2006). In current planning strategies, land-use dynamics is an important component of monitoring changes in the physical environment and managing natural resources (Samiullah, 2013). Globally, agriculture is one of the main dynamics of land-use change (McConnell et al., 2004). Different studies have shown that only a few landscapes on the Earth's surface are still in their natural state, whereas the rest has changed significantly (Opeyemi, 2006). Such changes in land use have greater implications for the natural environment and thus lead to new land-use patterns (Rahman, 2010).

Globally, 24% of the world's land is occupied by cultivated land (Saunders, 2010). It has been observed that in many Asian cities, there is a gradual decline in per-capita land availability (Oxfam, 2010). In this changing scenario, urbanization has been blamed as a major determining factor that is likely to further consume arable land, specifically in the periphery of existing urban settlements. This is mainly due to the engulfing of prime agriculture land for nonagriculture use (Khan, 2003). Such challenging issues have been multiplying day by day, resulting in the shrinking of farm size, specifically in countries with a rapid population growth rate. Urban expansion is not merely a cause of urbanization, but in most cities of the developing world, it has helped to trigger many urban problems.

Globally, the demand for food, fiber, and fodder is increasing, and as a result, farmers responded through land intensification by increasing the yield per unit area (Rahman, 2010). Similarly, this marginalized land, which has the potential for cultivation, has been intensively utilized, as reports from sub-Saharan Africa show (Cassman et al., 2005). Worldwide, in the past 40 years, intensification of farmland has contributed to an 80% enhancement in the output of primary products (Saunders, 2010). The United States, Japan, Australia, and Italy are the countries with the highest levels of productivity and stable population growth rates, but here the cultivated land is shrinking as part of the cost of urbanization (Samiullah, 2013). Likewise, in a populous country like China, area under agricultural production has also started contracting. Nevertheless, the opportunities for further expansion of cropland are gradually shrinking at a global level, most likely due to the ramifications of urban expansion over the arable land (Sudhira et al., 2004).

Pakistan is an agro-based country. Therefore, its farmland plays an important role in the national economy (Khan, 2003). The contribution of the agriculture sector to Pakistan's gross domestic product (GDP) is about 20% (Khan & Hye, 2010; Kandrika & Roy, 2008). However, the booming population is putting tremendous pressure on the amount of farmland, and there is an urgent need for agricultural land to expand further because the existing cropland is deficient in terms of food self-sufficiency. As a result, the population mainly relies on importing food grain. Additionally, to fulfill the shelter requirements for the increasing populations, both rural and urban settlements are horizontally expanding at an alarming rate, but most often, this horizontal physical expansion occurs at the cost of losing prime agricultural land (Rahman & Shaw, 2015). Khan and Hye (2010) pointed out that the trend of expansion in agricultural land has remained very slow compared to the country's increasing population rate.

In order to assess the impact of urban expansion on farmland, a microlevel case study in Peshawar, Pakistan, was undertaken. Peshawar is the capital city of Khyber Pakhtunkhwa province. In Peshawar capital city district, rapid urban growth has resulted in the growth in built-up areas, particularly in the countryside. Such urban expansion has brought changes in land use, and most often, this sprawl takes up the prime agricultural land within the city and on the urban fringe. However, this expansion has been very fast, consuming a good deal of fertile and productive agricultural land. This chapter particularly focuses on the menace of this silent disaster of reducing the "food basket" and increasing threats to food security and green hotspots. As economic growth is accelerating, urban centers are forced to grow beyond their boundaries and, as a result, land-use changes are occurring in big urban centers like Peshawar. It is, therefore, pertinent to detect and monitor spatial and temporal land-use changes over time, and highlight threats to food security, groundwater aquifers, and pluvial floods.

7.2 Spatial Pattern of Urban Land Use

The history of land utilization is as old as the history of agriculture and human settlement (McConnell et al., 2004). Scientific land use surveying was undertaken for the first time in Britain in the early 1930s (Stamp, 1962). The uses to which humans put land could include grazing, agriculture, physical development, and mining (Opeyemi, 2006). Broadly, land utilization can be classified into cultivated land, uncultivated land, and cultivable waste. Cultivated areas can be defined as farm areas that were sown at least once during the previous two years (GOP, 2004). Similarly, land use may also be classified into two categories: rural and urban.

Physical factors like rivers, plains, soil, climate, and mountains determine the direction of spatial growth (Nanda, 2005). Similarly, disasters like landslides, floods, earthquakes, soil erosion, fire, and waterlogging may also initiate changes in land-use patterns (Opeyemi, 2006). In fertile river valleys, there has been rapid physical development over good agriculture land, which may lead to further intensification of farmland. In a regional setup, socioeconomic and physical factors are important, interacting influences on land use that bring about changes in land use in terms of both time and space (Dramstad et al., 2001). The spatial pattern of land use reflects underlying human activities and processes that influence the ecology of the urban environment (Yeh & Li, 1999). Several researchers have discussed the possible causes and the underlying driving forces of land-use changes and their interactions with the environment (e.g., Fox & Vogler, 2005).

In an urban setup, several socioeconomic factors are responsible for spatial dynamics in land use. Population increases largely affect agricultural land use because people need to use land for residential, commercial, transportation, and industrial purposes, as well as for parks and playgrounds (Khan & Hye, 2010). Demographic factors such as changes in household composition, availability of labor, migration, and rural-urban distribution of land are the dominant factors causing physical urban growth (Lambin & Geist, 2007). Most urban growth models show that employment opportunities attract more and more people, who in turn need further physical development to accommodate their needs. Having road networks is also considered an important factor of rapid physical urban growth. Moreover, industrial establishment and commercial activities also boost economic opportunities, and as a result, they attract even more residents to a city, which influence the land-use patterns (Nanda, 2005).

Remote sensing (RS) and geographic information system (GIS) technologies have been recognized as powerful and effective tools for detecting the spatiotemporal dynamics of land use (Longa et al., 2009). It has been used often for the monitoring and management of resources (Sarma et al., 2001; Bradshaw & Muller, 1998). Satellite RS provides multispectral and multitemporal data that can be used to quantify the type, amount, and location of land-use change through GIS (Sarma et al., 2001).

7.3 Growing Urbanization Versus Growing Hazards

The concept of an urban area varies from time to time and country to country. Brockerhoff (2000) explained that an urban center is an area where the majority of people are engaged in nonagricultural activities with high population densities. In some countries, a lower limit is set for the population of an urban locality, and this limit also varies from country to country. For instance, the limit is 200 in Sweden, but 10,000 in Switzerland (United Nations, 2000). Apart from these statistics, a city can be defined on the basis of its functions and services, as an area with many economic activities, better educational facilities, and promising employment opportunities. In Pakistan, the term *urban area* has been defined on the basis of population and administrative setup. According to census figures for Pakistan, urban areas are confined to urban categories of town committee, municipal committee, cantonment board, or municipal corporations (GOP, 1999; Khan & Rahman, 2000). However, the boundaries of town committee, municipal committee and municipal corporations were abolished after the enactment of devolution legislation in 2000, where existing urban areas were turned into urban councils and urban centers were redefined as areas that remained under the administration of former municipal or town committees (GOP, 2004).

Urbanization is a process of transformation from a traditional society to a modern metropolitan society, associated with major changes in socioeconomic and land-use structures (Liu, 2002; Cheng & Masser 2003; Tan et al., 2005). It includes two phases: the movement of people from rural to urban places, where they primarily take part in nonrural occupations; and the change from a rural to an urban lifestyle, with its associated values, attitudes, and behaviors (Brunn et al., 2003). Globally, urban environments are a chief element of land-use change. Its importance will continue to increase as the majority of the world's population continues to migrate toward urban centers (Cai, 1995). In urban areas, the observable landscape is the urban expansion, increase in number of citizen, size of urban settlement and extent of commercial activities. Likewise, urbanization is a manifestation of historical, socioeconomic, geographic, and demographic forces (Pick & Butler, 1997).

Natural increase in population (birth-death) and rural-urban migration are the main components of urban growth. Reclassifications of settlements and area annexation also play a part. The term *reclassification* refers to the change of status of a settlement from rural to urban (or vice versa), or from one category of urban place to another. *Area annexation* means the extension of urban limits, which increases the urban population and becomes an important factor in the growth of individual cities (Khan et al., 1992). Sometimes deannexation produces a reverse effect. Urban dwellers are multiplying by a complex combination of all these factors.

The term *urbanization* means the increasing proportion of country's population living in cities (Khan, 2003). The shifting of rural populations to urban areas is mainly due to urban biases in terms of development and economic opportunities. It has been observed in developing economies that urban residents have a better standard of living, level of nutrition, and provision of services than rural dwellers. However, 30%–50% of urban populations in these nations live in dilapidated housing conditions, with limited access to urban services. The majority of these households are in illegal settlements that lack adequate access to water supply, educational institutions, health centers, and sanitation. Similarly, in low- and middle-income nations, it is hardly true that every urban dweller benefits equally from urban amenities and services.

The implications of these developments include the addition of urban dwellers to the existing population, urban sprawl, encroachment over farmland, threat to food security, reduction in carbon-sinking hotspots, decline in aquifers, traffic congestion, pluvial floods, and changes in land-use patterns. Most of the urban professionals consider urbanization as a cause of urbanization. However, no nation prospers without urbanization and all developed countries are predominantly urban. The world's poorest nations are also the least urbanized. Similarly, analyses show that a strong relationship between economic development and urbanization. The adverse effects of urbanization in most Asian cities include food insecurity, urban poverty, child mortality, slums and squatter settlements, limited access to health care, water, drainage, and sanitation. The same types of hazard have been observed in cities with exponential economic growth rate, such as Mumbai, India (Satterthwaite et al., 2010). China is a nation with rapid urbanization, high economic growth rate, and slow natural increase in population. On the other hand, in Afghanistan, Pakistan, India, Indonesia, and other developing countries have low economic growth and high birth rates.

Another negative aspect is the growth of slums and blighted areas in urban peripheries, which is mostly occupied by low-income dwellers and gradually become part of these cities. Such haphazard urban growth over the farmland is an urban sustainability issue. In developing nations, urban authorities are hard pressed to extend urban amenities and infrastructure to these newly built-up areas.

Urban growth is a global phenomenon. It may take the form of planned development or unplanned urban expansion over the farmland. The growth of cities can take various forms. If it is in the form of concentric expansion, it is known as *accretion.* Urban centers may also grow in a starlike shape if they have a radial road pattern, or by area annexation and the engulfing of villages into the city limits (Khan, 2003). There are two divergent views regarding forces behind urban growth in developing countries. Rogers and Williamson (1982) and Todaro (1984) stated that rural-urban migration is largely responsible for urban growth in developing countries. However, urban geographers and planners are of the opinion that natural increase is the main cause of urban growth (Rahman & Samiullah, 2005). To ascertain what is happening in Pakistan is almost similar in pattern to that of other south Asian countries, where all the abovementioned factors are equally responsible for urban expansion.

In major cities, population growth requires lateral expansion. Therefore, urban developers look into the neighboring natural territories to build needed dwelling units and other infrastructure. Growing populations in a city mean that it needs more housing units, and this causes urban land expansion. As the demand for housing increases, cities begin to expand over new peripheral areas. They might even begin to build on unlikely areas such as agricultural land, forests, and swamps (Brooks, 1999).

7.4 Impacts of Urban Expansion

Urban expansion is the physical growth over natural areas such as agricultural land, forests, and swamps (Brooks, 1999). Urban expansion is one of the indicators to measure the rate of urban growth and its spatial dimension (Deakin, 1989; Bourne, 1996) because it affects land cover both at the micro- and macro-level. Urban sprawl is a physical expansion of an urban center (Liu, 2002; Tan et al., 2005). Such processes includes the consistent transformation of open spaces into built environments. One of the prominent negative impacts of urban expansion on natural environment is the gradual loss of farmland by urbanization (Samiullah, 2013).

Land-use change has had an unprecedented impact on both physical and cultural environments, such as loss of prime agricultural land, reduction in agricultural production, loss of irrigation infrastructure, loss of agricultural jobs, and destruction of natural landscape (Firman, 2000). Conversion of farmland into built-up area has also caused pollution, destroyed green spaces, and increased traffic congestion (Nanda, 2005).

This process has substantially contributed to a city growing in almost all directions. The expansion of a city in all directions is and indicating new developments is called *leapfrog expansion* (Gordon & Richardson, 1977; Ottensmann, 1977), which is a typical feature of many South Asian cities. Policy choices have a great influence on the physical development of an area. Some of the available options to manage urban growth that meet the overall community goals include the preservation of farmland, expansion of economic development, and mitigation of regional congestion for physical development (Bradshaw & Muller, 1998).

7.5 Urbanization and Loss of Farmland

Population growth and economic changes are two important factors influencing land-use changes and the distribution and amount of available agricultural land (Turner et al., 1996; Helmer, 2000). The transformation of natural systems to farm production has been the primary foundation for the successful growth of populations (Samiullah, 2013). The urban expansion of cities has created numerous socioeconomic challenges, such as reductions in cultivable land, rural-urban migration, and rural-urban linkages (Deakin, 1989). Globally, urbanization is one of the significant human activities that has affected the local environment (Liu, 2002). The absence of appropriate land-use planning and the lack of any sustainable effort, urban expansion, and loss of farmland have posed serious threats to food security, including urban flooding and an expanding carbon footprint (Weng, 2001).

7.6 Urbanization and Food Security

Urbanization is a dynamic process that brings changes in citizen diets, agricultural products, and demand for food. Rapidly expanding urbanization and urban population growth require food security to be in place, which depends entirely on sustainable enhancement in yield and high-potential cereal production systems (Saunders, 2010). The UN Food and Agriculture Organization (FAO) defines the term *food security* as a condition when people have access to safe, sufficient, and nutritious food that meets their dietary requirements (FAO, 1996). It has been determined that in the future, the available agricultural land will be unable to feed the growing population in these areas. The booming population is resulting in the urbanization of the farmland and will put tremendous pressure on the remaining agricultural land, particularly in the urban periphery, to fulfill the residents' food requirements. For example, it has been found that in Accra, Ghana, 2600 ha of agricultural land is converted for urban purposes every year. Similarly, in Chilean city, over 3000 ha of arable land has been consumed by the built-up environment from 1975–2000. This is an irreversible negative consequence of urbanism prevailing over agricultural land and carbon-sinking hotspots.

Prior to industrialization, the loss of farmland was so rapid in a number of densely populated Asian sovereign states, including Japan, Taiwan, and South Korea (Heimlich & Brooks, 1989). Brown (1995) also revealed that initially, these countries were grain exporters, but now they are grain importers. The data further reveal that Japan's grain acreage declined by 52% in a 40-year period (1955–1994). Similarly, in South Korea, the cropland declined by 46% during 1965 to 1994 (30 year span). Likewise, Taiwan has lost 42% of its cultivated area in 32 years (1962–1994). China is another example of a reduction in grain-farming areas, from 90 million ha in 1990 to 85 million ha in 1994—an amazing 5.5% reduction in grain acreage in a mere 4 years. Weng (2001) stated that in China, encroachment over farmland has been a key feature of the urban periphery, and as a result, significant reduction in farmland has occurred due to rapid urbanization.

In order to feed the growing population of these areas, there is an urgent need for agricultural expansion. Every year, thousands of hectares of farmland is lost to physical growth (Bradshaw & Muller, 1998). In Japan, about 7% of the available agricultural land has been subjected to urban development over the past three decades. In India, 5.5% of the total land area is occupied for residential, commercial, industrial, and transportation purposes (as well as vacant land), and this share is multiplying day by day (Nanda, 2005). The same process is happening in Pakistan, where land resources are deficient in relation to its population and the per-capita availability of agricultural land has declined over the recent decades (Khan & Hye, 2010). In 1947, per-capita cultivated land was 0.4 ha, which decreased to less than 0.1 ha in 2009 (Khan & Hye, 2010). Its citizens require shelter and infrastructure, but that should not be at the cost of degrading agriculturally productive land; otherwise, this situation will pose a serious threat to the food security and environment.

Food scarcity and increasing loss of agricultural land have become issues of global concern (Bender, 1997). The continued increase in human populations, a decrease in freshwater supplies, loss of cropland, and the changing microclimate and macroclimate are all factors that may influence the ability of societies to produce sufficient food to feed the world's inhabitants. Global food security is further threatened by socioeconomic changes in China. Given its large and growing percentage of the global population, as well as the recent economic shift from agriculture to industry, China is fast becoming a major food importer. In a world economy, increases in food imports to China will affect all nations and could increase global food prices and create political and social instability, particularly in developing countries (Brown, 1995). It has been predicted that global demand for food will double by 2030 (Crosson, 1997). The prospects of increasing food production to meet future demands will require either more land and water or increase the efficiency of crop production on existing land.

The analysis of available literature revealed that with exponential economic growth and fast urbanization, there will be increasing demand for food items and other daily used goods and services for many countries. This will obviously influence the distribution of income and economic activities. In the present urban system, more people are working in the nonagriculture sectors, and very few are farm laborers. Similarly, the increasing number of nonlocal agricultural products has directly influenced the hikes in urban food prices. Therefore, in this research, attempts have been made to explore the impacts of urban expansion on agriculture in and around the city limits, and how this influences rural-urban linkages and the food supply chain.

Most often, the urban centers are expanding over the country prime agriculture land because the development cost on fertile land is comparatively less. Most of the world's major cities have been built on fertile land and the surrounding cultivable land fulfilled citizens' food and agricultural requirements for hundreds of years. Worldwide, urban expansion is unavoidably encroaching on prime agricultural land in the urban periphery. Such land-use changes occur due to the increasing population pressures within city limits. In addition, increasing land values in the peri-urban zones most often result into vacant land.

Cities gradually expand over the most productive agricultural land since the majority of these urban centers have been established over fertile agricultural land and support their residents. The declining amount of agricultural land around a city led to intensive cultivation of farmland. Similarly, the same process also poses challenges for rural and urban food security. It has been observed that urbanization will further extend her horizons. Unchecked urbanization persists and will continue, especially in low-income nations (Satterthwaite, 2007). The situation is comparatively more devastating for a growing number of megacities, while there are relatively few cities where deurbanization was noted in the last population census (United Nations, 2014). It will be interesting to see reduced pressures on urban agriculture and to quantify, monitor, and predict urban expansion over the agricultural land for each city or entire nation. Urban expansion over farmland is a serious threat to agricultural land and city air quality, in effect leading to a silent disaster. This situation needs government intervention to limit the loss of farmland to only built-up environments.

7.7 Farmland Conversion and Policy Issues

The rapid pace of city growth has resulted in competition with various land uses, such as residential, agricultural, industrial, commercial, and infrastructure, with little or no regard for the environment. Risks to humans has occurred as a consequence of growing incompatible land utilization. Similarly, high urban land prices seriously curtail urban renewal projects, and as a result, it affects project preparation and implementation. It is true in case of extending urban services and therefore most cities lack recreational spots and green spaces.

Real estate agents are playing another dark part in this silent disaster. Lack of urban land use planning and regulations has encouraged low-income dwellers to build their houses in high-hazard zones. Due to their unfortunate situations, these people are not able to settle in safer places. In middle- and low-income nations, there is an absence of urban land-use policy, programs, and plans to check haphazard urban expansion and guide physical urban expansion according to strategic planning (Satterthwaite et al., 2010). These land-use plans would provide a framework to determine physical expansion and where to locate residential, industrial, commercial, recreational, and physical infrastructure. In most cases, urban administrators lack any control over land use and urban expansion, and there has been a consistent conversion of farmland into nonagricultural uses. It is, therefore, crucial to devise city land-use policies, enforce those policies, and check encroachment over farmland.

7.8 Case Study: Peshawar, Pakistan

Urbanization in Pakistan has progressed gradually over time (Khan & Rahman 2000). In 1901, 9.8% of the total population lived in urban places. After the country gained its independence in 1947, it increased to 17.8% in 1951, while in 1981, it was 28.3% (Khan, 2003). In more recent years, the trend has continued, reaching 32.5% in 1998. This increase in urban population is substantial, but compared to the rest of the world, Pakistan has a low level of urbanization. For instance, in 1900, a total 14% of the world population lived in urban areas; in 1950, this percentage increased to 30%; in 1980, it increased to 41%; and in 1998, it increased to 45%. This comparative analysis indicates that Pakistan lags far behind the world, and the gap between the world's average urban population and Pakistan's urban population is increasing.

Despite this, Pakistan is experiencing urbanization at a rapid pace, but at the cost of degrading its natural environment. With urban growth, agricultural land and other natural environments have been transformed into built-up environments, which is irreversible. It is estimated that in 2005, almost half of the country's population was living in urban areas, and that same report estimated that by the year 2015, the total urban population will be 80 million (GOP, 2014). Of the country's total urban population, 40% will live in two cities: Karachi and Lahore. These developments will have a negative impact in the form of urban expansion on either agricultural land or open spaces, and give rise to complex economic, food security, and ecological problems for future generations.

In Pakistan, Peshawar is the fifth most populace city after Karachi, Lahore, Faisalabad and Rawalpindi and a capital city of Khyber Pakhtunkhwa province (Khan and Rahman, 2000). Physically, it is an integral part of fertile Peshawar valley (Khan, 2003; also see Figure 7.1). Being a provincial capital, Peshawar is a hub of financial, industrial, and administrative functions, and a rapidly growing urban center. From 1972 to 1981, the city experienced population growth of 4.69%, which was higher than the national average of 4.38% and provincial growth rate of 3.89% (GOP, 1983; Samiullah, 2013; also see Table 7.1). Similarly, between 1981 and 1998, Peshawar grew at an average annual rate of 3.3%, which is higher than the provincial (2.7%) and national (2.6%) averages (Samiullah, 1998). In addition to natural increases, people from surrounding areas migrated to Peshawar in search of employment, education, health facilities, and other benefits. The influx of hundreds of thousands of Afghan refugees further intensified the pressure on urban amenities and accelerated the haphazard sprawl of the city. According to the 1998 census, the population had reached 2 million, excluding Afghan refugees (GOP, 2000). As per Afghan refugee census, in Peshawar, total Afghan refugees were 611,501 during 2005. This process has substantially contributed to the growth of Peshawar, and as a result, unplanned development took place in all areas of the city.

In Peshawar, the urban growth pattern is similar to other major urban centers in Pakistan, where ribbon sprawl is prevalent along the radial road and leapfrogging is common between radial roads. Such urban expansion is rapidly engulfing prime farmland and poses a serious threat to urban food security. Both natural increases and rural-urban migration were blamed for the overwhelming growth of almost all Pakistani cities. This alarming urban sprawl has occurred at the cost of degrading prime agricultural land. In Peshawar, residential land use was the biggest consumer of agricultural land from 1991 to 2009, when 8750 ha of farmland was lost to residential uses (Samiullah, 2013). Similarly, large sections of agricultural land was also consumed by the brick kiln industries because factories were built on cultivable land.

In Peshawar, urban sprawl has been blamed for haphazard land utilization and encroachment over agricultural land. Urban areas are expanding at a rapid pace, resulting in large-scale physical growth and changes in land use. The spatial variations are more obvious in the neighboring fringe areas than in the city center. The analysis of the situation in Peshawar indicates that the city has been growing at a rapid rate in the recent past due to lack of planning control, and that there has been haphazard urban expansion over prime agricultural land. The rapid urban expansion harked back to the 1960s, when as the city presented a radial road pattern. Since then, Peshawar has grown in a leapfrog fashion, encroaching upon farmland with no or very few

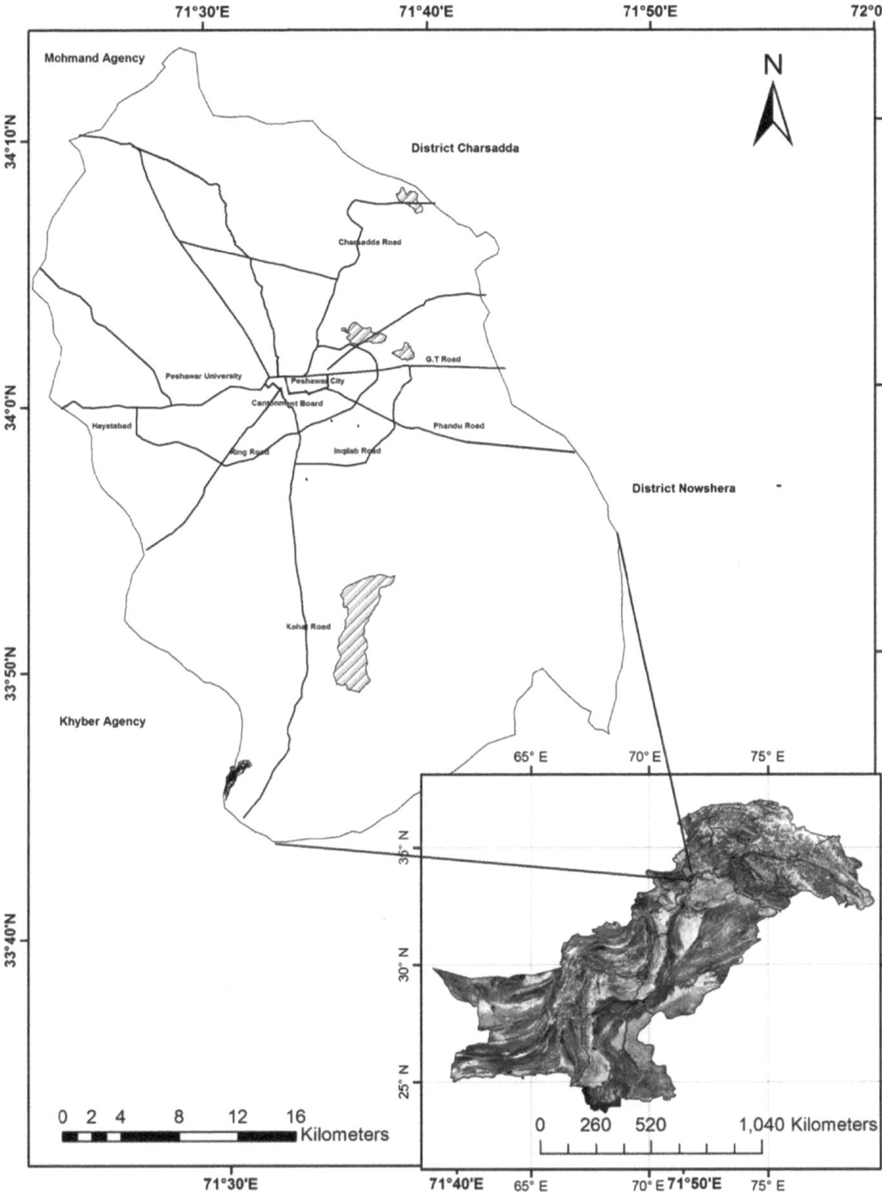

Figure 7.1 Peshawar district in Pakistan (modified after Samiullah, 2013).

urban services. Such unplanned urban expansion has resulted in slums and blighted residential clusters with a lack of amenities, a mixture of nonconformal land uses, and piecemeal commercial development (Samiullah, 2013). Economic development, population growth, and migration from surrounding areas are major driving forces behind urban growth. It is vital at this time to quantify and monitor spatio-temporal urban growth with sustainable land-use planning and management.

Table 7.1 **Peshawar Population Growth 1951–2009**

Census Year	Population Size (in thousands)	Intercensual Increase (%)	Annual Growth Rate
1951	391	-	-
1961	529	35.29	3.08
1972	807	52.55	3.7
1981	1113	37.92	3.89
1998	2019	81.4	3.56
2009 (Est.)	2765	34.94	3.56

Source: GOP (1983); GOP (2000); Samiullah (2013).

The city of Peshawar is rapidly growing in all directions, but the expansion toward the north and east is particularly noteworthy. Such development is haphazard and unplanned over the fertile farmland in the area, which was once the "food basket" for the city's residents (Samiullah, 1998). It has been observed that urban growth over farmland usually takes place along the radial roads. This phenomenon is referred to as *star-shaped city growth,* and this urban expansion is identified as *radial expansion* (Samiullah, 1998; Liu et al., 2005); it occurs in particular over prime agricultural land, posing serious threat to the "food basket." Once this prime agricultural land is consumed, the development is irreversible, and it may pose serious challenges. To check this overwhelming increase toward the north and east, the Peshawar Development Authority (PDA) has begun to plan Hayatabad Township in the western side of the city, and also recently planned Regi Lalma Township in the northwest. Despite all these planned initiatives, the unchecked and haphazard urban expansion over farmland still continues and needs a policy response and land-use regulations to stop further encroachment over agricultural land.

7.8.1 Spatial and Temporal Growth of Peshawar

Before 1866, Peshawar occupied only 193 ha of land, which is approximately 0.15% of the district area today (Table 7.2; Figure 7.1) and the city area increased to 126,710 ha in 2009. The spatial and temporal growth of Peshawar can be traced through five boundary extensions that occurred at different times. The first phase is the pre-1866 period, and it includes the area within the walled city. In the second phase of urban expansion (1866–1947), Peshawar was under British control (Table 7.2; Figure 7.2). The British colonial rulers did not add much to the indigenous city, other than a military cantonment in 1866 to the west. The third major expansion was made during 1947–1981, when the built-up areas doubled from 2853 ha in 1947 to 4635 ha in 1981 (Samiullah, 2013; also see Figure 7.2). This was a major boundary extension by the urban authority, and it gradually consumed over 10,000 ha prime agricultural land. The recent urban expansion took place during 1981–2009, and it was an explosive expansion over farmland. During the same period, another major

Table 7.2 Urban Expansion in Built-up Areas of Peshawar from Pre-1866 to 2009

Year	Area within Administrative Boundaries (ha)	Percentage of Present-City District Area	Size of Built-up Area Within City (ha)
Pre-1866	193	0.15	193
1866	919	0.72	919
1947	1678	1.33	2853
1981	11,100	8.83	4635
1991	11,100	8.83	7182
2009	126,710	100	16,986

Source: GoNWFP, 1987; GoNWFP, 2001; Samiullah (2013).

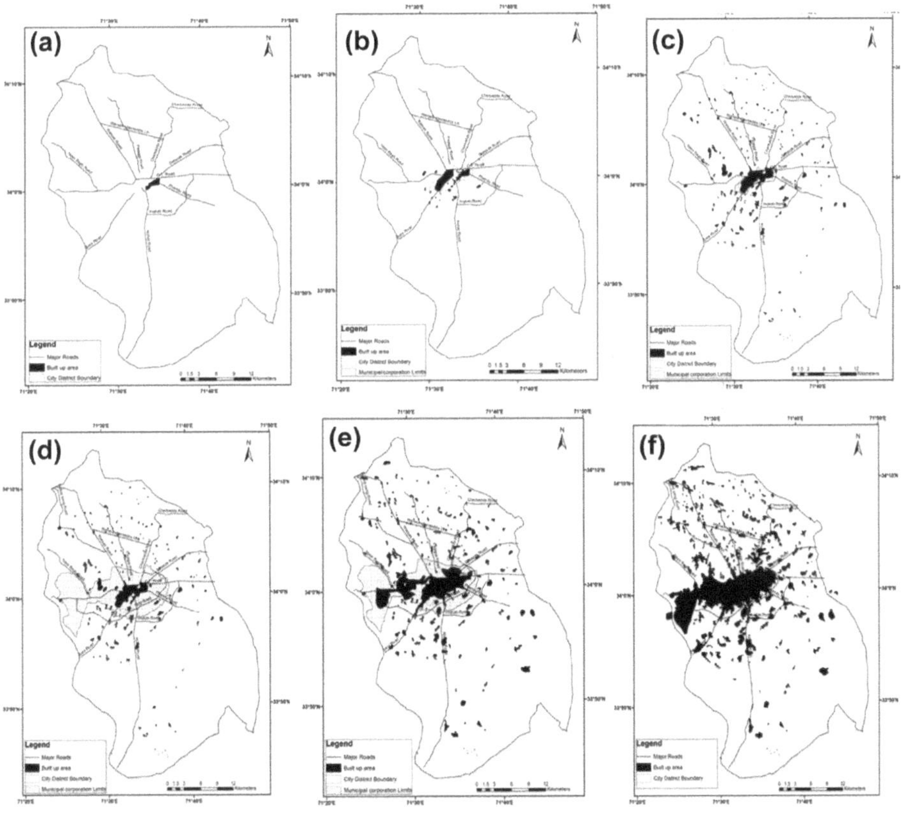

Figure 7.2 Spatiotemporal urban growth of Peshawar: (a) pre-1866 built-up area; (b) built-up areas (1866); (c) built-up areas (1947); (d) built-up areas (1981); (e) built-up areas (1991); (f) built-up areas (2009).
Source: Modified after GoNWFP, 1987; GoNWFP, 2001; Samiullah (2013).

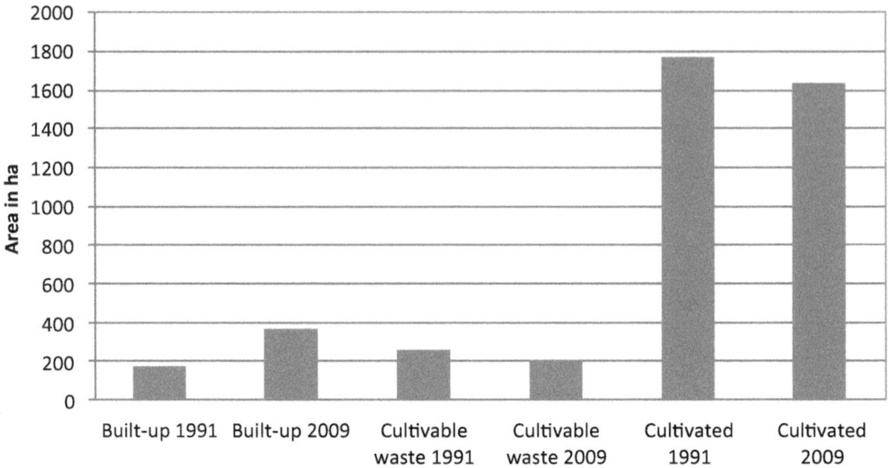

Figure 7.3 Change of land utilization from 1991 to 2009 (modified after Samiullah, 2013).

contributor was the influx of Afghan refugees into Peshawar, which further intensified the spatiotemporal growth. Similarly, it was estimated that from 1991 to 2009, an estimated 500 ha of prime agricultural land was consumed by the built-up environment (Samiullah, 2013).

In 2013, it was estimated that every year, about 700 ha is required for new housing for the Peshawar city dwellers (Samiullah, 2013). In Peshawar, residential land use was the biggest consumer of farmland during 1991 to 2009, followed by brick kilns (Figure 7.3). The farmland lost to built-up environment was mainly of the best quality (Figure 7.4). However, urban expansion over farmland and consumption of prime agricultural land were the major negative impacts. This also indicates the trend of declining farmland and a serious threat to food security and carbon sinking, leading to a creeping disaster for a fast-growing city like Peshawar. The urban encroachment over prime farmland is continuing at a rapid pace, with serious implications.

7.8.2 Trend of Declining Agricultural Production

In Peshawar district, various types of crops are grown to meet local demands. The data reveal that the total area under crop cultivation is fluctuating since the early 1980s, but overall, the decreasing trend has been noted (Samiullah, 2013). Microlevel analysis further confirms that urban areas are gradually consuming the prime agricultural land in the region. However, some new wasteland has been brought under cultivation between 2000 and 2009. Statistical analysis reveals that in Peshawar district, the land that has been lost to built-up areas falls mainly into class I (best agricultural land), followed by class VII, II, III, IV (Table 7.3). It was found from this analysis that residential land use was the biggest consumer of farmland during 1991–2009, and brick kilns were the second-largest consumer. This is an alarming trend: namely, the city population is increasing day by day, but "food basket" land is decreasing at a acceelerated pace. These results can safely be attributed to the silent disaster, and they

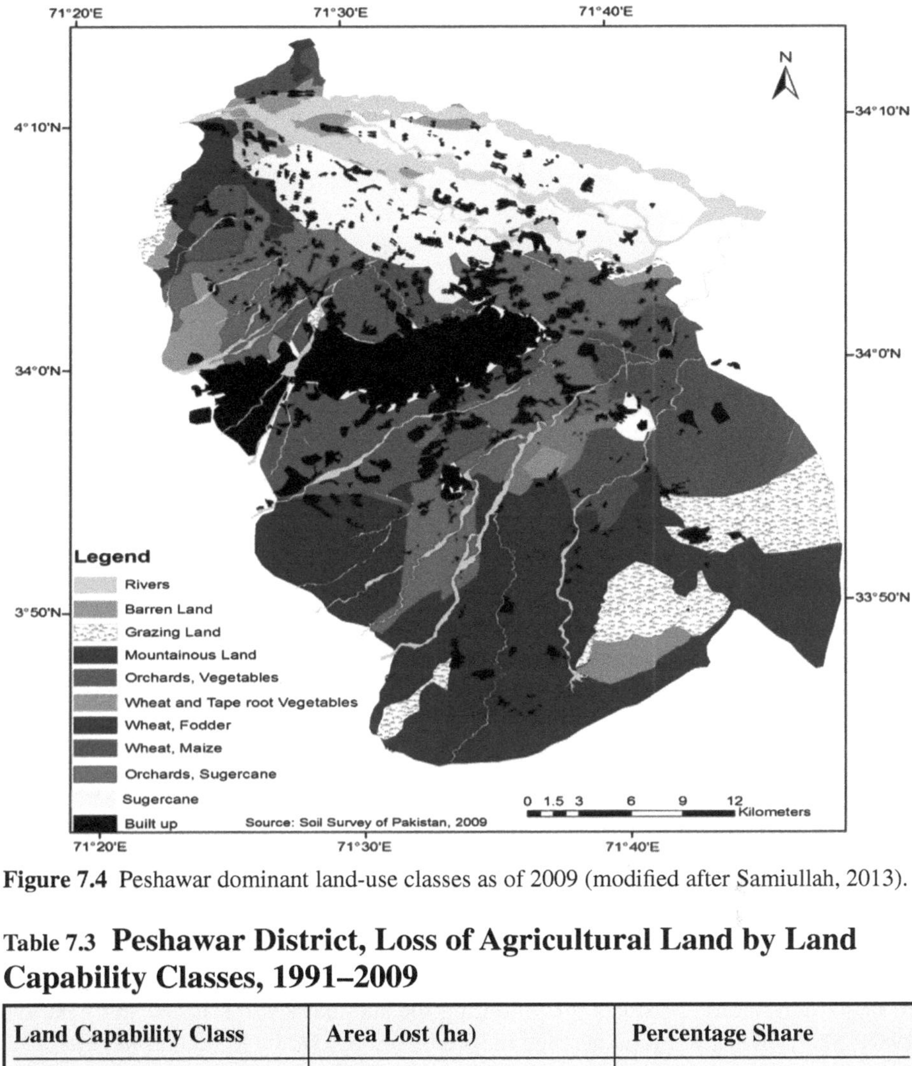

Figure 7.4 Peshawar dominant land-use classes as of 2009 (modified after Samiullah, 2013).

Table 7.3 **Peshawar District, Loss of Agricultural Land by Land Capability Classes, 1991–2009**

Land Capability Class	Area Lost (ha)	Percentage Share
Class I	8456	76
Class II	461	4
Class III	410	4
Class IV	246	2
Class VII	1620	14

Source: Modified after Samiullah (2013).

present a warning sign for policy makers to take preventive measures, devise policy regulations, and sensitize the community before it becomes irreversible.

Microlevel analysis revealed that at the village level, the changes are very alarming. It was observed that villages located close to the city centers present very disappointing positions than those located away from the city center (Samiullah, 2013).

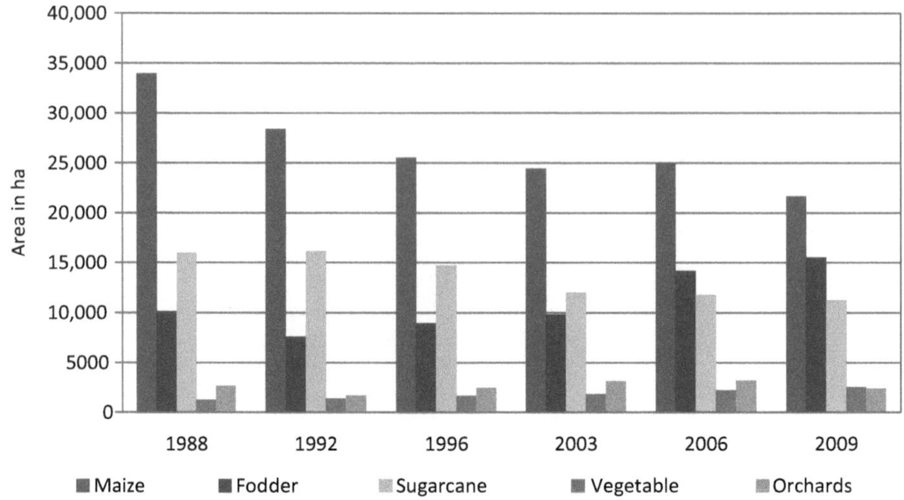

Figure 7.5 Peshawar district, summer cropping pattern, 1988–2009.

The assessment of land-use changes shows a consistent encroachment of built-up environment on farmland. The announcement of new housing schemes by individuals, groups, and private enterprises and real estate dealers has further aggravated this silent disaster, in effect posing a serious threat to food security and carbon-sinking hotspots. This emerging trend and serious urban issue need to be properly addressed and checked through effective land use regulation and enforcement.

The low productivity and yield that exists today is a major hindrance to the commercial viability of agriculture. There is a need to develop both crop and dairy farming along scientific lines to increase the yield. Factors such as traditional farming practices, inefficient irrigation methods, high input costs, lack of biosafety regulations, and insufficient institutional credit for poor farmers are the main causes of low yield. The analysis reveals that there is consistent reduction in agricultural production in Peshawar district. Figure 7.5 indicates that maize, fodder, sugarcane, vegetables, and orchards are major agricultural products grown during the summer season. In the study period (1988–2009), the agricultural acreage reduced from 63,000 ha to 53,000 ha. The data further reveal that the area under cultivation of maize, which is a major food grain, shrank from 34,000ha to 20,000ha during the study period. In 2009, the total farmland was 72,712 ha, with per-capita farmland of 262 m^2. It has been estimated that per-capita farmland is reducing at a rate of 5.18 m^2; if this trend continues, by 2060 there will be no farmland at all in Peshawar district.

7.8.3 Urbanization and Loss of Water Aquifers

In Peshawar, another negative but irreversible implication of the booming transformation of farmland into built-up environments is a drastic reduction in underground water aquifers. The expanding city population will face another challenge: maintaining its water supply. Historically, numerous cities have been ruined due to drought, desertification, and loss of water aquifers. The same could be true for Peshawar in the future because the collected data reveals that every year, the underground water level is dropping an average of 5 feet; and therefore, several tube-wells have dried up within the old walled city limits. In Peshawar, an underground aquifer is the major source of water for domestic and industrial use.

In Peshawar, which is experiencing an ever-rising demand for water, this situation has put tremendous pressure on aquifers and declines the water table. Analysis reveals that the pace of discharge from groundwater aquifer is faster than the recharge rate—another silent disaster. The aquifer is recharged through seepage and percolation of rainwater and irrigation. However, with urbanism and concrete structures, the process of seepage and percolation has reduced drastically and encouraged recurrent pluvial floods. This is one of the serious threats to the sustainability of Peshawar. In order to respond to this emerging issue, in 2008, the city government decided to supply water from the Warsak Dam and minimize the discharge from aquifers. So far, however, no effective step has been taken to put this project into operation.

7.8.4 Transformation of Farmland and Pluvial Floods

Pluvial floods occur when torrential or prolonged rainfall saturates and chocks the urban drainage system and excess water overflows the drains, causing sheet surface runoff. Recently, urban flooding probably has become the most significant hazard faced by the commercial hubs of any country, and Peshawar is no exception. In this city, rain-induced urban flooding is more serious than river and flash flooding because of its high economic and human exposures. In Peshawar, the rapid transformation of farmland into concrete surfaces has reduced the soil's imbibing capacity. Similarly, haphazard urban expansion has brought several issues to the fore, such as choking and lack of a drainage system. After every torrential and prolong rainfall, the existing drainage system is chocked and unable to carry the excess water. As a consequence, surface runoff occurs, leading to frequent pluvial floods. In addition, the concrete structures limit seepage and percolation, and it reduces the recharge of underground water aquifers.

Pluvial floods, which usually occur with little warning in areas that are not vulnerable to flood events, can be characterized as another silent hazard. In the face of climate change, the pluvial flooding is expected to increase and intensify further in Peshawar due to uncertain extreme weather phenomena. This situation seems inevitable to occur in cities of the developing world, and this is the case for Peshawar as well. For this city, increasing urban population, farmland transformation, loss of underground water aquifers, and climate change are the serious issues that need to be addressed properly. Nevertheless, the booming population is the root cause of almost all such hazards, and it has the potential to put urban dwellers at great risk.

7.8.5 Role of City Government

In Peshawar, there are several government agencies working on the process of effective land-use planning and enforcement. There is a need for local-level structural plans under the umbrella of a flexible master plan to foresee and guide the incremental development. Conversion of "food basket" land due to urbanization would cause greater shortage of foodstuffs and increase the need to import food items. It is currently necessary to formulate land-use policy and strictly prohibit future urban growth on agricultural land.

Spatiotemporal analysis of land use is the process of identifying variations in the state of land utilization through times (Khan, 2003; Khan & Hye, 2010). Even though humans have been modifying land for their use throughout history, the present rate of change has been unprecedented, with greater consequences than ever before. Change detection is an important process in monitoring patterns of physical growth and managing natural resources. It is also useful for understanding the conversion of farmland to other uses and provides quantitative analysis of human interaction with the environment and the consequences (Han et al., 2008). Since the 1990s, with increasing pressure on land resources, land-use change has become a major focus of the study of global change.

In addition, commercial, residential, and industrial pressure on agricultural land has to be solved through exploitation of various opportunities with available strengths. Similarly, farmers to take maximum advantage of economic opportunities at the urban periphery. The role of urban authorities is central in this regard. Urban and peripheral agricultural land need to be recognized as an integral component of urban land-use planning, and any city master plan must incorporate the same. Likewise, the initiation of urban environmental policy is one indispensable step toward smart city growth.

7.8.6 Peshawar and Urban Policy Initiatives

In Peshawar, despite numerous urban policy initiatives, the encroachment over the farmland has been unprecedented. Such increasing trends of built-up areas have generated multiple problems in Peshawar. In this regard, the government developed a master plan for the city in 1965. Similarly, a town planning act was passed in 1975 with the objective to check the haphazard urban expansion and guide the city's urban growth in an orderly fashion. Similarly, the Municipal Committee was upgraded to the status of Municipal Corporation in 1981. Another milestone occurred in 1986, when a structural plan was prepared for Peshawar to guide its planned development to the neighborhood level.

In order to implement the structural plan and work for the uplift of the Peshawar metropolitan area, the Peshawar Development Authority was established in 1987. Its main objective was to stop unplanned urban expansion, and particularly to save the prime agricultural land in the north and east and fill up the open spaces within the city limits. In this regard, Hayatabad Township in the west and Regi Lalma Township in the northwest were developed. In 2001, the Municipal Corporation was designated as the Peshawar Capital City District. Recently, an urban policy initiative called the Urban Policy Unit was established in 2013 to formulate a strategic urban policy, regulatory (legal) framework and urban management (institutional), and capacity building for

Peshawar, as well as providing technical support to line departments in managing urban affairs, and formulating economic regeneration and growth strategies for urban clusters. Despite government initiatives that include a national housing policy, a human settlement policy, a physical planning and housing sector, and a five-year plan, the encroachment into farmlands still continues around major urban centers like Peshawar.

References

Bender, W. H. (1997). How much food we will need in the 21st century? *Environment, 39*, 7–11.

Bhatt, B., Gupta, A. K., & Gogoi, G. (2005). *Application of Remote Sensing and GIS for Detecting Land Use Changes: A Case Study of Vadodara*. Accessed on April 6, 2014, from http://www.gisdevelopment.net/application/index.htm.

Bhatta, B. (2010). Analysis of urban growth and sprawl from remote sensing data. Springer, Heidelberg, p. 172.

Bourne, L. S. (1996). Reurbanization, uneven urban development and the debate on new urban forms. *Urban Geography, 17*, 690–671.

Bradshaw, T. K., & Muller, B. (1998). Impacts of rapid urban growth on farmland conversion: Application of new regional land use policy models and geographical information systems. *Rural Sociology, 63*(1), 1–5.

Brockerhoff, M. P. (2000). An urbanizing world. *Population Bulletin: A Publication of the Population Reference Bureau, 55*(3), 6–7.

Brooks, A. (2009). What is urban development? *Ehow editor (1999–2009)*. [online] Accessed on 2009 http://www.ehow.com/about_4728387_what-urban-development.html.

Brown, L. R. (1995). *Who will feed China?* New York, NY: W.W. Norton and Company.

Brunn, D. S., Williams, J., & Zeigler, D. J. (2003). *Cities of the world: world regional urban development*. Rowman and Littlefield. p. 548 on June 28, 2015.

Cai, F. (1995). Determinants, trend, and the policy of population mobility in China. *Chinese Demographic Science, 6*, 8–16.

Cassman, K. G., Wood, S., Choo, P. S., Cooper, H. D., Devendra, C., Dixon, J., et al. (2005). Cultivated systems. In R. Hassan, R. Scholes, & N. Ash (Eds.), *Ecosystems and human well-being: Current status and trends* (pp. 745–794). Washington, DC: Island Press.

Cheng, J., & Masser, I. (2003). Urban growth pattern modelling: A case study of Wuhan city, P.R. China. *Landscape Urban Planning, 62*, 199–217.

Crosson, P. (1997). Will erosion threaten agricultural productivity? Environment. *Science and Policy for Sustainable Development, 39*(8), 4–31.

Deakin, E. (1989). Growth controls and growth management: A summary and review of empirical research. In D. Brower, D. Godschalk, & D. Porter (Eds.), *Understanding growth management: Critical issues and a research agenda*. Washington, DC: Urban Land Institute.

Dramstad, W. E., Fjellstad, W. J., Strand, G. H., Mathiesen, H. F., Engan, G., & Stokland, J. N. (2001). Development and implementation of the Norwegian monitoring programme for agricultural landscapes. *Environmental Management, 64*, 49–63.

FAO. (1996). *Report of the World Food Summit*. Rome: Food and Agriculture Organization of the United Nations. Accessed on June 15, 2015 http://www.fao.org/docrep/003/w3548e/w3548e00.htm#date01.

Firman, T. (2000). Rural-to-urban land conversion in Indonesia during boom and bust periods. *Land Use Policy, 17*, 13–20.

Fox, J., & Vogler, J. B. (2005). Land-use and land-cover change in montane mainland, Southeast Asia. *Environmental Management, 36*, 394–403.

GoNWFP. (1987). *Peshawar Structure Plan and Programme*. Peshawar Development Authority.

GoNWFP. (2001). *District Gazetteer 2001*. Peshawar: Government printing press.

Government of Pakistan (GOP). (1983). *Peshawar District Census Reports*. Islamabad, Pakistan: Population Census Organization.

Government of Pakistan (GOP). (1999). *District Census Report of Mardan 1998. Statistics Division*. Islamabad, Pakistan: Population Census Organization.

Government of Pakistan (GOP). (2000). *Peshawar District Census Reports*. Islamabad, Pakistan: Population Census Organization.

Government of Pakistan (GOP). (2004). *Agriculture Statistics of Pakistan, 2003–2004*. Islamabad, Pakistan: Ministry of Food, Agriculture, and Livestock.

Government of Pakistan (GOP). (2014). *Pakistan Economic Survey, 2014-15. Population*. Islamabad: Labour Force and Employment, Government of Pakistan.

Gordon, P., & Richardson, H. W. (1977). Are compact cities a desirable planning goal? *Journal of the American Planning Association, 63*, 95–106.

Han, S. S., Li, H. T., & Gu, H. Y. (June, 2008). The Study of Land Use Change Detection Based on Object-Oriented Analysis. In Earth Observation and Remote Sensing Applications, 2008. EORSA 2008. International Workshop on (pp. 1–6). IEEE.

Heimlich, R. E., & Brooks, D. H. (1989). *Metropolitan Growth and Agriculture: Farming in the City's Shadow*. Washington, DC: U.S. Department of Agriculture, Economic Research Service. AER- 619.

Helmer, E. H. (2000). The landscape ecology of secondary tropical forest in montane Costa Rica. *Ecosystems, 3*, 98–114.

Kandrika, S., & Roy, P. S. (2008). Land use land covers classification of Orissa using multi-temporal IRS-P6 AWIFS data: A decision tree approach. *International Journal of Applied Earth Observation and Geoinformation, 10*(2), 186–193.

Khan, A. N., & Rahman, A. (2000). Urbanization: Trends and prospects in Pakistan. *Pakistan Journal of Geography, 10*(1–2), 31–43.

Khan, F. K. (2003). *A geography of Pakistan*. Karachi, Pakistan: Oxford University Press.

Khan, F. K., Wasi, B., & Rahman, S. (1992). Components of population growth of large towns in Pakistan. *Pakistan Journal of Geography, 2*(1), 22–30.

Khan, R. E. A., & Hye, Q. M. A. (2010). Agricultural land expansion in Pakistan: An empirical analysis. *Indian Journal of Agricultural Research, 44*(4), 235–241.

Lambin, E., & Geist, H. (2007). *Causes of land-use and land-cover change*. Accessed on October 18, 2015 Retrieved from http://www.eoearth.org/view/article/150964.

Lata, K. M., Rao, C. H. S., Prasad, V. K., Badarianth, K. V. S., & Rahgavasamy, V. (2001). Measuring urban sprawl: a case study of Hyderabad. *GIS development, 5*(12), 26–29.

Liu, J. Y. (2002). A study on spatial pattern of land-use change in China during 1995–2000. *Science China (Series D), 32*, 1031–1040.

Liu, J., Zhan, J., & Deng, X. (2005). Spatio-temporal patterns and driving forces of urban land expansion in China during the economic reform era. *AMBIO, 34*(6), 450–455.

Longa, H., Liua, Y., Wub, X., & Dongc, G. (2009). Spatiotemporal dynamic patterns of farmland and rural settlements in Su–Xi–Chang region: Implications for building a new countryside in coastal China. *Land Use Policy, 26*, 322–333.

McConnell, W. J., Sweeney, S. P., & Mulley, B. (2004). Physical and social access to land: Spatiotemporal patterns of agricultural expansion in Madagascar. *Agriculture, Ecosystems, and Environment, 101*, 171–184.

Nanda, T. (2005). *Urban sprawl and occupational change in Raipur City*. Unpublished M.Phil thesis submitted to the Department of Geography. India: University of Delhi.

Opeyemi, Z. A. (2006). *Change detection in land use and land cover using remote sensing data and GIS*. Unpublished M.Sc thesis submitted to the Department of Geography. Nigeria: University of Ibadan.

Ottensmann, J. R. (1977). Urban sprawl, land values, and the density of development. *Land Economics*, *53*, 389–400.

Oxfam. (2010). *Food security in India: Performance, challenges and policies*. New Delhi: Oxfam India Working Papers Series.

Pick, J. B., & Butler, E. W. (1997). *Mexico Megacity*. Boulder: Colorado, USA: Westview press.

Rahman, A., & Shaw, R. (2015). Urban risk and reduction approaches in Pakistan. In A. Rahman, A. N. Khan, & R. Shaw (Eds.), *Disaster risk reduction approaches in Pakistan* (pp. 295–314). Tokyo: Springer.

Rahman, A., Samiullah, & Khan, A. N. (2005). Land pooling system: A tool to control unplanned urban growth. *PUTAJ Science*, *12*(1), 11–16.

Rahman, A. (2010). *Ex post environmental impact assessment: A regional perspective*. VDM Verlag Publishing Co., Saarbrücken.

Rogers, A., & Williamson, J. G. (1982). Migration, urbanization and third world development: an overview. *Economic development and cultural change 30*, 463–482.

Samiullah (1998). *Urban sprawl in Peshawar city: A case study of eastern side of Peshawar city*. An unpublished M.Sc thesis submitted to the Department of Geography, Urban and Regional Planning. Pakistan: University of Peshawar.

Samiullah (2013). *Expansion of built-up area and its impact on urban agriculture: A case study of Peshawar-Pakistan. Ph.D thesis submitted to Institute of Geography, Urban & Regional Planning*. Pakistan: University of Peshawar.

Sarma, V. V. L. N., Muralikrishna, G., Malini, B. H., & Rao, K. N. (2001). Land use/land cover change detection through remote sensing and its climatic implications in the Godavari delta region. *Journal of the Indian Society of Remote Sensing*, *29*, 2011–218.

Satterthwaite, D., McGranahan, G., & Tacoli, C. (2010). Urbanization and its implications on food and farming. *Philosophical Transactions of the Royal Society B*, *365*, 2809–2820.

Satterthwaite, D. (2007). *The transition to a predominantly urban world and its underpinnings*. International institute for environment and development, London.

Saunders, D. (2010). *Arrival City: How the largest migration in history is reshaping our world*. New York: Random House.

Secretariat of the Convention on Biological Diversity (SCBD). (2012). *Cities and Biodiversity Outlook*. Montreal. 64 pages. Accessed on 18 October 2015 https://www.cbd.int/doc/health/cbo-action-policy-en.pdf.

Stamp, L. D. (1962). *Land of Britain: Its use and misuse*. London: Longmans Greens.

Sudhira, H. S., Ramachandra, T. V., & Jagadish, K. S. (2004). Urban sprawl: Metrics, dynamics, and modelling using GIS. *International Journal of Applied Earth Observation and Geoinformation*, *5*(1), 29–39.

Tan, M., Li, X., Xie, H., & Lu, C. (2005). Urban land expansion and arable land loss in China: A case study of Beijing-Tianjin-Hebei region. *Land use policy*, *22*(3), 187–196.

Todaro, M. (1984). Urbanization in developing nations: Trend, prospects and policies. In P. K. Ghosh (Ed.), *Urban development in third world*. Greenwood: Westport CT.

Turner, M. G., Wear, D. N., & Flamm, R. O. (1996). Land ownership and land-cover change in the southern Appalacian highlands and the Olympic peninsula. *Ecological Applications*, *6*, 1150–1172.

United Nations (UN). (2000). *World Urbanization Prospects: The 1999 Revision*. New York: United Nations. More developed countries include all of Europe and North America, plus Japan, Australia and New Zealand. Less developed" includes all of Africa, Asia (except Japan), Oceania (except Australia and New Zealand), and Latin America and the Caribbean.

United Nations (UN). (2014). *Department of Economics and Social Affairs Division, Population Division. World Urbanization Prospects; 2014 Revisions*. Accessed on July 8, 2015, from http://esa.un.org/unpd/wup/CD-ROM/.

Weng, Q. (2001). A remote sensing–GIS evaluation of urban expansion and its impact on surface temperature in the Zhujiang Delta, China. *International Journal of Remote Sensing*, *22*(10), 1999–2014.

Yeh, A. G. O., & Li, X. (1999). Economic development and agricultural land loss in the Pearl River Delta, China. *Habitat International*, *23*, 373–390.

Enhancing City Resilience Through Urban-Rural Linkages

8

Nitin Srivastava[1], *Rajib Shaw*[2]
[1]Graduate School of Global Environmental Studies, Kyoto University, Kyoto, Japan;
[2]Professor, Graduate School of Global Environmental Studies, Kyoto University, Kyoto, Japan

8.1 Introduction

Urban areas are considered to be areas that consolidate the safest infrastructure and vibrant economic opportunities in one location. Globally, a minimalistic estimation of economic concentration or density suggests that half of worldwide gross domestic product (GDP) was produced in just 1.5% of the world's land, most of that covered by cities (World Bank, 2008). These cities also face acute threats from many natural hazards because of their prosperity; the benefits of economic density come with a higher concentration of people and assets at risk from natural hazards. Consequently, the exposure to natural hazard risk in urban areas is large and increasing, largely driven by the greater economic opportunities in cities (Lall and Deichmann, 2011).

Statistics show that in 2011, around 370 million people lived in cities located in earthquake-prone areas, and 310 million of them lived in cities with high probability of tropical cyclones. The population under these threats is likely to double by 2050, thereby threatening these urban areas even more (Lall and Deichmann, 2011).The largest anticipated urban populations exposed to cyclones live in South Asia, where 246 million residents of large cities are estimated to be affected by severe storms by 2050 (World Bank, 2009).

Brown (2012) claims that in the 19th and 20th centuries, disasters disproportionately affected more families in rural areas than in urban areas. The basic needs of the affected families during this period were met easily through the distribution of food and water, shelter and health care, and basic sanitation. This was because these rural areas were characterized by sparse populations living in close proximity to their food sources. In the 21st century, with more and more urban populations getting affected and the urban nature of most of the disasters of recent years, the focus of assistance has shifted from the household to the neighborhood level (Brown, 2012). This is because only a small proportion of the urban population owns land; the majority of the rest rely on common resources. Disruption of these common resources affects a significant proportion of the population and hence poses challenges of a different nature.

8.2 Urban-Rural Linkages

Different authors attribute importance to different linkage elements to explain urban-rural interrelationships. Preston (1975) classified the interactions between the town and the country into five categories: (i) movement of people, (ii) movement

Urban Disasters and Resilience in Asia.

of goods, (iii) movement of capitals, (iv) social transactions, and (v) administrative and service positions. Tacoli (1998) differentiates the interactions between urban and rural areas into spatial (flow of people, goods, money, information, and wastes across space) and sectoral (across various segments such as agricultural, manufacturing, and services).

Urban-rural interrelationships and their related issues vary across various regions of the world. The relationship between the urban and rural areas in a developing country differs substantially from that in a developed country. With unprecedented growth in urban areas, developing countries have the challenge to address both urban and rural issues with the aim of all-encompassing regional growth.

8.3 Interdependency of Urban and Rural Areas

There is much evidence from developing countries proving that urban and rural areas are mutually dependent on each other for natural resources, raw materials, finished products, waste disposal, employment opportunities, and social interactions (Srivastava and Shaw, 2013). These interdependencies are defined in the next sections.

8.3.1 People

The flow of people between urban and rural areas provide for occupations, predominantly for the rural poor. It is an important factor in determining the well-being of the population during normal (nondisaster) scenarios, as "standard of living, disposable income, savings, education, health, and overall psyche of the society is determined by current levels of employment and unemployment" (Sharma and Krishna, 2007, p. 1). The flow of people between the spatial areas is the most widely considered element in interaction theories. Zewdu and Malek (2010) consider it as the 'key component in rural transformation' leading to industrial development in developing countries. This movement is usually irreversible as the migrating population is drawn to better economic, educational, and social opportunities (Brown, 2012). In developing countries, it might benefit the urban economy by the provision of an additional labor force. Simultaneously, this influence gives rise to a dual economy with the growth of an informal sector with negative environmental connotations (Gupta, 1993).

8.3.2 Natural Resources

The category of natural resources primarily includes sharing of natural resources such as water, land and energy. Zeleke and Trutmann (2006) highlights that there are negative effects on the environment between the urban and rural spheres, which can be minimized through harnessing urban and rural linkages. Rural incomes are predominantly agriculture-based in a developing country that is relying on natural resources; therefore, the impact on natural resources hurts the earning capacity of the rural population.

8.3.3 Products

The next factor to consider is a bidirectional flow of products. The rural-to-urban flow includes food for urban populations, raw and processed materials for urban manufacturing units, and labor flow for the nonfarm sector. On the other hand, manufactured and processed goods and inputs for the agricultural sector find market in the rural areas. Another dimension is that the peri-urban area may furnish parts or preliminary products for factories in urban areas, causing product linkages to be established.

8.3.4 Financial

The financial segment includes the flow of money in the form of remittances from urban to rural areas, rural savings channeled through urban institutions, and rent acquired by urban landlords. Johnson and Whitelaw (1974) established that the initial money flow is from rural to urban areas when the rural migrant becomes established in the urban setting; thereafter, there is a considerable reverse flow in remittances. Another important component is investment from urban areas into peri-urban and rural areas. The impact of greater investment in evidence-based, low-cost interventions at the rural level in developing countries, according to Buse et al. (2008), is through Millennium Village Projects being used to achieve the Millennium Development Goals (MDGs), especially poverty reduction.

8.3.5 Waste

Although waste is more of a physical linkage between the urban and peri-urban areas, it has spatial, sectoral, and natural implications. With less investment in waste disposal in major cities in developing countries, there is a greater reliance on waste disposal in the form of landfills on the periphery of the urban boundaries. Another component is the various forms of pollution affecting the water resources and air, both in urban and rural areas. This causes pollution of the land and water sources in the vicinity, which in turn affects the agricultural output of the neighboring areas and has indirect effects on the economy of these areas.

8.3.6 Information

Another category to consider includes the flow of information regarding resources or raw materials from rural to urban areas, markets and prices for these commodities, and the employment opportunities of urban and rural areas. It also includes the sharing of knowledge and technology between the two spheres, which is an important component. The usual situation with villages in developing countries is that they are undersupplied and underdeveloped. This can be rectified with an improvement in roads, rail service, and other vital communication linkages.

8.3.7 Social Interactions

Social linkages need to be considered under a separate category, as they include the intangible interactions between city dwellers and the villagers. Cultural exchanges, which play a vital role in strengthening ties, are also included in these social interactions.

Preston (1975, p. 172) says that this includes "promises of favor, the provision of protection, or the establishment of a formal relation," and these intangible elements may be more valuable than the tangible goods when it comes to establishing the linkages. These interactions improve the confidence of the rural migrants entering the urban society, and the consequential improvement in income brings about better social standing.

8.3.8 Governance

Governance forms the medium for implementating all policies, so it has an indispensable role to play in determining the flow of the abovementioned elements. Surprisingly, it has not been adequately represented in interaction theories other than the dilute reference to the administrative provision in Preston (1975). The government has the role of a facilitator in positive linkages, but it can also play a constricting role in negative exchanges. This element not only includes the formal setup of governments, but also includes civil society organizations and nongovernment organizations (NGOs), which share the responsibilities of governance in administering the policies of the government.

8.4 City Resilience

A city is resilient only if all its constituents can withstand and recover from the effects of a disaster. Similarly, a region is resilient if all its constituents (namely, those living in urban, peri-urban and rural areas) are able to face the negative consequences of disasters and return to their predisaster status in a minimal period of time. Before looking at what makes a city resilient, it is important to understand the specific characteristics that make a city less resilient.

- *Urbanization*—Urbanization tends to be associated with increasing incomes and better education, which generally reduce damages from hazards (Kahn, 2005). However, with the influx of the rural population into urban areas, the balance, and thus the resilience, of urban areas are disturbed. There is a greater population desnity per unit infrastructure, and also more dilapidated housing, unsafe working conditions, and denser environments. For example, three different disasters in 2011, a great earthquake and tsunami in East Japan; floods in Bangkok, Thailand; and an earthquake in Christchurch, New Zealand all hit densely populated areas.
- *Geography*—The physical geography of a city determines its vulnerability, and therefore its resilience. The presence of rural areas in the vicinity of a city also plays an important role in shaping its resilience. For example, in the event of an earthquake, the nonaffected rural areas might supply food to the nearby urban areas. This situation might also compel a large number of rural populations to work toward the reconstruction of urban areas. The interplay of economic and physical geography is one major reason for the high hazard risk in urban areas (Brown, 2012). Often these geographies overlap, with the most vulnerable areas in the cities such as low-lying areas, dilapidated housing, and the densest areas, being inhabited by the low-income population.
- *Land scarcity* —Local governments often exacerbate land scarcity by restricting high-density development. The desire to live close to jobs and amenities implies that even marginal city areas such as floodplains or steep slopes, which are the most vulnerable to hazards, often will be settled by poor people.

- *Varied income*—People with varying income levels have different resilient capacities against disasters (Surjan and Shaw, 2009). As economic sustenance for individuals holds paramount importance during a disaster, the human resources linkage needs special attention.
- *External factors*—Land scarcity leads to higher land prices and therefore, usually, to higher density occupation. The construction of larger buildings in cities can increase the damage and loss of life caused by severe earthquakes, especially where building codes are not enforced.

In addition, there are many other factors which are instrumental in determining a city's resilience such as the existing infrastructure, disaster management (including the disaster preparedness), city management systems (e.g. solid waste management, early warning mechanisms, rain water harvesting), enforcement of building bye-laws and urban-rural relations.

8.4.1 *The Case of India*

India has adopted the path of decentralized governance, which also had an influence over changing urban-rural linkages, with rural areas drafting their own policies. To give the appropriate significance to rural areas, the growing cities in India have resorted to regional statutory authorities such as the Mumbai Metropolitan Regional Development Authority (MMRDA), the Ahmedabad Urban Development Authority (AUDA) in Ahmedabad, the Hyderabad Urban Development Authority (HUDA) in Hyderabad, and the Delhi Development Authority (Figure 8.1). The macroeconomic and sectoral policies drafted by national governments must integrate urban and rural areas in their policies to achieve sustainable economic and social development, while local governments must go for overall development, taking care that the national policies are adequately altered for their local conditions.

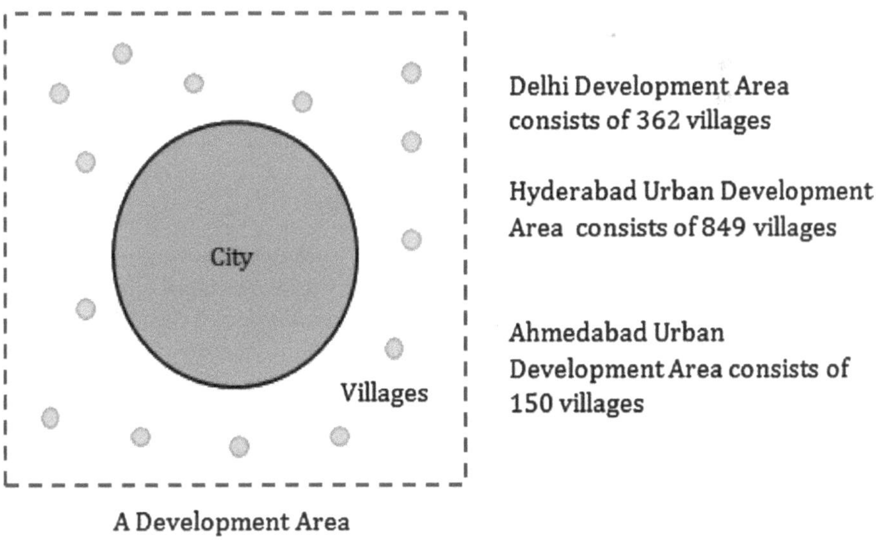

Figure 8.1 A typical development area consisting of villages under an urban development authority.

8.4.2 The Case of Pakistan

An example from Pakistan shows the comparative impact of a disaster on urban and rural areas. Kirsch et al. (2012) conducted a cross-sectional cluster survey of heads of households affected by the 2010 floods that occurred in Pakistan in January 2011. The survey was focused on the priority districts in the four affected provinces (namely, Balochistan, Khyber Pakhtunkhwa, Punjab, and Sindh) with 29 districts designated as "severely" flood affected by the government of Pakistan and the World Bank.

The flood adversely affected the income of 88% of households (90% rural, 75% urban). Kirsch et al. (2012) demonstrated that the rural settings were more adversely affected and had slower recovery after the flood. The impacts included greater impact on their income, sanitation, and electricity supply and less economic recovery at six months. Rural residents were also more likely to have moved to an entirely different geographic area and less likely to have been able to return to their original homes after six months. This postdisaster phenomenon undermines the natural resilience of these cities due to increased dependency and abbreviated self-reliance.

The Indian example shows that the development model integrates both urban and rural areas under the same authority. However, disaster management systems and initiatives often overlook the interdependency of these areas. This coexistence of urban and rural populations can be used as an opportunity to create beneficial systems that can not only enhance a city's resilience, but also reduce the bias created by the current model of development, where most of the initiatives are meant for urban populations. The Pakistan example reinforces the notion that disasters in rural areas are bound to affect the urban areas, and vice versa.

8.5 Discussion

In section 8.2, the interdependency of the urban-rural areas was discussed. This section expands this idea to how these linkages behave in fast-growing developing countries in the event of a disaster.

8.5.1 Urban-Rural Linkages and Disasters

Global disaster damage statistics are not classified by urban versus rural. The extent to which urban areas or rural areas are affected depends on the country's overall urbanization rate, the type of natural disaster studied, and the measure of damage considered (Lall and Deichmann, 2011). An urban earthquake or hurricane may be more damaging than a flood or drought in a rural area, but it also may equally disrupt livelihoods in both types of location.

However, these urban and rural livelihoods are very different. Brown (2012) aptly differentiates between the two. According to him, urban livelihoods, in varied forms, have the primary objective of earning money to enable families to purchase those household items most urgently needed. In contrast, rural communities, which are primarily agriculture- and land-based, strive to produce sufficient food for its families and only then sell the surplus. Therefore, disruption to urban livelihoods, even for short durations, affects families' ability to survive. This might lead to price increases for food, water, and essential commodities (Brown, 2012). The urban poor, particularly migrants, are hit the hardest and often lack the financial, social, and physical assets to rely upon when there is a reduction of or interruption in income (Table 8.1).

Certain strategies that might be useful in enhancing a city's resilience include the following linkage elements:

- *Provision for utilization of natural resources spread over a region*—Proper systems can be created to utilize natural resources, like water and land, that are spread over a region, including both urban and rural areas, in the aftermath of a disaster. There can be an increased demand for water due to the collapse of distribution systems. In the postdisaster period, rural areas can channel their resources to bail out urban residents.
- *Risk communication*—Information in the affected cities about shelter, rehabilitation, employment opportunities, distribution of food, and other postdisaster welfare activities can be disseminated to rural areas by utilizing strong urban-rural ties between the inhabitants from both domains. Also, the strong ties between urban and rural areas, especially for families that are spilt between these areas, can help foster stronger recovery processes.

Table 8.1 Urban-Rural Linkage Elements

S. No.	Linkage Elements	Probable Disaster Impacts
1	People	A disaster in rural areas can trigger an influx of population from rural to urban areas in search of new employment opportunities.
2	Natural resources	A disaster in an urban area might put pressure on its limited natural resources. In these cases, urban areas might rely on rural areas for survival.
3	Products	Food and other rural products fall short of the demand during urban disasters and the city's catchment area increases in size considerably.
4	Financial	Disasters might create new jobs in cities affected by disasters, but they also take away a lot of economic opportunities, especially the ones in the informal sector, bringing the microeconomy to standstill.
5	Waste	Disaster waste is a major problem in cities, often with impacts beyond the administrative boundaries. For example, earthquakes produce a lot of debris, which is often disposed of in peri-urban and rural areas. During floods, excess water flows to the downstream villages.
6	Information	Much information—both right and wrong—regarding relief efforts and preparedness (training and evacuation) activities is communicated through word of mouth if there are no proper mechanisms for doing so.
7	Social Interactions	The social interactions descend to minimum with huge gap between urban and rural, rich and poor and their specific needs.
8	Governance	Local governments concentrate solely on the urban core and its periphery to provide relief, even though the model of development is regional in nature.

- *Poverty reduction through employment generation*—It has been widely recognized in the recent past that urban-rural linkages provide opportunities and constraints for poverty reduction and regional development (Zeleke and Trutmann, 2006; Braun, 2007; Cali and Menon, 2009; Srivastava and Shaw, 2013). Sustainable rural development and urbanization can be achieved only by integrating poverty alleviation efforts with urban-rural dimensions. The eradication of rural poverty requires access to social facilities like education and health, but mere social welfare schemes will not help eradicate chronic poverty. Fundamental transformations of the society, through a mix of capitalist and social paths of development, are needed, thereby requiring a partnership between private and public sectors. Poverty has a two way relationship with the linkages; it contributes to the flow of people, while urban-rural linkages can contribute to the poverty by biased policies.

  A diversification of the economic base would result in the increased job creations in nonfarm sector, thereby arresting rural to urban migration. The provision of assistance to small-scale enterprises, which have the capability to sustain themselves on their own once they get a helping hand, would allow rural populations to be involved in the mainstream economy. Also, the informal sector is playing an important role in rapidly growing cities in developing and transition countries. This role is a consequence of growth in the labor force without a matching response in the level of formal employment opportunities. The informal sector has great freedom of action, being free of government interference, and tends to deliver labor resources to productive areas of the economy. With higher incomes, better-quality housing can be built, mitigation measures can be made more affordable, and better institutions lead to enforcement of rules and regulations aimed at reducing vulnerability.
- *Regional growth*—Rural prosperity will enhance urban resilience. This is influenced by the structure of the settlement system, which has immense potential to be a positive force in the process of economic development (Funnell, 1988). Conversely, growth can regulate the linkages between the urban, peri-urban, and rural areas. Therefore, the linkages should be connected to the wider government vision of regional growth. Authorities often fail to distinguish the differentiated needs of urban and rural areas and aim for overall growth, assuming that the effects of such economic growth would trickle down to the poor, with more employment opportunities and higher wages. The growth-oriented approach must be replaced with a "growth with social justice" approach.
- *Decentralization of urban markets*—Urban populations rely almost exclusively on buying whatever goods they need from the market. In an urban setting, markets are integral to survival. Often urban and rural domains face market failures and stagnation due to the failure of one or more of the following: the labor market, services market, or goods market. The linkages between urban and rural areas can provide markets to rural products achieving "Regional Economic Resilience" (Hill, Wial and Wolman, 2008). In Guinea and Zimbabwe, Action Contre la Faim (ACF) found that urban-rural linkages had a major effect on food security and nutrition. The provision of access to markets would help improve the market linkages for the benefit of both the urban and rural poor.

8.6 Conclusion

The latest urban population estimates from the United Nations suggest that, globally, urban populations exceeded rural populations for the first time in 2008 (UN, 2008). In less-developed regions, this threshold is expected to be reached by 2019. That is the part of the world that presents an opportunity to utilize urban-rural linkages for enhancing the resilience of cities. It is agreed that the loss of life is much less in rich countries. Additionally, the economic damages tend to be larger, but when measured as a share of exposed wealth, they are less than in poor countries. The same is true for progressive urban areas as opposed to the lagging rural areas. The impact caused varies between the two areas; however, the total loss to a region is constant and can be nullified.

Urban-rural linkages provide an opportunity to plan for the enhanced resilience of cities through policies that can contribute to both urban and rural resilience. However, this resilience enhancement should not be achieved at the cost of other areas. Instead, urban areas should alleviate the inadequacies of rural areas for their own good as well as the good of the rural areas. The villages are becoming vulnerable due to their lack of infrastructure, scattered population, lack of disaster management capabilities, and limited livelihood opportunities. However, these can be tackled through initiatives planned and formulated in urban areas.

A city is resilient if all its inhabitants are resilient to the impacts. And city's resilience is the basis for rural resilience and consequently regional resilience. A well-oiled symbiotic system between urban and rural areas is bound to be beneficial for the urban areas. The adjoining rural area's prosperity would certainly enhance the resilience of the cities in more than one ways.

Acknowledgments

The first author acknowledges the support from the MEXT scholarship to undertake this research in Kyoto University.

References

Buse, K., Ludi, E., & Vigneri, M. (2008). *Beyond the village : The transition from rural investments to national plans to reach the MDGs, Sustaining and scaling up the Millennium Villages.* Overseas Development Institute, Synthesis Report, London, UK. Available at http://www.odi.org.uk/sites/odi.org.uk/files/odi-assets/publications-opinion-files/3288.pdf, last Accessed on 13.01.13.

Braun, J. V. (2007). *Rural-Urban Linkages for Growth, Employment, and Poverty Reduction, Ethiopian Economic Association Fifth International Conference on the Ethiopian Economy June 7–9, 2007, Keynote, Plenary Session I.* Retrieved on August 5, 2011 from http://www.ifpri.org/sites/default/files/publications/20070607jvbruralurban.pdf.

Brown, C. (2012). *The 21st-century urban disaster. CHF International (Global Communities).* Retrieved on March 27, 2015, from www.alnap.org/pool/files/2012-chf-urban-disasters.pdf.

Cali, M., & Menon, C. (2009). *Does Urbanisation Affect Rural Poverty? Evidence from Indian Districts.* Spatial Economics Research Centre (SERC), Discussion Paper 14. Retrieved on July 11, 2011 from http://eprints.lse.ac.uk/33205/1/sercdp0014.pdf.

Funnell, D. C. (1988). Urban-rural linkages: Research themes and directions, Geografiska Annaler, Series B. *Human Geography*, *70*(2), 267–274.

Gupta, M. R. (1993). Rural-urban migration, informal sector, and environmental problems in a small, open, less-developed economy: A theoretical analysis. *Journal of Development Economics*, *41*(1), 137–151.

Hill, E. W., Wial, H., & Wolman, H. (2008). *Exploring regional economic resilience*. University of California, Berkeley: Institute for Urban and Regional Development Working Paper Series, no. 2008-04.

Johnson, G. E., & Whitelaw, W. E. (1974). Urban-rural income transfers in Kenya: An estimated-remittances function. *Economic Development and Cultural Change*, *22*(3), 473–479.

Kahn, M. E. (2005). The death toll from natural disasters: The role of income, geography, and institutions. *Review of Economics and Statistics*, *87*(2), 271–284.

Kirsch, T. D., Wadhwani, C., Sauer, L., Doocy, S., & Catlett, C. (2012). Impact of the 2010 Pakistan floods on rural and urban populations at six months. *PLOS Currents Disasters*. http://dx.doi.org/10.1371/4fdfb212d2432 Aug. 22 Edition, 1.

Lall, S. V., & Deichmann, U. (2011). *Density and disasters: Economics of urban hazard risk*. World Bank Research Observer, http://dx.doi.org/10.1093/wbro/lkr006. http://wbro.oxfordjournals.org/content/early/2011/07/06/wbro.lkr006.short.

Preston, D. (1975). Rural-urban and intersettlement interaction: theory and analytical structure. *Area*, *7*(3), 171–174 Wiley Publishers on behalf of The Royal Geographical Society (with the Institute of British Geographers).

Srivastava, N., & Shaw, R. (2013). Establishing Urban-rural linkage elements for Growth and Poverty Reduction in India. *Asian Journal of Environment and Disaster Management*, *5*(1).

Sharma, B. K., & Krishna, N. T. (2007). *Employment-unemployment situation in Million Plus cities in India: A critical analysis* Directorate of Economics and Statistics government of NCT of Delhi. Retrieved from http://delhi.gov.in/DoIT/DES/Publication/paper61.pdf. Accessed on September 18, 2011.

Surjan, A., & Shaw, R. (2009). Essentials of urban disaster risk reduction. In R. Shaw, & R. R. Krishnamurthy (Eds.), *Disaster management: Global challenges and local solutions* (pp. 543–555). University Press (India) Private Limited, Hyderabad, India.

Tacoli, C. (1998). Rural urban interactions: A guide to the literature. *Environment and Urbanization*, *10*, 147–166.

United Nations (UN). (2008). *World Urbanization Prospects 2007 Revision*. New York: United Nations. Population Division.

World Bank. (2008). *World Development Report 2009*. Reshaping Economic Geography. Washington, DC: World Bank.

World Bank. (2009). *Current and future estimates of exposure to natural hazards in the world's large cities*. Background analysis for the Economics of Disaster Risk Reduction report. Washington, DC: World Bank.

Zeleke, G., & Trutmann, P. (2006). *Fostering new development pathways: Harnessing rural-urban linkages (RUL) to reduce poverty and improve environment in the Highlands of Ethiopia*. Working Paper 1, Working Paper Series on Rural-Urban Linkage Theme of the Global Mountain Program (GMP).

Zewdu, G. A., & Malek, M. (2010). *Implications of land policies for rural-urban linkages and rural transformation in Ethiopia*. Ethiopia Strategy Support Program (ESSP) II, Working Paper No.15. Washington, DC: International Food Policy Research Institute. Available at http://www.ifpri.org/sites/default/files/publications/esspwp15.pdf.

Urban Disaster Risk Reduction in Vietnam: Gaps, Challenges, and Approaches

9

Huy Nguyen, Phong Tran
Institute for Social and Environmental Transition, Vietnam

9.1 Introduction

As of December 2013, Vietnam has about 770 cities and towns, with an urbanization rate of over 33% (according to a report by Vietnam Ministry of Construction, 2014). As estimated by the United Nations Department of Economic and Social Affairs (UNDESA) in its World Urbanization Prospect report (UNDESA, 2011), urbanization will continue to expand in Vietnam and is estimated to reach nearly 56% in 2050 (Nguyen, 2014). Urbanization poses many potential risks to both existing and new urban areas and will make urban residents more vulnerable to the growing pressures related to urban planning, urban services, environment, and climate change.

Disaster risk reduction (DRR) and climate adaptation has become the priority issues for Vietnam nowadays. Because of its topography, Vietnam is susceptible to several types of natural disasters, such as typhoons, floods, landslides, erosion, drought, and some other climate-related impacts such as salinity intrusion and sea-level rise (SLR). Almost cities of Vietnam are located in the coastal area and in the two main deltas that face those type of disasters (CCFSC, 2014).

Disaster risk management (DRM) and reduction are not new topics in Vietnam. In fact, they have been widely implemented for many years, with government agencies at the national to local levels, nongovernmental organizations (NGOs), and civil society all playing active roles. However, in most cases, efforts have focused only on rural areas, and most of the activities involve preparedness, response, and recovery. The approach taken by government agencies and NGOs includes four main procedures of DRM (preparation, response, recovery, and mitigation) and highlights five key components (capacity building, preimpact, emergency, restoration, and reconstruction).

Most cities in Vietnam choose a single frame of analysis through which to interpret vulnerabilities such as specific hazards, geographical locations, and urban sectors (Tyler et al., 2010). This frame of analysis helps ensure that proposed actions respond to the relevant vulnerabilities in the respective framework and should make transparent any gaps in the proposed plan of action. Using a single frame of analysis also brings limitations, however. For example, cities that focus on geographical vulnerabilities may be more likely to miss an important intervention related to a citywide sectoral issue such as water management or energy systems.

Urban Disasters and Resilience in Asia.

Due to the complexities of urban populations and systems, as well as the nature of disasters that urban areas face, it is much more challenging to plan DRR for urban communities. In the context of a growing urban population, it is essential to develop and organize a system of suitable approaches and tools for DRR and climate change adaptation (CCA) in urban areas, with an initial focus on the concept of urban resilience.

This chapter summarizes the experiences, difficulties, and challenges with the current approach to DRR and CCA in Vietnam, especially in applying these traditional approaches and tools for vulnerability assessment in urban and peri-urban areas in the context of urbanization and climate change. The text also introduces tools used by the Institute for Social and Environmental Transition (ISET) for vulnerability assessment and building urban climate resilience; and provides suggestions on how to tailor these tools for the urban context.

9.2 Methodology

This chapter focuses on multiple approaches, including desk study and two main consultation workshops with input data from relevant stakeholders who work in the fields of DRR and urban resilience.

Most of the information given here is what was discussed during a technical workshop entitled "Shared Learning Workshop on Vulnerability Assessment Tools and Approaches for Disaster Risk Reduction in Rural and Urban Areas," organized by ISET, the International Federation Red Cross (IFRC), the Vietnam Red Cross (VNRC), and the American Red Cross (ARC). In December 2013, the workshop was attended by representatives of various entities, including ARC; Da Nang; the provinces of Thua Thien–Hue, Quang Nam, Quang Ngai, and Ha Tinh; the Disaster Management Center (DMC)–Ministry of Agriculture and Rural Development (MARD); the Central and Central Highlands Flood and Storm Control Center, and other stakeholder organizations in Da Nang, including the Da Nang Climate Change Coordination Office and Da Nang Women's Union. The meeting explored the challenges involved in conducting vulnerability capacity assessment (VCA) in urban areas and aims to provoke further discussion on how to address them.

A second workshop (in July 2014) was on "DRR in the Urban Context and Urban Resilience Approach," with the participants coming from MARD, RCs, representative of nine cities in Vietnam, and practitioners from relevant NGOs. It explored the challenges, gaps, and approaches for urban DRR in Vietnam.

9.3 Policies for Disaster Management in Urban Areas in Vietnam

9.3.1 Vietnam's Disaster Profile

Because of its topography, Vietnam is susceptible to several types of natural disaster, as shown in Table 9.1. Typhoons and floods are the most frequent and most devastating of these events. The storm season lasts from May to December. In most years, storms hit

Table 9.1 Relative Frequency for Specific Disasters in Vietnam

High	Medium	Low
Flood Typhoon	Hail rain/tornado Drought	Earthquake Accident (technology like chemical accidents, fire)
Inundation	Landslide Flash flood Erosion Salinity intrusion	Frost Damaging cold

Source: Vietnam Disaster Law (2014).

the northern part of Vietnam from May through June, and then move gradually south from July to December. Typhoons and floods often cause a heavy loss of life and damage to infrastructure and economic activity because of the concentration of people living along the coastline and in low-lying deltas, particularly those of the Mekong and Red rivers. Natural disasters cause an average of 750 deaths yearly and result in annual economic losses equivalent to 1.5% of gross domestic product (GDP). Because loss data is chronically underreported, the actual totals may be much greater (GFDRR, 2011)

Dangers to these urban populations have increased with the spread of economic activity and development into only marginally suitable areas such as floodplains, coastal swamps, drainage channels, or other natural buffers. The result is that most of Vietnam's cities are now located in low-lying river basins and coastal areas, so they are exposed to risks from multiple natural hazards.

An estimated 80%–90% of the total population of the country is affected by typhoons, according to the Ministry of Agriculture's Central Committee Flood and Storm Control. Vietnam experiences an average of 6–8 typhoons or tropical storms of varying intensities each year, with the northern and central coastal regions being hardest hit in the early months of the storm season (May–December). Communities along the coast are directly affected, but so too are communities in upland areas, which can experience flash floods resulting from the heavy rains of typhoons. River-plain flooding is extensive and prolonged throughout the wet season in large deltas. Most of Vietnam's 2,360 rivers are short and steep, so that heavy rainfall in their basins produces intense floods of short duration.

Sizable portions of the country—especially the Central Highlands and Central Coast—are subject to heavy rainfall. Table 9.2 listed the main disasters that have hit Vietnam in recent years, which have occurred in the urban areas and resulted in significant loss of life and property.

9.3.2 DRR Policy Framework

Vietnam has a long history of being prepared for and actively responding to natural disasters. Its extensive system of dikes and seawalls is evidence that citizens and leaders over the centuries have recognized the country's vulnerability to the ravages

Table 9.2 Recent Disaster Events in Urban Areas in Vietnam

Year	Event	People Killed	People Injured	People Missing	Economic Loss (Billion Dong)	Cities Affected
2014	Flood	11	N/A	3	N/A	Lang Son, Bac Can, and Lao Cai
2013	Typhoon Nari	38	N/A	N/A	1491	Cities in the central and north central (Da Nang, Hue, Quang Nam) of Vietnam
	Typhoon Wutip	65	N/A	N/A	11,000	Hue, Quang Tri, Dong Ha, and Quang Binh
2012	Typhoon Son Tinh	27	N/A	N/A	N/A	Cities from Quang Binh, Ha Tinh, Vinh and Thanh Hoa
	Flood in the Mekong Delta	83	N/A	N/A	N/A	12 cities in the Mekong Delta
2011	Flood in the Central of Vietnam	28	N/A	N/A	N/A	Cities in the Central of Vietnam
2010	Typhoon Mindulee	20	N/A	N/A	N/A	Quang Binh, Ha Tinh, and Vinh city
	Flood	100	N/A	N/A	N/A	Quang Binh (Figure 9.1) and Ha Tinh city
2009	Typhoon Marine	120	N/A	N/A	N/A	Cities in the south central part of the country
2008	Storm Kammuri	133	91	34	1939	North-central (Quang Tri, Quang Binh and Ha Tinh) cities
	Floods	144	N/A	N/A	3000	Hanoi and north-central (Hanoi, Nam Dinh, Thai Binh) cities
2007	Storm Lekima	88	180	8	3216	17 north and central cities
2006	Storm Xangsane	72	532	4	10,402	15 central and south-central cities
2005	Storm No. 7	68	28	N/A	3509	12 north and central cities
2004	Storm No. 2	23	22	N/A	299	5 central cities
2003	Floods	65	33	N/A	N/A	9 central cities
2002	Floods	171	N/A	N/A	457	12 cities in the Mekong Delta
2001	Floods	393	N/A	N/A	1536	12 cities in the Mekong Delta
2000	Flash flood	28	27	2	43	5 northern cities (Son La, Lao Cai, Dien Bien, Lai Chau and Bac Can)
1999	Floods	595	275	N/A	3773	10 central cities (Quang Ngai, Quang Nam, Hoi An, Da Nang, Hue, Quang Tri, Quang Binh, Ha Tinh, Vinh, Thanh Hoa)

N/A: data is not available
Sources: Compiled from (GFDRR, 2011); CCFSC website; and data from DMC.

Figure 9.1 Flooding in Dong Hoi city, Quang Binh, in October 2010. (*Source:* Vietnam Television Corporation News)

of typhoons and other tropical storms. The government of Vietnam (GoV) is actively engaged in preparing for what could be even greater challenges of climate change adaptation and DRR in the future. The following sections will explore these activities.

9.3.2.1 Less Urban DRR Focus

Most of the laws and regulations on natural hazards in Vietnam have dealt with high-risk events like floods and storms, while other hazards were dealt with in other laws and regulations. The legal framework on floods and storms dates back to 1946, when the Central Dyke Protection Committee, the forerunner of the present Central Committee for Flood and Storm Control (CCFSC), was established by President Ho Chi Minh. This framework included the Ordinance on Prevention and Control of Floods and Storms of 1993, the decree of 1996 providing details for its implementation, the 1996 Decision on the Establishment of the CCFSC, and the amendments of the 1993 ordinance in 2000. The low-to-moderate risk of earthquake and tsunami was regulated in the PM Decisions (on Disaster forecasting) of 2006 and 2007. The Decree on Civil Defence of 2008 dealt with all natural and manmade hazards and specified the structure and role of civil defense forces at the national, provincial, district and commune levels (IFRC, 2014).

These laws and regulations defined the functions and responsibilities of different agencies for disaster response, but there was little information and guidance about urban DRR. The disaster prevention and mitigation measures laid out in the abovementioned laws and regulations mainly involved short-term actions, while building a resilience strategy calls for long-term actions. They dealt with measures to prevent or mitigate the impact of forecasted hazards by taking actions such as issuing early warnings for sea crews and communities, evacuating people and property from at-risk areas, closing disaster-prone areas, preventing ships from going out to sea, directing ships to safe shelters, and guarding dikes. However, the 1993 ordinance identified some long-term disaster prevention actions. In particular, the law required the development of prevention plans for each area, construction of flood and storm prevention and mitigation infrastructure, planning for resettlement, and the promotion of DRR awareness within the population. The separate law concerning dikes also provided more details on the infrastructure (Dams, Irrigation system). These ordinances mostly focused on disaster management in rural areas.

In 2007, the center government issued the National Strategy for Natural Disaster Prevention, Response, and Mitigation, which outlines Vietnam's strategy for disaster mitigation and management and focuses on floods, storms, and drought. In addition to setting policies for disaster response, the National Strategy provided long-term strategic orientations. The strategy provided a very concrete and practical action plan that included programs on improving legislation and policies, consolidation of organizational structures, community awareness-raising, forestation and protection of upstream forests, strengthening of disaster management capacities through science and technology, structural measures, and programs on strengthening warning and forecast capacities. Again, this National Strategy did not mention urban areas, and there were no action plans for DRR in cities.

A Strategic National Action Plan (SNAP) on DRR was released in 2009, along with a national program for community-based disaster management (CBDRM). After the approval of the National Strategy, all of the ministries, provinces, and cities in Vietnam had to make their own strategic action plans and base these action plans on the CCFSC. The DMC has been implementing this program with support from donors and NGOs throughout Vietnam since then, on the scale of 6000 communes in rural areas.

From 2005 to 2014, the GoV made progress in mainstreaming DRR within national, sectoral, and provincial socioeconomic development planning frameworks and also passed a new Law on Natural Disaster Prevention and Control, which went into effect in May 2014. The majority of sectoral development plans and Socio-Economic Development Plans (SEDPs) at the national and provincial levels for 2011–2015, as well as the master plans for 2011–2020, integrate elements of DRM.

9.3.2.2 Overlapping on Disaster Management

The Ministry of Natural Resource and the Environment (MONRE) has been designated as the leading agency for climate change coordination in Vietnam, while MARD maintains the overall responsibility for natural disaster mitigation and response. In addition, the Ministry of Construction (MoC) is responsible for drainage systems and major public works, the Ministry of Planning and Investment oversees the issues of land use and master planning, and the Ministry of Science and Technology is involved in climate forecasting.

The Central Committee for Flood and Storm Control (CCFSC), chaired by the MARD minister, provides a coordinating umbrella for DRM in Vietnam. MONRE is a member of the committee, and other members include representatives from the ministries of planning and investment, finance, fisheries, transportation, science and technology, construction, health, industry, labor and social affairs, telecommunications, and foreign affairs. The VRC, Vietnam TV, Voice of Vietnam, Department of Dyke Management and Flood, Storm Control (DDMFC), and National Hydro Meteorology Center (NHMC), are also represented (see Figure 9.2).

Each sectoral ministry has its own Committee for Flood and Storm Control, which cooperates with the national central committee and offices at the provincial level. According to their own functions and duties, ministries, sectors, and local entities are responsible for effectively implementing relevant contents, objectives, duties, and solutions stated in the National Strategy.

MARD and the CCFSC preside over the implementation of the National Strategy up through 2020. The implementation of the strategy is urged by regulation in other ministries, sectors, and localities. CCFSC acts as the national focal point with international communities in the field. Specific implementation responsibilities have been assigned to ministries, sectors, and local entities, as follows:

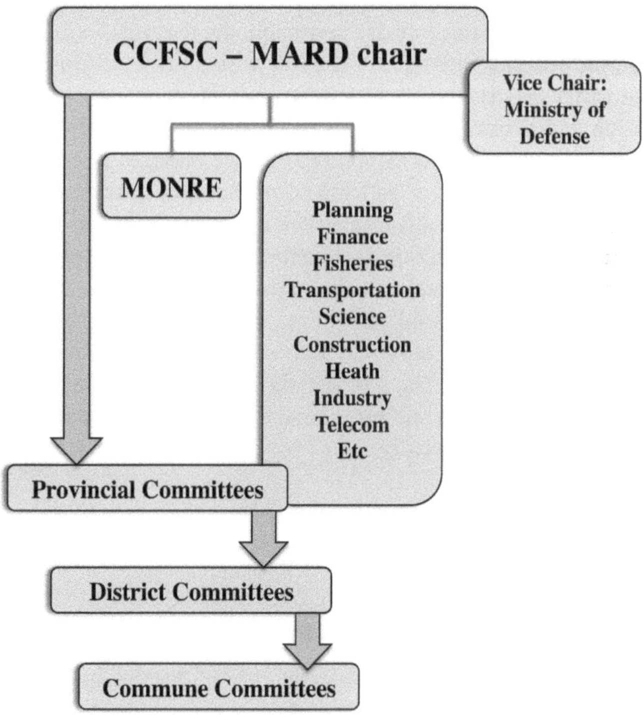

Figure 9.2 Disaster management system in Vietnam.
(*Source:* DMC)

Ministry of Planning and Investment takes the lead, working in collaboration with the Ministry of Finance, Ministry of Agriculture and Rural Development, CCSFC, National Committee for Search and Rescue, and other relevant ministries and sectors to balance and arrange annual investment resources in accordance with the Law on State Budget and other funding resources to implement the Strategy on DRM. MARD is in charge of inspection and assessment of the strategy implementation by ministries, sectors, and local entities. It conducts a review of the strategy implementation every year, and every five years, it draws out experience, and recommend to the prime minister suitable adjustments.

At the provincial level, the disaster management structure is followed as it is with the national level. Each province is requested to establish a provincial Steering Committee for Flood and Storm Control (CFSC). Each province have 1 CFSC. Vietnam has 63 provinces so it has 63 CFSCs. This committee is chaired by the vice chairman of the People's Committee, with the members coming from all relevant sectoral departments. The CFSC has quarterly meetings, and then they hold other meetings in emergencies. Almost all the activities of provincial CFSCs focus on making plans for disaster preparedness, disaster response, and disaster recovery. Only the CFSC of Da Nang city takes on additional DRR activities, such as awareness-raising and building resilience by improving early warning systems.

The prime minister tasked MONRE with preparing a National Target Program (NTP) on climate change, which was approved in December 2008. The NTP defines a set of actions at the central level and adaptation efforts at the provincial, district, and commune levels following a three-phase approach: the first phase (2009–2010), which was the startup; the second phase (2011–2015), which involved implementation; and the third phase (after 2015), which was development. It is a clear responsibility of MONRE to guide and assist the ministries, sectors, and provinces in developing and implementing their action plans to respond to climate change. The Ministry of Investment and Planning (MPI) takes the lead and coordinates with other ministries, sectors, and provinces to develop a standard framework and guidelines for mainstreaming climate change issues into socioeconomic development strategies, programs, and planning. Finally, People's Committees of provinces and major cities should develop and implement action plans to respond to climate change in their respective areas. By 2015, after 7 years from its start, only very limited activities have been implemented in urban areas, and there has been a lack of collaboration between MONRE and MARD on DRR and CCA. At the sector and provincial levels, climate change responses have not yet been addressed systematically.

9.3.2.3 The Roles of Ministry of Construction and Ministry of Planning and Investment

Influencing the Urban Form and Resourcing Implementation. The MoC is a key player in influencing the form that cities in Vietnam are taking, in terms of providing guidelines and regulations related to building codes and master planning. A Danida-sponsored sustainable urban development (SDU) component is assisting the MoC with the preparation of a handbook on urban planning and design with reference to climate change mitigation and adaptation considerations. Recently, the ministry issued Decision 2623/BXD, which requests cities to mainstream urban climate change and urban DRR into master planning and urban planning. The Decree of the Government No. 61/2003/ND-CP, released on June 6, 2003, prescribes the functions, tasks, powers, and organizational structure of the MPI. The MPI is an agency of the government that performs the functions of state management of planning and investment, including the provision of general advice on strategies and plans on national socioeconomic development; on mechanisms and policies for general economic management and some other specific fields; on domestic and foreign investment, industrial parks, and export-processing zones; on the management of official development assistance (ODA), bidding, enterprises, and business registration throughout the country; and perform state management over public services in the fields under its management as prescribed by law.

9.4 Challenges on Urban DRR Practice and Tools for Urban DRR

9.4.1 Challenges on Urban DRR Practice

The experiences of practitioners who have been working in urban areas in DRR and CCA through pilot programs in several cities indicate that implementation has involved the following challenges:

- Urban communities have a higher level of demographic complexity. The bonds that connect urban community members are weaker, as are those between communities and the areas where they live. This is because people living in urban communities often come from many different places, do not have long-established connections, and often do not know one another or have a strong understanding of the history of natural disasters in their neighborhoods.
- There are multiple difficulties in encouraging participation due to the differences in income level, standard of living, and lifestyles of urban and rural residents. Unlike rural dwellers, people living in cities often have to travel longer distances to work and have stricter working hours, whether they are administrative employees or workers.
- Urban systems are more complicated than rural ones, and more sensitive to factors outside the direct administrative control of the ward/commune being assessed. For example, when looking at the water supply system of a downstream city of a river basin, we need to consider influences at the watershed scale.
- There was a lack of tools that can assess the impact of local institutions, including laws, regulations, social rules, level of decentralization, and access to information.
- There were difficulties in mobilizing local government support. Urban local governments are under more pressure and have multiple priorities, including socioeconomic development. Local government leaders might not agree with adaptive measures suggested because they do not consider them their top priorities.
- The impacts of urban-specific processes such as industrialization play an important role in DRR preparation. Urbanization impacts tend to outweigh those of natural disasters or climate change alone. For example, construction and urban development might be the direct cause of serious flooding when a storm happens. It is very difficult for DRR plans to keep pace with this rapid urbanization process.
- There are not enough practitioners and facilitators that have experience in urban DRR planning.

9.4.2 Current Tools for Urban DRR: Limitations and Solutions

Practitioners, especially the VNRC and many other similar organizations in Vietnam, often employ a set of key tools in their work. The decision about which tools to use and how to use them depends on the specific needs of each locality, as well as the available time and human, financial, and physical resources. Importantly, local actors should be supported in implementing the follow-up actions suggested by the VCA findings.

In addition to the general difficulties in conducting VCA in urban areas (as described in the previous section), each VCA tool also presents specific limitations. Specific VCA tools and their limitations are described in Table 9.3.

9.5 Urban Resilience Approach for Urban DRR

Urbanization is bringing many new pressures to these areas. For example, disasters pose major risks to dikes and embankments along rivers; energy demand is driving the development of dams and hydroelectricity systems; greenhouse gas emissions are rising; there are more frequent and intense extreme weather events and continuing SLR; and disasters are having a greater impact on safety and infrastructure. It is not feasible or realistic to simply apply rural approaches in urban areas; and key issues besides urbanization and SLR include subsidence, upstream flooding, rainstorms and other kinds of disasters. Major challenges and considerations include large-scale structural interventions, where urban planning is often based on outmoded data; and a limited focus of risk mitigation/structural interventions (e.g., building dikes), where one failure can paralyze the whole system. These investments can create a false sense of security. There are many institutional challenges, especially shortcomings in monitoring and evaluation methods and implementation. In addition, the activities are often solely focus on emergency response.

The Asian Cities Climate Change Resilience Network (ACCCRN) is an innovative program with the goal of assisting medium-sized Asian cities to adapt to climate change. The program has taken an action research approach that has catalyzed city-level entities to assess key climate stresses and potential vulnerabilities and propose measures to respond to them rather than commissioning external experts or national agencies to prepare such plans (Tyler, 2010). In Vietnam, ACCCRN has been implementing activities in Da Nang, Quy Nhon, and Can Tho. The ACCCRN process was originally designed to operate for a relatively short period of time. At the beginning, ISET and other regional partners underestimated the time required to fulfill the key aspects and objectives of the plan: for local partners to absorb new and challenging concepts; for the translation of complicated technical materials into local languages (and training translators to undertake this work); for rigorous data collection, climate, and vulnerability assessment; and for the resolution of conflicting perspectives among stakeholders on key issues. ISET has developed a conceptual framework for urban climate resilience. This framework focuses on the following areas:

- Understanding cities' vulnerability to the impacts of natural disasters and climate change based on three key aspects: systems, agents, and institutions. More detail on this point is given by Moench and Tyler (2011).

Table 9.3 Current VCA Tools: Limitations and Solutions for Urban DRR

Tool	Limitations	Solutions
Review of secondary data	Although urban areas have an advantage over rural ones in terms of information and data, human resources, and education levels of the community, the review of secondary data still faces many challenges because of the following factors: • There is no practitioner group with experience in doing research, assessments, and analysis in urban contexts. • No clear mechanisms exist for mobilizing data sources. Thus, information is abundant but very difficult to access and explore, especially highly sensitive information such as details of land-use planning, construction planning, and urban development. On the other hand, the large amount of information from many different sources, sectors, and levels is time-consuming to collect and difficult to synthesize and analyze. • There is limited information on flood and storm control, nor are there instructions for long-term flood and storm control. Available information about climate change scenarios is limited and not detailed enough for community-level research.	Building the capacity of facilitators and the assessment team; coordination and sharing among sectors; publish the climatic disaster statistics
Historical profile	When researching historical profiles, there are usually more available data in urban areas than in rural ones. However, when doing surveys, interviews, or group discussions, it is difficult to identify the right group to target due to the greater mobility of urban populations. Urban communities also know very little about the history of disasters in their areas due to their citizens' short duration of residence. It is also difficult to verify their information.	Research should be based mainly on secondary data, verified by local people through in-depth interviews with the longest-term dwellers.
Mapping	Although information and data in urban areas (e.g., administrative maps, topographic maps, and detailed construction plans) are more available, detailed, and well organized, mapping tools still encounter difficulties when applied in urban areas. This is because cities have higher construction densities, with many view-blocking structures. Landscapes in urban areas are less stable than in rural areas, and can alter very rapidly in the urbanization process. Urban people rarely know their areas well, so it is difficult to identify vulnerable areas.	Engaging an assessment team with a good range of skills, including map analysts; using scientific data with verification from local people

Continued

Table 9.3 Current VCA Tools: Limitations and Solutions for Urban DRR—cont'd

Tool	Limitations	Solutions
Seasonal calendar	So far, this tool has been applied in rural areas only, and is based on crop schedules. It can still be used for VCA in peri-urban areas. However, for highly urbanized areas, it should be adjusted to focus on specific aspects of livelihoods only, because urban areas are much more complex in terms of labor market structure and types of livelihood, which makes it challenging to cover every aspect in depth. Urban livelihoods are less dependent on seasons, so this analysis is not useful in describing livelihoods.	Using reports on labor market structures to learn about livelihoods; developing a new tool for analyzing complex urban livelihoods structures, or focusing on major areas to avoid thinning out the analysis
Transect walk	Due to the high density and complexity of urban infrastructure, it is difficult to produce the transect diagram.	Using urban plans and geographic information systems (GISs)
Direct observation	Similar to the transect walk, a limitation in urban areas is that it is difficult to have an overview of these areas by observation from a single direction, while the density of urban population and infrastructure is also high and unstable. There are also issues related to safety for assessment staff as the rate of road accidents and crime in cities can be high.	Studying the urban master plan; engaging an assessment team, including people working on urban infrastructure, transportation, and civil engineering
Venn diagram	The difficulty in analysis using Venn diagrams is in how to identify roles of the political or social organizations of interest because the amount and complexity of information in urban areas makes it difficult to collect and analyze that information. This will lead to differences in people's viewpoints and opinions. Venn diagrams assess people's viewpoints only regarding a specific area, and they can be challenging to use when dealing with the complicated and complex urban systems and sectors.	Improving this tool to assess agents in the urban areas, clarifying roles and linkages between organizations and individuals

Focus group discussions	It is difficult to verify information provided and to identify groups because the selection and classification of groups are not the same as for rural areas. Urban people have busier lives and work schedules, which might creates time pressures and can cause assessment to be hasty and ineffective.	Working closely with leaders of residential areas
Interviews	First, it is difficult to meet with targeted interviewees in urban areas due to the nature of their work and lifestyles. The consumption habits of urban people might also prevent the development of effective adaptive measures. In addition, there is currently no questionnaire developed for urban areas, with consideration of their special circumstances.	Raising public awareness while doing assessment; working closely with leaders of residential areas
Strengths, weaknesses, opportunities, and threats (SWOT) analysis	This tool is based mainly on available secondary data and therefore faces similar challenges to those faced with secondary data reviews. However, SWOT analysis can also derive from primary data sources, such as focus group discussions. Although there are many sources of information, the level of consistency can be low.	Improving the capacity of assessment staff
Livelihood analysis	Urban populations are dense and highly diversified in terms of livelihoods. They lack stability across time and space. The nature of livelihoods is also different. Therefore, the quantity and accuracy of information is lower and the analysis more complicated in urban areas than in rural ones.	Using reports on labor market structures to learn about livelihoods groups; developing a new tool for analyzing complex urban livelihood structures

- Shared learning dialogues that try to fully mobilize local knowledge and the in-depth scientific knowledge of all stakeholders involved.
- Intervention actions that build the resilience of these systems, agents, and institutions based on their own distinctive characteristics.
- The innovative and interactive nature of these processes.

The challenge is how to deal with uncertainty in climate information and extreme events for planning. A key issue in CCA and DRR practices is how planners can make effective use of available climate information and disaster profiles despite great uncertainty and unfamiliar or unhelpful data formats that they must work with. In the early stages of the ACCCRN program, many partners expressed frustration at the lack of probabilistic data regarding future climate impacts, which they had hoped to use to design infrastructure standards. Here are some aspects could be considered for applying urban DRR (Tyler et al., 2010):

- *Focus on existing vulnerabilities:* All of these strategies have a strong focus on current challenges and vulnerabilities, which are very likely to be exacerbated by climate change. These issues included waterlogging, water scarcity, flood protection, ecosystem degradation, and solid waste management.
- *"No-regrets" planning:* No-regrets strategies are actions that yield positive outcomes regardless of climate conditions and across a wide range of futures. Examples of such strategies include awareness building, early warning systems, improved service delivery, drainage systems, and wastewater management. In many cases, no-regrets strategies respond to current problems, guaranteeing benefits that can be felt immediately but will also strengthen resilience to future climate hazards.
- *Further studies on potential interaction of climate and key urban systems:* All cities identified the need for better local data and detailed scientific studies of plausible local climate impacts that would allow them to plan with more sophistication in the future. A common interest across most cities was learning more about impacts on key hydrological parameters and water management systems in relation to planned urban development.
- *Avoiding maladaptation:* Strategies recognize the increasing risks of further development in exposed sites, of overexploitation of key resources (e.g., groundwater), and the vulnerability of particular sectors (e.g., fishing, agriculture). They propose approaches that would redirect "business as usual" efforts away from these sensitive sectors or areas.
- *Awareness:* Several cities have prioritized awareness-raising among different groups, from the general public to private businesses and elected officials, in order to generate broad support for resilience actions and to build the capacities for behavioral change and autonomous adaptation.

9.6 Good Practices on Implementing Urban DRR in Vietnam

9.6.1 Practice 1: Storm and Flood-Resistant Housing for the Poor, implemented Through the Women's Union

The city of Da Nang is considered highly vulnerable to the effects of climate change due to its geographical exposure and limited socioeconomic ability. Poorer households typically live in poorly constructed and maintained housing and frequently struggle to recover from storm damage due to low incomes, limited compensation payments, and rising costs. With loans provided by the Rockefeller Foundation and technical support from ISET, a storm- and flood-resistant housing project has been implemented in Da Nang that helps vulnerable households to increase their climate resilience. This project, implemented through the city's Women's Union, has enhanced the capacity of vulnerable wards in Da Nang through a revolving loan fund for storm-resistant housing.

The project has so far achieved the following:

- Provided credit to improve the homes of poor households in eight disaster-prone wards, as well as professional technical assistance in constructing and monitoring construction progress
- Established a revolving loan fund for small businesses, which helped at least 30 poor households improve their credit rating
- Helped 20 extremely poor families to build new houses that withstand typhoon and floods
- Enhanced the capacity of program participants and community members to undertake CBDRM, manage finances, and maintain houses

In the first three years of the program (2011–2013), 320 households have been reinforced and rebuilt to withstand typhoons (Tran, 2014). This number is expected to increase to 376 households by 2016, following the second round of the credit. The project contributes to the following resilience characteristics:

- *Resourcefulness*—The feasibility study provided an opportunity for several organizations to work together in identifying problems and solutions, mobilizing their collective resources and developing a shared vision for reducing vulnerability to climate change.
- *Responsiveness*—The capacity to organize or reorganize is enhanced through forming a network with vulnerable communities and other stakeholders.
- *Learning* —Training on climate change impacts, city resilience measures, and storm-resistant housing for the urban poor and the Women's Union, and stakeholder workshops support the internalization of learning within partner organizations.

The main beneficiaries of the project are households that acquire affordable loans for housing reinforcement, as well as gain new skills and experiences, which were promoted within their communities to build awareness of storm-resistant housing and to generate additional demand for credit access. The Women's Union of Da Nang is another key beneficiary, as it will develop skills for managing and scaling a fund for storm-resistant housing and emerge as a key agent of change in climate change resilience for poor and vulnerable households. And it has shown results: During Typhoon Nari, which hit Da Nang in October 2013, none of the houses developed or reinforced as part of this project were damaged.

The Da Nang Women's Union is currently working with the Da Nang Climate Change Coordination Office (Da Nang CCCO) and Central Vietnam Architecture Consultancy (CVAC), which is the consulting firm providing technical support in storm-resistant building techniques, to provide instructions on this model to a larger number of households. The Women's Union has also proposed to the City People's Committee to develop policies and mechanisms to integrate storm-resistant standards into housing development procedures in Da Nang.

9.6.2 Practice 2: Comanagement for Riverbank Stabilization and Erosion Control in Can Tho

The city of Can Tho is situated in the center of the Mekong Delta, where the entire urban population (approximately 600,000 people) lives and works at an elevation of less than 3 m above sea level. In recent years, during high-water season, tidal flooding has become more common in the city's many canals and drainage channels. Commercial shipping has also increased along the main canals. The combination of these factors has led to increased riverbank erosion and loss of bank stability in parts of the city where urban development is expanding rapidly. In 2014, Can Tho, with the support of the Rockefeller Foundation and ISET, has launched a project called "Community-Based Urban Flood and Erosion Management for Can Tho City." We applied an Urban Resilience Framework (Moench & Tyler, 2011), tool called Shared, Learning, and Dialogue (SLD) to bring deferent stakeholders in contact with the local community to develop a comanagement capacity to help negotiate conflicts and overcome existing limitations.

In an experimental approach taking place in An Binh ward of Can Tho, the local government and the community are collaborating to test a number of low-cost bank stabilization measures. These measures include several innovations:

- They are based partly on the community's knowledge of bank vulnerability and stabilization measures.
- They involve the selection of appropriate biological materials for bank stabilization.
- They involve the community in planning, decision making, construction oversight, and contribution of labor and finance with the approval of government authorities.

These measures together constitute a model for the comanagement of adaptive infrastructure measures. The most emphasis result is that the contribution of the local people who are living in the both sites of the river. The local people understand that no one could take care for the riverbank better than them. The riverbank is their property and they have responsibility to protect it. Inspired by the key message "My riverbank, my responsibility," the local people contributed to the project 680 working days and $7700 in cash to build and protect 3086 m of the riverbank by biological measures. The community has also established a community board to maintain the riverbank after the project has ended. Key issues that were resolved included establishing the roles and responsibilities of various stakeholders, such as riverbank dwellers, community members living farther from the river, local government, and technical advisors. The community's plan for riverbank stabilization was more ambitious than originally anticipated, and their financial and labor contributions enabled the project to be enlarged. Roles of different stakeholders and community are listed in Table 9.4.

This result shows how low income peri-urban communities can develop new financial and institutional mechanisms for adaptive infrastructure investments to build climate resilience. Co-management of some kinds of urban infrastructure may be a practical solution to reduce climate risks.

Table 9.4 **Roles of the Stakeholders and the Community**

Local Government	Local Community	Scientists/Experts	NGOs and Donors
• Cofunding • Policy guidance • Comonitoring and comanagement • Negotiating with households as needed • Organizing and monitoring construction • Materials management, asset and investment for the project • Ensuring the settlement of the time prescribed	• Comanagement • Participating in group meetings to provide ideas on the construction plans and methods of contributions • Launching initiatives • Contributing materials and labors if as possible • Participating the construction • Discussing and proposing regulations and conventions in order to operate and maintain the projects • Managing operations and maintenance in accordance with regulations	• Conducting studies on riverbank erosion and flooding • Conducting cost-benefit analysis (CBA) • Working with local people to define solutions • Supporting local people to develop construction plans	• Cofunding • Supporting the establishment of a model of comanagement in urban areas • Technical assistance • Management during the project period

References

Central Committee for Flood and Storm Control (CCFSC). (2014). Progress report on the National Strategy for Disaster Prevention, Mitigation, and Response to 2020. *Ministry of Agriculture and Rural Development of Vietnam.* Hanoi, Vietnam.

Global Facility for Disaster Reduction and Recovery (GFDRR). (2009). *Disaster Risk Management Programs for Priority Countries: East Asia and Pacific* Vietnam, 107. GFDRR Secretariat, Washington, DC, USA. Available at: http://www.unisdr.org/files/14757_6thCGDRMProgramsforPriorityCountrie.pdf.

Global Facility For Disaster Reduction And Recovery (GFDRR). (2011). Annual Report: Building Resilience to Disasters — Delivering Results, GFDRR, Washington, DC 20433, USA.

Government of Vietnam. (2014). Law on natural disaster prevention and control. Hanoi, Vietnam.

International Federation of Red Cross and Red Crescent Societies (IFRC). (2014). *Viet Nam: Country Case Study Report, How Law and Regulation Support Disaster Risk Reduction.*

Ministry of Agriculture and Rural Development (MARD). (2006). Decision No. 312/QD-PCLBTW on regulations on reporting and information sharing for dealing with floods and storm. Hanoi, Vietnam.

Moench, M., & Tyler, S. (2011). *Catalyzing urban climate resilience: Applying resilience concepts to planning practice in the ACCCRN program (2009–2011).* Bangkok: ISET Boulder.

Nguyen, H., Tran, P., & Nguyen, A. T. (2014). *Applying Vulnerability and Capacity Assessment (VCA) Tools in the Urban Context: Gaps, Challenges, and New Approaches.* Hanoi, Vietnam: ISET. Available at: http://i-s-e-t.org/resources/working-papers/vca_urban.html.

Tran, P. (2014). *Lesson from the Typhoon Nari, 2013.* Vietnam: ISET.

Tyler, S., Reed, S. O., MacClune, K., & Shashikant, C. (2010). Planning for Urban Climate Resilience: Framework and Examples from the Asian Cities Climate Change Resilience Network (ACCCRN). *(Climate Resilience in Concept and Practice Series Working Paper No. 3). Boulder, CO: Institute for Social and Environmental Transition (ISET).*

United Nations Department of Economic and Social Affairs (2011). World Urbanization Prospect: the revision, United Nation, New York.

Vietnam Ministry of Construction (2014). Annual report on urban development in Vietnam, Hanoi Vietnam.

Urban Disasters and Microfinancing

10

Gulsan Ara Parvin[1], *Rajib Shaw*[2], *Kazi Farzana Shumi*[3]
[1]Researcher, Graduate School of Global Environmental Studies, Kyoto University, Kyoto, Japan; [2]Professor, Graduate School of Global Environmental Studies, Kyoto University, Kyoto, Japan; [3]Department of Business Administration (DBA), International Islamic University Chittagong (IIUC), Dhaka, Bangladesh

10.1 Introduction

The world is steadily becoming more urban. It is estimated that more than half of the world's population now lives in cities, and two-thirds of the population will by 2030 (World Economic Forum, 2014; Sharma et al., 2011; Srinivas et al., 2009). World Economic Forum (2014) forecasted that the urban population will nearly double, to an estimated 6.4 billion, by 2050. Further, it is estimated (World Bank, 2011) that about 70 million people in the developing world move to urban areas, and about 1 billion people already live in slums (a number that is projected to double by 2050). Intensive economic activities and lifestyles of this large number of people greatly contribute to urban environmental problems and disaster risk. It has already been proved that rapid and unplanned expansion of urban settlements face severe risk (including climate-induced disasters) in the light of climate change. Therefore, urban populations will increasingly need to cope with increased incidents of flooding, air and water pollution, heat stress, and vector-borne diseases (Tanner et al., 2009). UNHABITAT also has explored that urban growth and disaster risk are statistically correlated (UNHABITAT, 2007). It is said that "Urbanization affects disasters just as profoundly as disasters can affect urbanization" (Pelling, 2003, cited in Tran et al., 2009). In this cycle of rapid urbanization and urban exposure to disasters, poor and socially disadvantaged groups are the most vulnerable to and affected by natural disasters, since they have a disproportionate impact on the poor. There is no doubt that the poor suffer most and are more adversely affected by natural disasters (GAR, 2009; UNISDR, 2007, 2013; Parvin & Shaw, 2013; Johnson, 2006, Yodmani, 2001). In particular, the urban poor who usually live in hazard-prone areas are at particularly high risk from the impacts of climate and natural hazards (World Bank, 2012).

Poverty often migrates from rural to urban areas (Tran et al., 2009). After migration, these poor people often live in the lowest-quality housing in the most hazardous and disaster-prone locations, which lack basic facilities, a healthy and hygienic environment, and basic infrastructure. After migrating from their rural homes, large numbers of urban poor embrace the enormous economic

Urban Disasters and Resilience in Asia.

opportunities offered by cities, but nevertheless, they trade that opportunity with the associated risks of poor infrastructure and utilities, unhealthy, and unsanitary conditions.

Both governments and non-government organizations (NGOs) try to avoid investing in the areas where the poor live, since these areas have no secured tenure and the potential for disaster is high (World Bank, 2012). Nonetheless, responding to criticism from human rights organizations donors, government and NGOs have introduced different approaches like slum up gradation projects and health and sanitation programs in order to improve the living conditions of the urban poor. These approaches are sometimes directly or indirectly aiming to reduce the vulnerabilities of the urban poor from various shocks and risks, including disasters. But financial services or supports like microcredit/microfinancing to rebuild livelihoods, strengthen community bonds, and protect the poorest urban residents from income shortfalls after a shock or natural disaster are not yet widely practiced. This is because slum dwellers are not always considered creditworthy, especially due to their lack of legal tenure and absence of strong social capital, which are the prime preconditions of creditworthiness. Here, it should be noted that microfinancing is a mechanism that offers loan, savings, insurance, and other financial services to the poor on a small scale so that they can generate income opportunities, build their asset bases, stabilize consumption, and protect themselves against risk (Hammill et al., 2008). Therefore, this approach benefits poor community by providing support for economic and social rejuvenation after a natural disaster by providing financing to rebuild their livelihoods, strengthen community bonds, and protect them from income shortfalls (Poston, 2010). Research findings also reveal that in many cases, microfinancing beneficiaries can quickly recover their income and rebuild their houses after a disaster, and their disaster risk reduction (DRR) ability also has improved (Parvin & Shaw, 2013). In addition, there are numerous possibilities for strategic collaboration between microfinancing institutions (MFIs) and disaster management by concentrating on their core competencies, for the ultimate benefit of affected communities (Mathison, 2003). However, this approach of collaboration between MFIs and disaster management entities mostly exists in rural areas, to help the rural poor. Nevertheless, studies show a close link between poverty and urban vulnerability and DRR in a context that is as much about poverty reduction as any other measure (Tran et al., 2009). Unfortunately, there are almost no studies exploring how the most vulnerable group (i.e., urban poor) in urban disasters can be assisted via the most applicable tool, which is microfinancing for poverty alleviation. Bangladesh, India, and Indonesia, which are some of the world's most populous countries, are embracing rapid urbanization, having large numbers of urban poor, and have been working for more than three decades on intensive microfinancing for poverty alleviation. Therefore, this chapter gives an overview of the availability, accessibility, and role of microfinancing for DRR and recovery of the urban poor by focusing on the disaster risks and urban poverty of three cities in Bangladesh, India, and Indonesia.

10.2 Urban Disaster and Microfinancing—An Overview from the Perspective of Urban Poor in Dhaka, Bangladesh

Numerous research studies have addressed different challenging issues in Dhaka, Bangladesh, such as urbanization, disaster risk, and helping the poor in Dhaka. Some studies have also attempted to integrate these issues and explored how one issue is exaggerating another one. The World Bank, World Vision, several other different organizations, and a number of researchers have investigated the vulnerabilities of Dhaka's urban poor and their disaster risks. Nonetheless, no studies have focused on urban microfinancing in Dhaka and how it is alleviating poverty and reducing the vulnerabilities of Dhaka's poor. Studies addressing urban microfinancing are rare, even though Bangladesh is the home of modern microfinancing and also has experienced rapid urbanization and in increasing numbers of urban poverty, especially in Dhaka. Considering the vast research gap in this area, this section aims to address the vulnerabilities and disaster risks of Dhaka's urban poor and what microfinancing approaches are doing to reduce their disaster risk.

10.2.1 Vulnerabilities and Disaster Risks of Dhaka's Urban Poor

The main natural hazards affecting Dhaka include floods, earthquakes, and tornadoes. Apart from these, fire is another serious manmade hazard in Dhaka. By virtue of its geography along with rapid urbanization, Dhaka is considered to be one of the cities most susceptible to climate-related disasters, especially floods, which are associated with river water overflow and rainwater stagnation. Dhaka City experienced heavy flooding at least 10 times from 1954 to 2007. Among those, the last four in 1988, 1998, 2004, and 2007 floods were catastrophic (Rabbani, 2009). Besides flooding, according to United Nations IDNDR-RADIUS initiatives, Dhaka and Tehran are the cities with the highest relative earthquake risks in the world (Jahan et al., 2011). Further, the Earthquake Risk Index (EDRI) for Dhaka ranked it first in its list of the 20 high-risk cities in the world (Khan & Phibbs, 2005).

In Dhaka, the disaster risks are most extreme amongst communities inhabiting low-lying slum areas. There are about 3.4 million people—perhaps more than 35%–40% of Dhaka's population—living in the slums and squatter areas (Ahmed, 2014). These slum and squatter residents often lack security of tenure and are predominantly characterized by very poor housing, very high population density, poor environmental services, and low socioeconomic status (Islam et al., 2009; Roy & Abdullah, 2005). Furthermore, according to the International Organization for Migration (IOM), 70% of the slum dwellers of Dhaka city went there after facing environmental hardships in their own homes. But their hardship does not end when they arrive in the city, since they live in unhygienic conditions with few utilities and higher risks of hazards and disasters (Figures 10.1 and 10.2). It is evident that this existing physical, economic, and social vulnerabilities of slum residents heighten their poverty, marginality, and disenfranchisement, making them highly

Figure 10.1 Slum located along the riverbank.

Figure 10.2 Garbage disposal in a slum.

susceptible to disasters and their lasting effects (Stott & Nadiruzzaman, 2014) (Figures 10.3 and 10.4).

According to Baker (2007), housing structures of the poor in Dhaka tend to be of poor quality, and access to basic infrastructure services is extremely limited. For the poorest quintile, only 9% of households have a sewage line, and 27% obtain water through pipes (compared with 83% of the wealthiest population). Spatial mapping shows that only 43 of the 1925 identified slums have a public toilet within 100 meters. An estimated 7600 households live in slums that are within 50 meters of the river,

Figure 10.3 Vulnerable business conditions of Dhaka's poor face a number of difficult business conditions (for instance, they cannot continue business on rainy days).

Figure 10.4 Roadside houses of Dhaka's poor on a rainy day.

increasing their risk of being flooded. Dhaka's slums are mostly located on relatively low-lying terrain areas. Therefore, flooding and subsequent waterlogging are common. In addition, Stott and Nadiruzzaman (2014) reported that many slums in Dhaka are located close to the epicenter of liquefaction zones. Although the simple, lightweight infrastructure of slum settlements in Dhaka shields them from the immediate risk of building collapses, the liquefaction of surrounding buildings can cause much destruction. In addition to flood and earthquake risks, fire is a frequent hazard in Bangladesh's urban areas, especially in slums and squatters, and such events cause great loss of life and assets every year. In the urban context, slum areas are at the greatest risk from fire (Maniruzzaman & Haque, 2013). Damage caused by fires is exacerbated in slum areas due to their crowded conditions and limited mechanisms for fire response. Firefighters have reported that the narrow streets and lack of hydrants or other sources of water cause delays in responding to and extinguishing slum fires (CBS News, 2010). The evacuation of slums is also very difficult due to the compact alignment of houses and dense population. In Dhaka, the mean duration of fires was significantly higher in slums, with an average of 68 minutes, than outside slums, at 28 minutes (Maniruzzaman & Haque, 2013).

In spite of having poor environmental services, low socioeconomic status, risk of frequent eviction, and most important, risk of numerous hazards and disasters, these poor people have limited alternatives in terms of what to do and where to live, due to their economic vulnerability. Since microfinancing has the potential to strengthen the risk management capacity of the poor, we will highlight its use for DRR of Dhaka's urban poor in the next sections.

10.2.2 Microfinancing for Dhaka's Urban Poor and DRR

In the policymaking domain, urban poverty is becoming critical due to climate change and the potential displacement of some 20 million poor into urban areas. Despite the urgency of this matter, Bashar and Rashid (2012) claimed that their study was the first to look at urban microfinancing in Bangladesh. Therefore, knowledge of urban microfinancing in Bangladesh is still almost completely lacking. The Institute of Microfinance does not provide a separate category for urban microfinancing (Bashar & Rashid, 2012). It should be noted here that microfinancing is mostly operated by NGOs. Therefore, these MFIs are abbreviated as NGO-MFIs.

The Microcredit Regulatory Authority (MRA) was established by the national government in August 2006. Between that time and the end of June 2010, it issued licenses to 482 NGO-MFIs operating microfinancing programs for the poor through more than 17,200 branches. According to the authors, among these 482 organizations, about 121 were working for the urban poor in Dhaka city (including Savar, Dhamrai, and Kranigonj). But there is no available data related to the number of beneficiaries of microfinancing institutions in Dhaka. According to the Institute of Microfinance, urban microfinancing institutes have helped over 2 million borrowers (according to Bashar & Rashid, 2012) throughout Bangladesh. However, Dhaka alone has more than 3 million poor residents. Furthermore, according to Bashar and Rashid (2012), among the total borrowers in the microfinancing industry, 12% were the urban poor, and they received one-seventh of the total amount of microfinancing disbursed. It is estimated that the overall demand for microfinancing is almost three times greater than the supply available to the borrowers (Bashar & Rashid, 2012). The demand for capital and average loan size were higher in Dhaka compared with other cities, which indicates a substantial demand for microcredit. Since there are no detailed statistics about Dhaka's microfinancing demand and supply, and the role that the city plays, the abovementioned facts related to microfinancing in overall urban Bangladesh can provide a clear indication of the widespread scarcity of microfinancing access to the urban poor of Dhaka.

The prime role of urban microfinancing is to give the urban poor access to credit, especially those who have stability and a definite identity for their income generation, and thus had concrete ways to fight poverty. The most common microenterprises in Dhaka are petty shopkeeping, and selling fish, vegetables, clothing, and tea. These microenterprises operated by the urban poor of Dhaka (among other cities) have very high profit margins—sometimes more than 100%. In spite of this high profitability, due to inadequacy of start-up capital, sudden loss of uninsured businesses, social insecurity, larceny, fires, natural hazards and other disasters, many urban poor often fail to achieve any gains from microfinancing. However, most microfinancing institutes working in the urban areas of Bangladesh have no insurance or risk management mechanisms.

10.2.3 Urban NGO-MFIs in Dhaka

In spite of having more than three decades of success in various issues of vulnerabilities and poverty, MFIs in urban Bangladesh, including those working in Dhaka, have not yet been intensively involved in housing, water supply, health- and sanitation-related programs, which directly and indirectly reduce the disaster risks and vulnerabilities of urban poor. MFIs in urban areas have prioritized income-generating activities as their main focus. However, a recent study by Stott and Nadiruzzaman (2014) provides information related to the efforts of a few international and national NGOs working for the improvement of living conditions and DRR of Dhaka's urban poor. These efforts are discussed in the next section.

10.2.3.1 Concern Worldwide

Concern Worldwide is currently focusing on four main areas in Bangladesh: livelihoods, education, health and nutrition, DRR, and climate change adaptation (CCA). In Dhaka, prioritizing the urban poor, Concern Worldwide is working to solve the problems of waterlogging, flood, fire, waste management, air pollution, and a lack of clean water, which are crucial issues for disaster preparedness and risk reduction.

10.2.3.2 Dushtha Shasthya Kendra (DSK)

Dushtha Shasthya Kendra (DSK) is a Dhaka-based national NGO. Along with microcredit, health services, water, sanitation, hygiene promotion, education, and training are the main concerns of this organization. DRR in urban areas has been a recent focus as well. DSK is focusing on fire and flood and other natural hazards. In Dhaka, it is working to raise awareness of slum residents to these important issues, providing training on different issues related to eviction, fire incidents, flooding, and waterlogging. This NGO also provide emergency packages of rice, dhal, oil, and water to those affected by disasters. However, only 25,000 beneficiaries are fortunate enough to have this training and emergency packages.

10.2.3.3 Habitat for Humanity (HFH)

Habitat for Humanity (HFH) is not an MFI; rather, it is an international housing organization that works for universal housing security in order to secure environmental and health benefits for communities. Housing programs of HFH aim to improve disaster mitigation and disaster response. HFH has started a pilot project in Mirpur Bihari, a slum area of Dhaka, where it has completed a hazard assessment and is mapping community development for future projects.

10.2.3.4 Islamic Relief Worldwide (IRW)

By addressing poverty, illiteracy, and disease, Islamic Relief Worldwide (IRW) tries to promote sustainable economic and social development. IRW is currently implementing a DRR project in Sylhet district and planning to initiatives in Dhaka districts and its urban areas. It is adapting community risk assessment methods to tailor an urban risk assessment strategy. The main focuses of the organization's urban DRR project are fire and earthquake, including training on earthquake resilience and landslides.

10.2.3.5 Plan International Bangladesh

Plan International Bangladesh supports disadvantaged children and their families so that they can become active citizens in their communities and society at large. It undertook an Urban DRR program in Dhaka city, with support from European Commission Humanitarian Office (ECHO) under DIPECHO (Disaster Preparedness-CECHO) action plan V and VI for South Asia starting from 2009 to 2012. Currently, Plan International Bangladesh is implementing another three-year (up to 2015) DRR and CCA program in five wards under the Dhaka City Corporation with the aim to build safe, resilient communities in which children and young people contribute to managing and reducing their disaster risk.

10.2.3.6 Islamic Bank Micro Financing Program (IBMFP) for Urban Poor

Islamic Bank Micro Financing Program (IBMFP) in urban areas formally started operating on May 14, 2012, in Dhaka city. Up to December 2014, the total number of **IBMFP** beneficiaries in Dhaka city is 5000. Before starting Islamic Micro Finance these beneficiaries were mostly borrowing/receiving loan from conventional NGO-MFIs. In addition, most people used to seek loan from informal sources because that was the easiest, least formal way to get it. However, the interest rate they paid would be extremely high (according to the borrowers, sometimes 50%–100%). IBBL operates its Urban Poor Development Scheme (UPDS) program very inclusively, visiting respondents door to door to see their practical lifestyles. In this way, these people become convinced by the attitude and behavior of the staff of IBBL that UPDS would rid them of high interest rates and give them a way to acquire significant financing on the basis of profit and loss sharing with the capital provider. Thus, the beneficiaries shifted their previous mode of financing to UPDS.

Even though different NGO-MFIs and national and international NGOs have started to work with the urban poor, the supply compared to the demand remains negligible. There are some problems stemming from this negligence. NGOs, governmental organizations, and donors have agreed to concentrate on three main problems: (i) lack of government policy giving slum residents rights and access; (ii) eviction of slum residents; and (iii) the role of *mastaans* (Extortionist or hoodlum or Hooligan) in the absence of formal government structures.

10.3 Disaster Vulnerability of Mumbai, India, and Its Urban Poor—The Role of Microfinancing

Mumbai (formerly known as Bombay) used to be a congregation of seven tiny islands in India, mostly lived in by fishers and farmers (Mehrotra & Dwivedi, 1995, cited in Nallathiga, 2009). Now Mumbai is considered one of the largest megacities in the world and one of the world's top 10 centers of commerce (Surjan et al., 2009; Patankar et al., 2010). In terms of population, Mumbai is currently ranked fourth in the world, after Tokyo, Mexico City, and New York (Patankar et al., 2010). It is the capital of the leading industrial state and second-leading urbanized state of Maharashtra in India. In spite of being the industrial, commercial, and financial hub of India, Mumbai faces severe environmental crisis and disaster risk, which have collectively made the city vulnerable to both manmade and natural disasters.

The city is surrounded on three sides by water: the Arabian Sea and Harbour Bay to the west and Thane Creek to the east, and it is just 10–15 meters above sea level. Being a coastal city, Mumbai is prone to cyclones and heavy winds. It falls in seismic zone III, which is a moderate damage risk zone. As per the 2001 census, Mumbai has over 276,000 dwellings (residential, industrial, and commercial), of which 9% are made of reinforced concrete, 31% are engineered constructions, and around 60% are nonengineered constructions, which correspond with the large presence of slum settlements (Patankar et al., 2010).

Greater Mumbai Disaster Management Action Plan (GMDMAP, 2007) reveals Mumbai's high risk from flooding, cyclones, earthquakes, landslides, traffic accidents, industrial and chemical accidents, and fire. Furthermore, Mumbai's flat topography, geology, wetlands, and flood-prone areas, projected sea-level rise (SLR), building conditions, slums and squatters, poor sanitation and waste treatment, and low income were defined as "bundle of stresses." All these stresses reduce the city's disaster preparedness ability and thus enhance its vulnerabilities (Sherbinin et al., 2007). Considering the extent of exposure and vulnerability to climate extremes, the Organisation for Economic Co-operation and Development (OECD) has ranked Mumbai as first among 136 cities with populations of more than 1 million (Nicholls, et al., 2007).

10.3.1 Disaster Risks of Mumbai's Urban Poor

The Mumbai metropolitan region contains about 17.7 million people, with the majority (10 million) living in slums and squatters (Surjan et al., 2009). In fact, more than half of Mumbai's population (54.5%) live in slums (Bhatia & Chatterjee, 2010). Further, according to the Mumbai City Development Plan (MCDP) in 2006, 36% of the households in this city lack hygienic sanitation facilities, and these households are connected with old stormwater networks and poor drainage facilities that can carry only a maximum of 50% of the potential sewage it needs to handle. All these facts clearly indicate the disaster risk and vulnerabilities of Mumbai, especially those of that majority who live in the slums.

Researchers have found that the most important factor of the risks and vulnerabilities of Mumbai is its ever-growing population. However, slums are nonexistent on the city's developmental plans, with none of the 2335 settlements being recognized. Regular displacements prevent slum dwellers from settling in safe and secure locations, and they react by moving toward landslide-prone areas (Figure 10.5), low-lying regions, and unsafe dumping sites, among other dangerous areas, which are breeding grounds for hazards in changing climate conditions (Patankar et al., 2010).

10.3.2 Urban Microfinancing to Address Disaster Risks—Focus on Mumbai

Since its independence almost 70 years ago, India has continued its great efforts toward the inclusion of large segments of rural people, especially the poor, into the financial sector. Therefore, during the period 1951–1991, financial institutions' total share of rural household debt increased from 8.8% to 53.3%, and the role of moneylenders declined significantly (Arun et al., 2006). It is argued by scholar (Imai et al., 2010) that in spite of having different banking and cooperative finance institutions and strong microcomponents in various programs, the formal financial sector still fails to adequately reach out

Figure 10.5 Slum in a hazardous area in Mumbai.

to or reflect and respond to the requirements of the poor. Due to this failure of the formal financial sector, MFIs became increasingly important in India in the 1990s, mainly due to their better access to local knowledge and information at the community level and their use of peer group monitoring (Imai et al., 2010). However, in spite of the huge outreach and success of microfinancing in rural India, the urban microfinancing approach has not been embraced by MFIs for long time. It has been reported that despite the prospects of huge loan demand, larger average loan size, and higher population density, urban microfinancing was left unexplored due to the greater mobility of and lack of social cohesion among the urban poor (Ghate, 2008). This is about a decade since urban microfinancing began receiving attention and interest in India. However, its growth and role is not the same in all parts of India. Although the reasons for this are not clearly identified, the growth of urban microfinancing has been relatively fast in Bangalore, Hyderabad, and Kolkata, but less so in Mumbai and Delhi (Ghate, 2008).

As noted by the Institute for Financial Management and Research (IFMR, 2006), in contrast with the development of microfinancing in rural areas, there has been a near-absence of urban MFIs in India. Due to the lack of access to financial services such as bank branches and automatic teller machines (ATMs), the urban poor in need of credit rely on street lenders, who often charge up to 10% per day. However, recently, this picture has been changing, especially among some of the large MFIs such as SPANDANA, Slum Development Society (SDS), and BASIX. These MFIs are either increasing or planning to increase their activities in urban lending. In 2006, the Centre for Microfinance at IFMR has done a study of urban microfinancing in India, in which organizations like SEWA Bank, Ujjivan, Working Women's Forum, Indian Bank's Microstate Branch, Village Welfare Society, and Sharada's Women's Association for Weaker Sections have been identified as prominent participants.

Among other examples of urban microfinancing in India, the efforts of ACCION are especially remarkable. ACCION is a global nonprofit organization that supports microfinancing institutions in their work to provide financial services to low-income clients. In 2005, ACCION partnered with Unitus to form the Unitus-ACCION Alliance for India, and two years later, it joined with India's YES Bank to create a microfinancing subsidiary, YES Microfinance India. Both Unitus-ACCION and Yes Microfinance India provide financial services to the country's urban poor (Ghate, 2008). The study cites four new and one existing urban MFIs (two in Bangalore and one each in Ahmedabad, Kolkata, and Mumbai); all are negotiating with equity investors to enable them to expand rapidly as Non-Bank Financial Institutions (NBFIs) (Merrill & Suri, 2007).

In urban India, especially in Mumbai, there are a few organizations that are playing a remarkable role in different critical aspects of disaster vulnerability, such as housing, water and sanitation, and health and income generation through microfinancing. The following sections describe these groups.

10.3.2.1 Society for the Promotion of Area Resource Centers (SPARC)

The Society for the Promotion of Area Resource Centers (SPARC) is one of the largest NGOs in India. It opened in 1984 with the aim to support the most vulnerable and invisible of Mumbai's urban poor. Its prime focus is dealing with housing and infrastructure issues of the urban poor. SPARC usually works with its alliance of women's cooperatives (such as *Mahila Milan)* formed by pavement dwellers and the National Slum Dwellers Federation. With their relentless efforts, SPARC and its alliance have established some good guidelines for poverty reduction. It has showed how different risk-associated critical sectors of life, such as community-based savings and credit, housing construction, development of toilet blocks, and the management of resettlement, can contribute to poverty reduction. (Patel & Mitlin, 2001).

These microfinancing institutions are creating awareness about health and hygiene for the CBOs of Community toilets. It has started a new program called Energy Justice in order to provide the urban poor better access to modern energy and also reduce their energy costs (http://www.sparcindia.org/lnews.php#energy).

10.3.2.2 Ujjivan

Samit Ghosh, founder of Ujjivan, started the microfinancing institution in 2005. He stated his ideology as follows: "The urban poor, the fastest-growing segment of our population who live in abject kind of surroundings, has nobody to look at them. I said, why don't I focus on the urban poor." Following the same model of Grameen Bank, Ujjivan serves 1.3 million customers with 350 branches in 23 states of India. Beneficiaries of Ujjivan started to claim that microfinancing from Ujjivan only helped them to earn a livelihood, not to get them out of poverty. Therefore, apart from income-generating loans, Ujjivan has started to provide education loans, too. Furthermore, Ujjivan has recognized the opportunity of financing for affordable housing and has launched a pilot urban housing microfinancing scheme. It is found that 35% of the nonbusiness loans of Ujiivaan in past years were for housing (Merrill & Suri, 2007; Ghate, 2008).

In spite of the emergence of different highly motivated urban MFIs, urban microfinancing is still in the rudimentary stages in India. Therefore, direct involvement of urban microfinancing in DRR and recovery of the poor is not yet observed. However, since urban microfinancing also has the similar approach of livelihood improvement of the poor through giving access to small amount of money, it is assumed that, like rural microfinancing, it has the potential to protect the poor against risk and to help them recover after a natural disaster.

10.4 Urban Disaster, Urban Poor, and Microfinancing—The Case of Jakarta, Indonesia

Jakarta is a city with an average growth rate of 3.08% between 1980 and 1990. Based on the estimation of Jakarta National Socio-Economic Survey 2007, there were 9.06 million people living in an area of 662.33 km^2. It gives Jakarta's population density as 13,700 per square kilometer, which makes it the most populous region in Indonesia. Another estimation shows that Jakarta contains 28 million people within its 5897-km^2 area. Further, about 250,000 people are emigrating there from different parts of Indonesia each year (Rahardjo et al., 2014; Baker, 2012). As a result of this overshot carrying capacity, anthropogenic disasters such as floods and landslides are increasing in terms of quantity, intensity, and distribution (*http://indonesiaurbanstudies.blogspot.com*). Residents of Jakarta are forced to struggle with floods every year. Different studies have noted the increasing woes for Jakarta residents between 2000 and 2013, both in terms of area under inundation and people that must be required to evacuate due to floods. In addition to the effects of flooding, the exploitation of groundwater has made land subsidence a major problem. The number of villages experiencing landslides in the region surrounding Jakarta also increased from 2000 to 2008 (*http://indonesiaurbanstudies.blogspot.com*). There is no doubt that the urban poor of Jakarta are more vulnerable to these increased disaster risks. Like all cities in Asia, these urban poor and vulnerable communities in Jakarta live close to riverbanks, canals, and drainage areas (Figures 10.6 and 10.7). Most of their houses are in flood-prone areas (Baker, 2012).

Figure 10.6 Slum close to a dumping ground and railroad in Jakarta.

Figure 10.7 Slum close to a riverbank in Jakarta.

About 40% of Jakarta city is below sea level (Baker, 2012). According to the poverty line established by the Central Statistics Agency (BPS) in 2011 (Rp 233,740/capita/month, which is $26.80 in US dollars), about 4.97% of the total population in Jakarta is defined as "Poor" (*http://indonesiaurbanstudies.blogspot.com*). Other classifications include "Almost Poor" (7.6%) and "Vulnerable" (12.5%). The first two categories include the number of people whose incomes are below the poverty line, while the rest include the number of people whose incomes are slightly above the poverty line. For these poor residents of Jakarta, the main disaster risks are associated with water management and flood control (Baker, 2012). Nowadays, different city governments, NGOs, and world aid organizations have started to devote more and more attention to the improvement of the local urban life of Jakarta. Therefore, issues such as congestion, urban poverty, slums, and environmental risk have been the focus of efforts to resolve them. Although different NGOs are taking a leading role for resolving different challenges, the involvement of MFIs in these allied initiatives of environmental and DRR is not much observed in Jakarta. Nevertheless, microfinancing have been a standard element of Indonesian governmental poverty alleviation programs for a long time, and unlike in Bangladesh and India, the government in Indonesia plays a key role in microfinancing. In spite of this, the Bank Rakyat Indonesia (BRI), which is one of the world's most successful commercial MFIs, does not serve the poorest residents, but rather the better-off poor. Further, the role of BRI in DRR for the urban poor is not well documented. One exception is the Urban Poor Consortium (UPC), which is an NGO that deals with various urban issues, including community organizations for housing rights and eviction, urban poverty, and the urban environment (http://www.globalhand.org/en/organisations/23436).

Overall, this discussion in the previous sections has given a snapshot of the disaster risks faced by Dhaka, Mumbai, and Jakarta, and especially the risk faced by the urban poor of these cities. The extent and nature of urban MFIs for the poor have also been highlighted.

10.5 Discussion

It is already been observed that the pace of urbanization has been very fast in recent decades, and consequently risks to the urban population are increasing. Lack of income opportunity, poverty, natural disaster, and climate change are some of the principal influences of urbanization in different Asian cities (Jaysawal & Saha, 2014). Through this rapid urbanization, rural poverty is transforming into urban poverty. As mentioned earlier, the urban poor often embrace better income opportunities after entering the wider urban market. However, the urban poor need to face numerous challenges associated with housing, water supply, sanitation, health, education, and social cohesion (Figure 10.8). All these challenges increase the disaster risks and vulnerabilities of urban poor (Baker, 2012). Previous discussions in this chapter related to the disaster risk faced by the urban poor in Dhaka, Mumbai, and Jakarta describe this reality.

In recent decades, both poverty reduction and disaster management approaches have experienced a paradigm shift. The poverty reduction approach has changed its

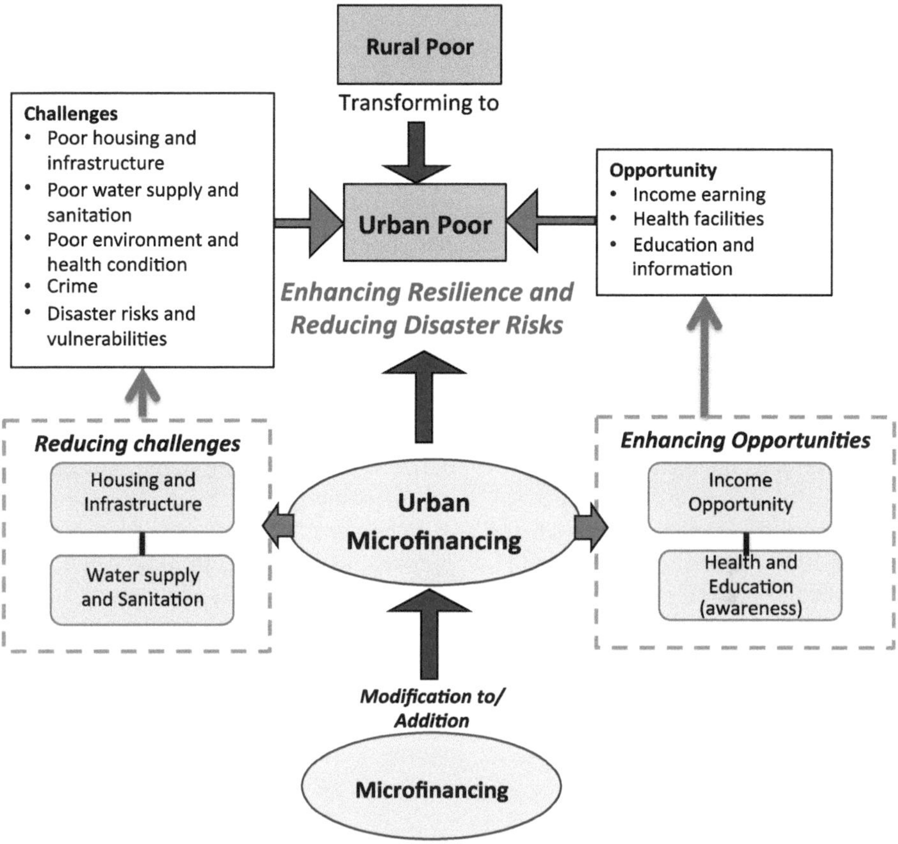

Figure 10.8 Urban microfinancing—A pathway for enhancing the resilience and reducing disaster risks of the urban poor.

focus from income poverty to human poverty, and disaster management has shifted from top-down relief and response to intersectoral risk management. The present approach of DRR has more application to the poor (Strange & Bayley, 2008). Nonetheless, the poor in urban areas still have not been given much accommodation due to their lack of secure tenure. Governments, NGOs, and entities in the private sectors are reluctant to work for the urban poor, since most of the slums have no tenure right.

Due to these efforts to fight poverty and vulnerabilities, it has been proved that microfinancing can support economic and social rejuvenation after a natural disaster by providing monies to rebuild livelihoods, strengthen community bonds, and protect the poorest from income shortfalls (Poston, 2010). But highlights from the three megacities described in this chapter show that in spite of having large numbers of urban poor and a high risk of natural disaster, the role of urban microfinancing institutions has not been so widespread. In the case of each of these cities, a few MFIs have launched programs directed at helping the urban poor. However, these programs are either in the pilot stage or are addressing only a small percentage of their respective city's poor. As mentioned before, in spite of having over three decades of success and growth, microfinancing is in a rudimentary stage in these populous and rapidly urbanized cities. Lack of legal tenure, lack of social cohesion, and mobility of the urban poor have discoursed MFIs to deal with urban poverty and vulnerabilities.

However, this study implies that during the last decade, rapidly urbanized developing countries in Asia like Bangladesh, India, and Indonesia are starting to adapt urban microfinancing. Although the numbers of MFIs are very few, even in different megacities of Asia, there is the hope that MFIs and their efforts will flourish. A traditional microfinancing approach must be modified for the urban poor, and these urban microfinancing is recommended to address all the crucial sectors of vulnerabilities and risk, including income, housing, water supply, sanitation, health, education, and awareness.

It is said that financial resilience and vulnerability are two sides of the same coin (Jacobson, 2009). Therefore, risk reduction strategies for the poor should work toward reducing economic vulnerability (supporting income opportunities). Nevertheless, at the same time, there is a demand to capitalize on the inherent social and cultural capacities of the poor communities. A microfinancing approach for the urban poor should take this holistic approach for risk reduction, and thus resilience building (Figure 10.8). In addition to the holistic approach, as Pantoja (2002) advised, urban MFIs need to be well informed about the potential consequences of disasters at three levels: the clients, the institution, and the sector. Many studies have found that urban microfinancing must be developed as a tool for reducing the risks and vulnerabilities of the urban poor and thus to enhance their resilience.

Acknowledgments

The support of Sumimoto Foundation and SATREP project JICA-JST are gratefully acknowledged by the authors of this chapter.

References

Ahmed, I. (2014). *Factors in Building Resilience in Urban Slums of Dhaka, Bangladesh.* Procedia Economics and Finance 18(2014). Elsevier, 745–753.

Arun, T., Imai, K., & Sinha, F. (2006). *Does the microfinance reduce poverty in India? Propensity score matching based on a national-level household data.* Manchester, UK: *Economics Discussion Paper*, the University of Manchester.

Baker, J. L. (2007). *Dhaka: Improving living conditions for the urban poor.* Bangladesh Development Series, Paper No. 17. World Bank, Dhaka.

Baker, J. L. (Ed.). (2012). *Climate change, disaster risk, and the urban poor: Cities building resilience for a changing world.* World Bank Publications, International Institute of Environment and Development (IIED). London, UK.

Bashar, T., & Rashid, S. (2012). Urban microfinance and urban poverty in Bangladesh. *Journal of the Asia Pacific Economy*, *17*(1), 151–170.

Bhatia, N., & Chatterjee, A. (2010). Financial inclusion in the slums of Mumbai. *Economic and Political Weekly*, *45*(42), 23–26.

CBS News. (2010). *Bangladesh Fire Rips through Building, 116 Dead.* Available from: http://www.cbsnews.com/news/bangladesh-fire-rips-throughbuilding-116-dead/. Accessed on June 15, 2015.

Citywatch India, A SPARC, NSDF, Mahila Milan Publication. Published on February 28, 2014. http://www.sparcindia.org/lnews.php#energy. Accessed on May 26, 2015

Ghate, P. (2008). *Microfinance in India: A State of the Sector Report, 2007.* India, New Delhi: SAGE Publications.

Global Assessment Report (GAR). (2009). *Global Assessment Report on Disaster Risk Reduction, Risk, and Poverty in Changing Climate.* Geneva, Switzerland: United Nations.

Greater Mumbai Disaster Management Action Plan (GMDMAP). (2007). *Greater Mumbai Disaster Management Action Plan, Maharashtra Emergency Earthquake Management Programme, Govt. of Maharashtra.*

Hammill, A., Matthew, R., & McCarter, E. (2008). Microfinance and climate change adaptation. *IDS Bulletin, 39*(4), 113–115.

Indonesia's Urban Studies. (2014). The Megacity of Jakarta: Problems, Challenges and Planning Efforts. http://indonesiaurbanstudies.blogspot.jp/2014/03/the-megacity-of-jakarta-problems.html. Accessed on May 29, 2015.

Institute for Financial Management and Research (IFMR). (2006). *Reaching the Other 100 Million Poor in India: Case Studies in Urban Miocrofinance.* India: Centre for Microfinance at Chennai.

Imai, K. S., Arun, T., & Annim, S. K. (2010). Microfinance and household poverty reduction: New evidence from India. *World Development, 38*(12), 1760–1774.

Islam, N., Mahbub, A. Q. M., & Nazem, N. I. (2009). Urban slums of Bangladesh. *Daily Star* June 20, 2009.

Jacobson, K., Marshak, A., & Griffith, M. (2009). *Increasing the Financial Resilience of Disaster-Affected Populations.* Washington, DC: Office of Foreign Disaster Assistance (OFDA), U.S. Agency for International Development (USAID).

Jahan, I., Ansary, M. A., Ara, S., & Islam, I. (2011). Assessing social vulnerability to earthquake hazard in Old Dhaka, Bangladesh. *Asian Journal of Environment and Disaster Management, 3*(3), 285–300.

Jaysawal, N., & Saha, S. (2014). Urbanization in India: An impact assessment. *International Journal of Applied Sociology, 4*(2), 60–65.

Johnson, J. D. (2006). *Natural Disaster and Vulnerability, OECD Development Center Policy Brief No. 29.* Paris, France.

Khan, A., & Phibbs, P. (27–30 October 2005). Housing and education: An example of a non-shelte outcome. In *The refereed 46th annual conference of American Collegiate Schools of Planning (ACSP).* Kansas City: MO.

Maniruzzaman, K. M., & Haque, Q. (2013). Fire Hazard in Dhaka City: A Case Study of the Service Area of Mohammadpur Fire Station. *Bangladesh Institute of Planners.* Available from *http://www.bip.org.bd/SharingFiles/journal_book/20130718121437.pdf.* Accessed on Feb 17, 2015.

Mathison, S. (2003). *Microfinance and Disaster Management.* South Brisbane, Australia: Foundation for Development Cooperation.

Merrill, S. R., & Suri, D. A. (2007). *Low-Income Shelter Finance in Slum Upgrading: India Urban Initiatives.* United States Agency for International Development, Washington DC: Urban institute.

Mumbai City Development Plan (MCDP). (2006). Available in http://jnnurm.nic.in/wp-content/uploads/2010/12/CDP-Mumbai.pdf. Accessed on May 7, 2015.

Nallathiga, R. (2009). From master plan to vision plan: The changing role of plans and plan making in city development (with reference to Mumbai). *Theoretical and Empirical Researches in Urban Management*, *4*(13), 141–157.

Nicholls, R. J., et al. (2007). *Ranking of World's Cities Most Exposed to Coastal Flooding Today and in the Future*. OECD Environment Working Paper No. 1 (ENV/WKP(2007)1). France: Organisation for Economic Co-operation and Development (OECD).

Pantoja, E. (2002). *Microfinance and disaster risk management: Experiences and lessons learned.*

Parvin, G. A., & Rajib, S. (2013). Role of microfinance institutions (MFIs) in coastal community's disaster risk reduction, response, and recovery: A case study of Hatiya, Bangladesh. *Disasters Journal*, *37*(1), 165–184.

Patankar, A., et al. (2010). *Mumbai City Report. Presented at International Workshop on Climate Change Vulnerability Assessment and Urban Development Planning for Asian Coastal Cities*. Bangkok, Thailand, August 2010.

Patel, S., & Mitlin, D. (2001). *The Work of SPARC, the National Slum Dwellers Federation, and Mahila Milan* (Vol. 5). London: International Institute for Environment and Development (IIED).

Pelling, M. (2003). *The vulnerability of cities*. London: Earthscan.

Poston, A. (2010). Lessons from a microfinance recapitalization programme. *Disasters*, *32*(2), 328–336.

Rabbani, G. M. D. (2009). Environmental risks in Dhaka: Present initiatives and the future improvements. In R. Shaw, H. Srinivas, & A. Sharma (Eds.), *Urban risk reduction: An Asian perspective* (pp. 319–338). Bingley, United Kingdom: Emerald.

Rahardjo, H. A., Suryani, F., & Trikariastoto, S. T. (2014). Key success factors for public-private partnership in urban renewal in Jakarta. *International Journal of Engineering and Technology*, *6*(3), 217–219.

Roy, S. G., & Abdullah, A. Q. M. (2005). Assessing needs and scope of upgrading urban squatters in Bangladesh. *BRAC University Journal*, *2*(1), 33–41.

Sharma, A., Srujan, A., & Shaw, R. (2011). Chapter 1, Overview of urban development and associated risks. In R. Shaw, et al. (Ed.), *Climate and disaster resilience in cities (Community, Environment and Disaster Risk Reduction)* (Vol. 6) (pp. 1–16). Bingley, UK: Emerald Group Publishing Limited.

Sherbinin, A., Schiller, A., & Pulsipher, A. (2007). *The vulnerability of global cities to climate hazards, environment, and urbanization* (Vol. 1). Sage Publications, on behalf of International Institute for Environment and Development.

Srinivas, H., Shaw, R., & Sharma, A. (2009). Chapter 1, Introduction to urban risk reduction. In R. Shaw, H. Srinivas, & A. Sharma (Eds.), *Urban risk reduction: An Asian perspective Community, Environment, and Disaster Risk Management* (Vol. 1) (pp. 3–12). Bingley, UK: Emerald Group Publishing Limited.

Stott, C., & Nadiruzzaman, M. (2014). *Disaster Risk Reduction in Dhaka City—From Urban Landscape Analysis to Opportunities for DRR Integration*. World Vision International, Asia Pacific HEA Community Resilience Team, Singapore.

Strange, & Bayley. (2008). *Sustainable Development—Linking Economy, Society, Environment*. France: Organization for Economic Cooperation and Development (OECD).

Surjan, A., Redkar, S., & Shaw, R. (2009). Community-based urban risk reduction: Case of Mumbai. *Urban Risk Reduction: An Asian Perspective*, *1*, 339–354. Bingley, UK: Emerald Group Publishing Limited.

Tanner, T., Mitchell, T., Polack, E., & Guenther, B. (2009). Urban governance for adaptation: Assessing climate change resilience in ten Asian cities. *Institute of Development Studies, Working Papers*, (Vol. 315) (pp. 1–47). Brighton, UK.

Tran, P., Kaneko, F., Shaw, R., Victoria, L. P., & Oi, H. (2009). Urban disaster risk analysis, action, and implementation management. In R. Shaw, H. Srinivas, & A. Sharma (Eds.), *Urban risk reduction: An Asian perspective* (pp. 13–36). Bingley, UK: Emerald Group Publishing.

United Nations International Strategy for Disaster Reduction (UNISDR). (2013). *Hyogo Framework for Action (HFA) Asia Pacific. (2011–2013).* Kobe, Japan: United Nations Office for Disaster Risk Hyogo office.

UNHABITAT. (2007). *Disaster Risk: Conditions, Trends and Impacts, Chapter 7, Enhancing Urban Safety and Security, Global Report on Human Settlement, 2007.* London: United Nations Human Settlement Programme, UN Habitat, Earthscan, 168–192.

United Nations International Strategy for Disaster Reduction (UNISDR). (2007). *Hyogo Framework for Action, 2005–2015: Building the Resilience of Nations and Communities to Disasters.* Geneva, Switzerland: United Nations International Strategy for Disaster Reduction (UNISDR).

World Economic Forum. (2014). *Global Risks 2014* (9th ed.). Geneva, Switzerland: World Economic Forum.

World Bank. (2011). *Climate Change, Disaster Risk and the Urban Poor, Cities Building Resilience for a Changing World,* The International Bank for Reconstruction and Development/ The World Bank, Washington, DC.

Yodmani, S. (2001). *Disaster Risk Management and Vulnerability Reduction—Protecting the Poor* Paper delivered at Social Protection Workshop 6: Protecting Communities—Social Funds and Disaster Management, at the *Asia and Pacific Forum on Poverty* held at the Asian Development Bank, Manila, February 5–9, 2001.

http://www.globalhand.org/en/organisations/23436, accessed on June 9 to get information related to Urban Poor Consortium (UPC).

Urban Food Security in Asia: A Growing Threat

11

Ranit Chatterjee[1], Atta-ur-Rahman[2], Tho Tran[1], Rajib Shaw[3]
[1]Graduate School of Global Environmental Studies, Kyoto University, Kyoto, Japan; [2]Associate Professor, Institute of Geography, Urban and Regional Planning, University of Peshawar, Peshawar, Pakistan; [3]Professor, Graduate School of Global Environmental Studies, Kyoto University, Kyoto, Japan

11.1 Introduction

A rapidly urbanizing Asia is overburdening its land, water, and energy resources. The 2012 UNAPCEAM policy brief suggests that, by 2050, Asia will see a rise in cases of urban poverty due to urban population growth and the consequent demand for agricultural products will increase by 70%. A considerable number of this future urban population will include productive adults moving to cities because of the inadequacy of income opportunities, lack of rewarding employment sources, and issues of poverty in the rural areas (Afshar, 2003). In Asian cities, especially in developing countries, open spaces and agriculture lands are being transformed into metropolitan spaces to accommodate the increasing urban population. The urban food insecurity is an immediate consequence of this phenomenon. Njungbwen et al. (2011) points out that, as a result of urban expansion, available agricultural land in the urban areas will be impacted, resulting in a reduced food supply to cities. The 2008 world food crisis is a stark reminder of the importance of increasing crop yields or reducing consumption of crops and meat products in order to manage the food demand and supply for an ever-increasing population. This scenario worsens in the case of a disaster because the supply chain and local markets will be severely impacted. Thus, securing food for urban areas in terms of availability, access, and affordability is one of the many challenges for increasing urban resilience globally.

Bailkey et al. (2000) discusses urban agriculture as the growing, processing, and distribution of food and other products through intensive plant cultivation and animal husbandry, in and around cities. Burton et al. (2013) suggest that one of the major factors contributing to food insecurity is the loss of valuable agricultural land, especially those lands on the peripheries of cities that have been lost to development. This loss of agricultural land can be attributed to weak planning and policies that lead to unregulated urban growth and a weak governance system. Acharya et al. (2004) states that the disparity between the availability of land and the actual need for that land leads to a loss of open spaces and prime agricultural land, among other implications for the environment. As a result, the costs of storing, purchasing, and locating in the urban areas is increased (Bernstein, 1994). The negative impact on the access and availability of food will be borne directly by the urban population and will be more prevalent among the urban poor.

Urban agriculture at the household level contributes to the population's income and gives more stable access to healthy food. In addition, urban agriculture adds to the overall production of daily consumable food items such as milk and vegetables at the local level (Zezza and Tasciotti, 2010). Hence, diminishing agricultural land in the urban areas will play an important role in urban poverty and food insecurity in Asia. In order to counteract this trend, various measures have been initiated in Asia aimed at reducing the potential effect of urban poverty, strengthening food security, and supporting sustainable city growth. The peri-urban areas are an important resource for introducing reform in planning and policies to manage and secure food for the urban population. In order to make food more accessible and available, the supply chain management plays an important role in determining the cost, storage, and distribution of food in the urban areas.

This chapter will attempt to understand and explain the impact of urban expansion on agricultural land in Asia's urban and peri-urban areas. The loss of agricultural land will worsen Asia's already alarming food security situation, especially in its developing countries. This study considers the urban centers of India, Pakistan, and Vietnam and observes the trends of urban expansion in recent years and their impact on agricultural land and food security. The analysis of these case studies will lead to the identification of different factors, which can improve the food security conditions in the urban centers and, in turn, improve urban resilience.

11.2 Impacts of Urban Expansion on Farmland in Asian Cities

The global urban population is multiplying daily (Wu et al., 2006). This rapid urban expansion has increased the built-up area in and around Asian cities (UN, 2005), but this development comes at the cost of losing farmland to physical infrastructure (Wu et al., 2006; Samiullah, 2013). With an increase in population and economic growth, cities have grown horizontally. It is important to monitor the impact of urban expansion over farmland, particularly in cities of the developed world (Thompson & Prokopy, 2009). Numerous urban challenges emerged when the agricultural lands in and around cities are engulfed by urban land uses. Prompt government attention is necessary in order to take the appropriate measures and guide the urban expansion, while protecting the prime agricultural land.

It was estimated that cities had utilized over 400,000 km^2 by the year 2000, which is approximately 0.3% of the world landmass (Taubenbock et al., 2009; Samiullah, 2013). The same area under urban territory was 3% of the total arable land (UN, 2005). It has been calculated that, with increasing urban population, the area occupied by cities will further increase 2.5 times by the year 2030 (Samiullah, 2013). This development will mostly be around cities. In Asia, a predominant majority of cities are surrounded by prime agricultural land (Rahman & Shaw, 2015). There is competition in land utilization between urbanization and agriculture.

There is close association between rapid urbanization and the threat to food security, which requires urban planners and decision makers to seriously think and address this emerging issue in their planning and policies. Conceptually, food security aims to explore various means to enhance access to food and to accelerate agricultural production. With the expanding urban population, there is an increasing pressure on the agricultural businesses to meet the food demand in a sustainable way, while also assuring the poor access to food. The rural poor have been pushed towards urban areas in search of jobs and assured access to food. Urbanization has the potential to minimize inequality and poverty.

11.3 Case Studies

In this section, we will discuss three cases from the developing Asian countries that have predominately agrarian economies: Pakistan, Vietnam, and India rank 13th, 14th, and 17th, respectively, in the 2015 Asia Pacific food security index. These three countries have also shown a contrasting trend of agricultural land use in the past decade; Pakistan has shown a decreasing trend, Vietnam shows an increase in agricultural land holding, and India has been more or less constant in this respect. In order to understand the issue in detail, the case studies are examined at different levels, ranging from national, state, and city for Vietnam; Maharashtra (with a focus on the megacity of Mumbai); and Mardan, respectively. In addition, all three countries are exposed to climate change induced hydrometeorological disasters.

11.3.1 Case Study from Vietnam

Vietnam is considered a food-secure country because its rice production matches its population's needs, and the country ranked 65th in 2015's global food security index ranking. In Vietnam, food security is dependent on the natural disasters induced by climate change and the prices and poverty levels for specific groups (Thanh, 2013). Vietnam is divided into eight regions, as shown in Figure 11.1.

In Vietnam, land use and land cover patterns have undergone a major change due to rapid economic development supported by legislation and policies for economic reforms. According to result of the national total land inventory in 2010, Vietnam experienced an upward trend in agricultural land and nonagricultural land areas and a downward trend in unused land during the period from 2000 to 2010. Agricultural land expanded considerably from 21,532 ha (65%) to 26,226 ha (79.2%). Similarly, nonagricultural land area increased from 2850 ha to 3705 ha. These trends will continue in the future. Vietnam's national land use plan through 2020 was approved by its national assembly on November 22th, 2011. According to this plan, the national target aims at reducing nearly 2 million ha of unused land for agricultural and nonagricultural land. More specifically, agricultural land will be increased to nearly 500,000 ha by 2020, half of which will be reached by 2015. Nonagricultural land will also be expanded over 1 million ha, according to the national land use plan through 2020. During the period from 2011-2015, the land use plan will expand agricultural and nonagricultural land areas to 26,732 and 4448 ha, respectively (Figure 11.2).

At regional level, from 2000 to 2010, the northern highlands had the largest area of agricultural land, with 7.7 million ha occupying 76.2% area of this area in 2010. The region's agricultural land spreads 2197 ha from 5.5 million ha in 2000 to 7.72 million ha in 2010. Similarly, the upward trend of agricultural land area occurred in the north central, south central coast, and central highlands, with an increase of 1 million and 0.6 million ha, respectively. In contrast, agricultural land in the Red River delta reduced by 36,000 ha, and by 37,000 ha in the southeast region. In Mekong delta, agricultural land area increased from 3.3 million ha to 3.44 million ha between 2000 and 2005, however

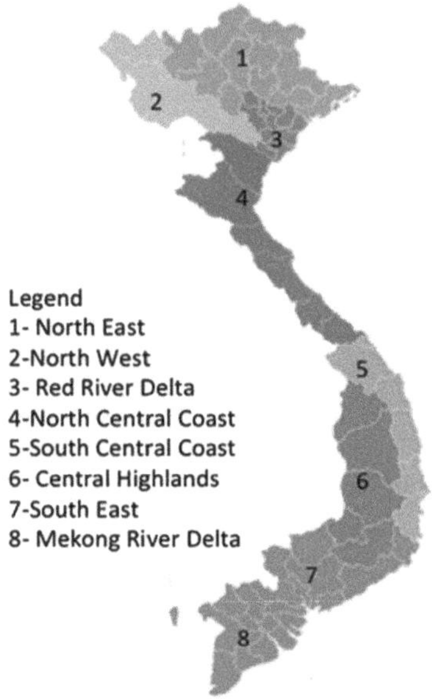

Figure 11.1 Key map of eight regions in Vietnam.

Figure 11.2 Vietnam land use structures from 2000 to 2020.
Source: VNG (2011)

this area of land decreased to 3.4 million ha in 2010. According to the land use plan through 2020, the upward trend of agricultural land will continue in the northern highlands, the north central, the central highlands, and the south central coast. However, the downward trend of agricultural land in the period from 2010 to 2020 will occur in three regions: the Red River delta, the Mekong delta, and the southeast (Figure 11.3).

Agricultural land in Vietnam is mainly classified into three types: rice land, aquacultural land, and forest land. The diagram in Figure 11.4 shows that forest land has expanded greatly from 2000 to the present day. For example, production forest land increased significantly from 1.1 million ha in 2000 to 7.4 million ha in 2010. This increase indicates that forest land will spread only slightly during the next decade. Aquaculture land also developed rapidly from only 30,000 ha in 2000 to 6.9 million ha in 2010. Aquaculture land is expected to increase to 7.9 million ha by 2020.

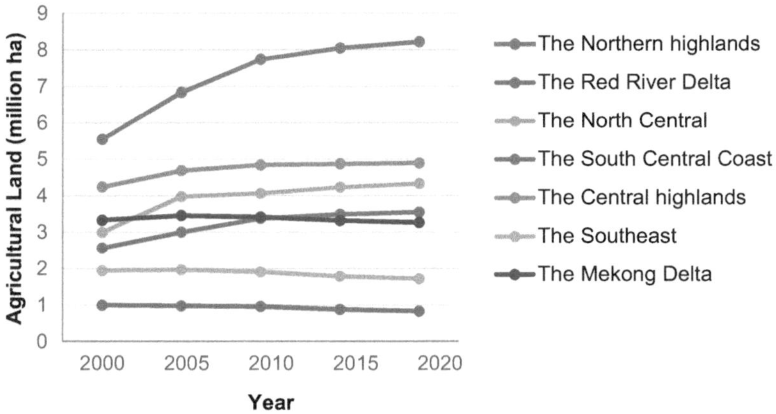

Figure 11.3 Vietnam agricultural land use change in regions between 2000 to 2020.
Source: VNG (2011)

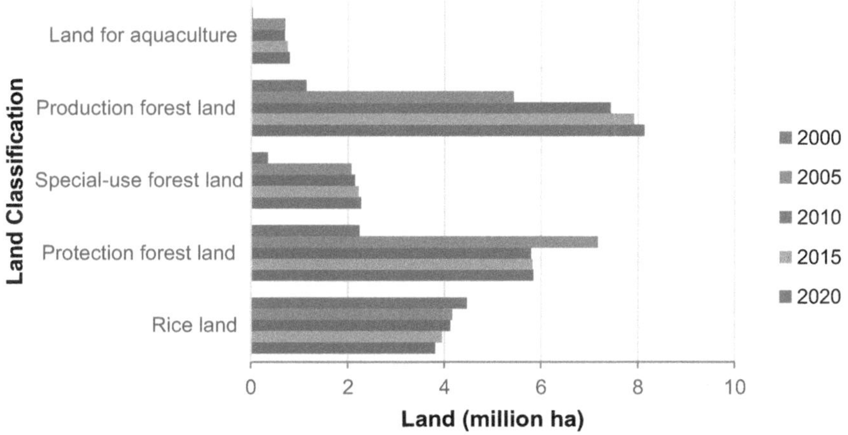

Figure 11.4 Vietnam's agricultural land use change in various categories between 2000 and 2020.
Source: VNG (2011)

Paddy fields play the most important role in Vietnam's agricultural land use. Rice production ensures Vietnam's food security and the country is one of the largest exporters. In 2010, the total rice land area was 4.1 million ha, occupying 12.45% of the national area and 15.71% of agricultural land. Rice land area decreased considerably from 4.4 million ha in 2000 to 4.1 million ha in 2010. It is expected to reduce gradually to 3.8 million ha by 2020. Although rice-sown area is reducing, the rice yields increased from 42.4 kg per ha to 53.2 kg per ha, and rice production increased from 32.5 million tons to 39.9 million tons during the period from 2000 to 2010. During the past decade, the Mekong delta had the greatest reduction of rice land with only 0.16 million ha, down from 2 million ha in 2000 and 1.9 million ha in 2010 (Figure 11.5). The rice land mainly transitioned into aquaculture and fruit land. The Red River delta and the southeast provinces also experienced high decreases of rice land. In the Red River Delta, the reducing ratio in

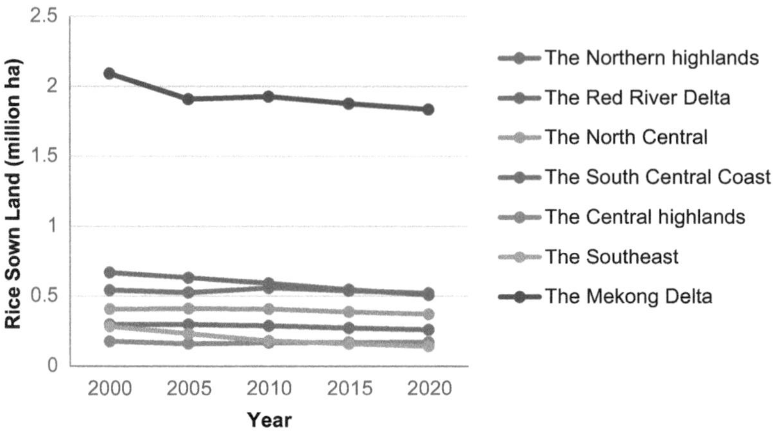

Figure 11.5 Vietnam's rice land change in different regions between 2000 and 2020.
Source: VNG (2011)

Hai Duong is 1400 ha / year, Vinh Phuc's is 1200 ha / year, Hanoi's is 1100 ha, and Hung Yen's 1000 ha / year). The reduction rates in the southeast includes Ho Chi Minh City's 2700 ha / year, Tay Ninh's 3100 ha / year, Binh Duong's 1600 ha / year, and Dong Nai's 1700 ha / year). The two regions have been primarily converted into industrial parks.

11.3.2 *Case Study from India*

The rapid urbanization in India poses a threat to the urban agriculture landholdings. The urban population in India has increased from 27.8% to 31.2% of the total population the in decades ending in 2001 and 2011. In the next 50 years, it is expected that India will transform from a largely rural and agricultural society to a predominantly urban society. The country ranks 68th in the 2015 global food security index. Factors such as income, urbanization, population growth, and the distribution rate of income play an important role in managing India's food security situation (Radhakrishna, 2004).

In a highly urbanized states like Maharashtra, the agriculture output contributes to 13% of the total sectoral outputs and is mainly concentrated in the rural areas (IIHS 2012). According to the 2011 census, 45.22% of the population in Maharashtra lives in urban areas. The contribution of the three major urban centers (Mumbai, Thane, and Pune) to the growth process of the country is 48%, with far less contribution coming from the rural districts faring well in the primary sector, as per the Maharashtra human development report 2012. In addition, the nonagriculture sector has 88% of the income generation compared with 12% of the income generated by the agriculture sector. The employment pattern shows a downward trend of the population working in a primary sector both in the rural and urban areas as shown in Figure 11.6. The average size of agricultural landholding has also decreased from 4.28 ha in 1970 to 1.44 ha in 2011. The total number of operational holdings in the Maharashtra state has increased from 2000-01 to 2010-11 by 12.86%, but at the operated area has decreased by 1.67% during this same period. The overall area and production of Kharif and Rabi crop has

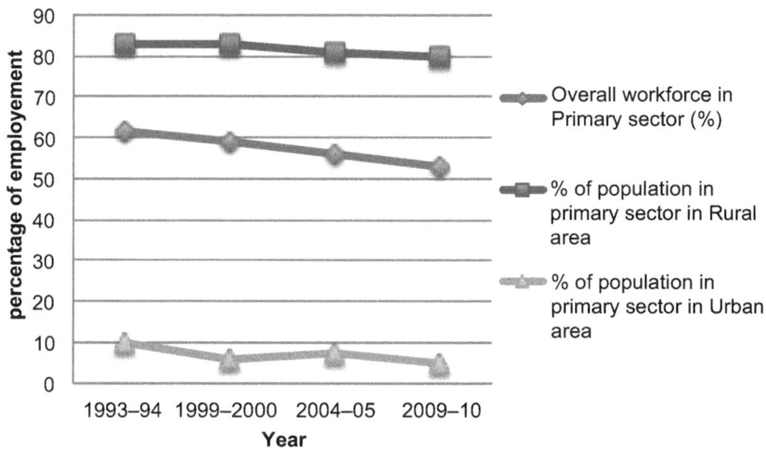

Figure 11.6 Employment trends across sectors in Maharashtra.
Source: MHDR (2012)

Table 11.1 Change in Agriculture and Built-up Area for the Mumbai Metropolitan Area

	Existing Land Use in 1971		Existing Land Use in 1991		Existing Land Use in 2011	
Sector	**Area (km^2)**	**% of total area**	**Area (km^2)**	**% of total area**	**Area (km^2)**	**% of total area**
Agriculture	2098.3	54.4	1445.9	37.5	1381.3	35.8
Built up Area	149.1	3.9	351	9.1	1074.04	27.8
Forest, wetland, industry, water body, and other	1607.6	41.7	2058.1	53.4	1399.66	36.4
Total	**3855**	**100**	**3855**	**100**	**3855**	**100**

Source: Acharya et al. (2004)

shown a decrease by 4% from 2011–12 to 2012–13 (Commissionerate of Agriculture, GoM as cited in MSME development institute report, 2014).

In Maharashtra, the Konkan division has the highest urbanization rate that comprises the Mumbai Metropolitan Region (MMR). Mumbai has no agricultural land and its agricultural landholding in Thane has decreased by 6.56% from 2001 to 2011 (Agriculture Census 2010–11). Table 11.1 shows the change in agricultural and built up land-use in MMR from 1971 to 2011. Acharya et al. (2004) suggests that the growth of built up area in Mumbai and surrounding areas comes at the cost of agricultural land and wetland. In particular, the eastern part of Mumbai focuses less on agriculture due to its rapid urban expansion and increase in real estate business (Vazhacharickal et al. 2011). Figure 11.7 suggests a recent trend of moving from the core to the periphery

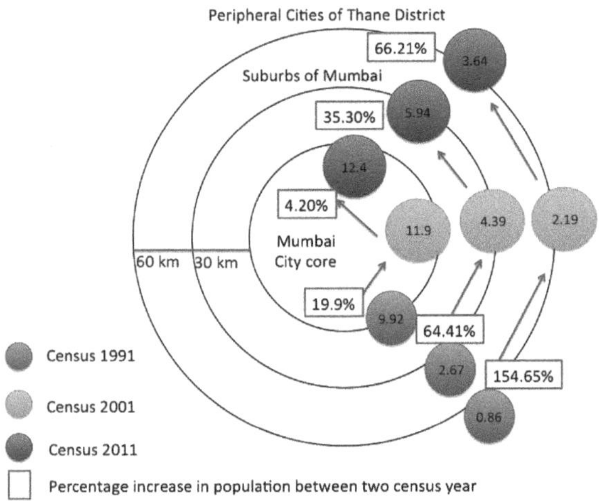

Figure 11.7 The scattering of population from the core to the periphery in MMR.
Source: Bhagat and Jones, 2013

due to availability of land, cheaper cost living and land, and closer workplaces. This trend has been supported by the development of rail and road infrastructure in the MMR. Kim (2012) states that, on the basis of land use change analysis, approximately 23% of agricultural land in Mumbai suburban areas and 6% in the peripheral areas have been repurposed for urban infrastructure development.

Mumbai has various types of urban and peri-urban agriculture (UPA), but considering that very few households engage in these and cater to the local market, the supply to meet the city's food demand comes from outside MMR. Considering that Maharashtra is a drought-prone state, the food supply variation fluctuates annually based on the rainfall conditions (Dev et al. 1996).

11.3.3 Case Study from Pakistan

In Pakistan, the urban population has been increasing at a rapid pace. The World Bank has estimated that half of the country's total population will become urban in the near future (Rahman & Khan, 2015). This growth in urban population is manifested in the encroachment of prime agricultural land by housing and other infrastructure, including for industrial, educational, and commercial uses (Abul, 2001). Change in urban land uses occurs through two processes; the reorganization of land uses, and the outward expansion of urban areas (Lean & Goodall, 1977). In the first process, the pockets of urban agricultural land inside the city are converted into built-up areas. In second process, the urban sprawl or expansion has a direct impact on the available prime agricultural land surrounding the urban areas, which are important for food production and supply (Bender, 1997). The problem is aggravated by the enhanced conversion of agricultural land to built-up area (Engelman & Leroy, 1995). The problem is also compounded by land degradation (Li & Sun, 1997). Soil erosion and salinity are considered the most important threats to agricultural land in Pakistan. With direct effects on the country's

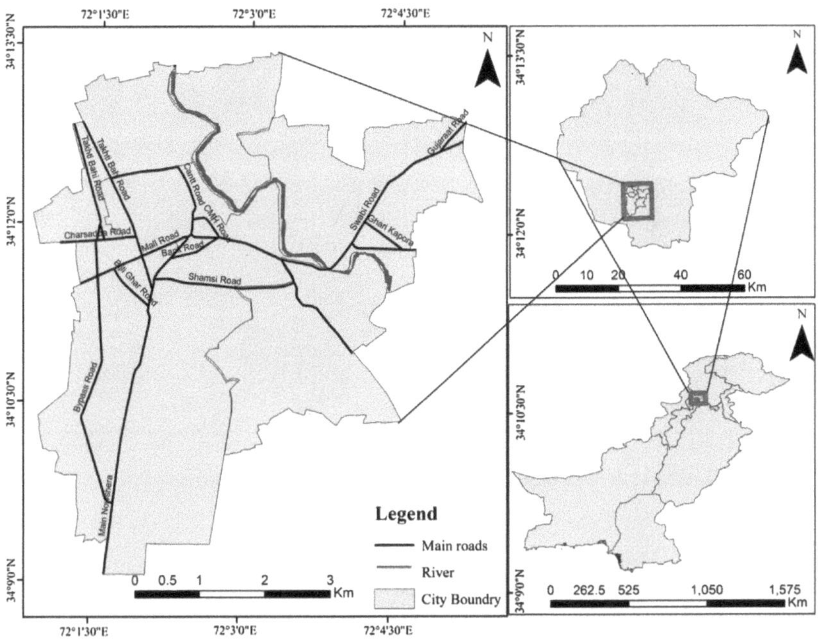

Figure 11.8 Location of Mardan, Pakistan (Yar, 2014).

food production (Liu, 1991), soil erosion that leads to siltation is also harmful to irrigation systems (World Bank, 1992) because it reduces the lifetime of dams.

Global urban food security is a growing threat and Pakistan is no exception; the country ranks 77th on the 2015 global food security index. In order to analyze the farmland conversion into built-up areas, Mardan was selected for microlevel assessment. Mardan is one of Pakistan's fastest growing cities. It is the district headquarters and is also the hub of administrative and financial functions for the region. Geometrically, Mardan stretches from 34°9' 77″ to 34°13' 83″N latitude and 72°3' 11″ to 72°14' 48″E longitude, covering an area of approximately 32 km^2 (Figure 11.8). It is located 64 km from the provincial capital, Peshawar, and 148 km east of Islamabad, the federal capital of Pakistan.

The population of Mardan is increasing at a rapid pace. As a consequence, the city is horizontally expanding in all directions haphazardly, and is engulfing the prime agricultural land, which once served to grow and supply food for the city dwellers. Based on the spatial data analysis, the growth of residential land use is very rapid towards the east, north, and west (Yar, 2014). This development is mostly in an unplanned manner over the farmland. Farmland's conversion to an urban environment would negatively impact the area and would serve as a threat to farmland and urban food security.

The data was extracted through the analysis of multitemporal satellite images of Landsat 1990 and SPOT 2010 (Figure 11.9a,b; Table 11.2). Figure 11.9 illustrates the overall land use change that occurred during 1990–2010. The land utilization has been classified into built-up area, agricultural land, and wasteland (Figure 11.9a,b). Figure 11.9 illustrates the conversion of farmland and cultivable wasteland to different land uses. The analyses clearly depict that there is rapid but consistent conversion of farmland into built-up

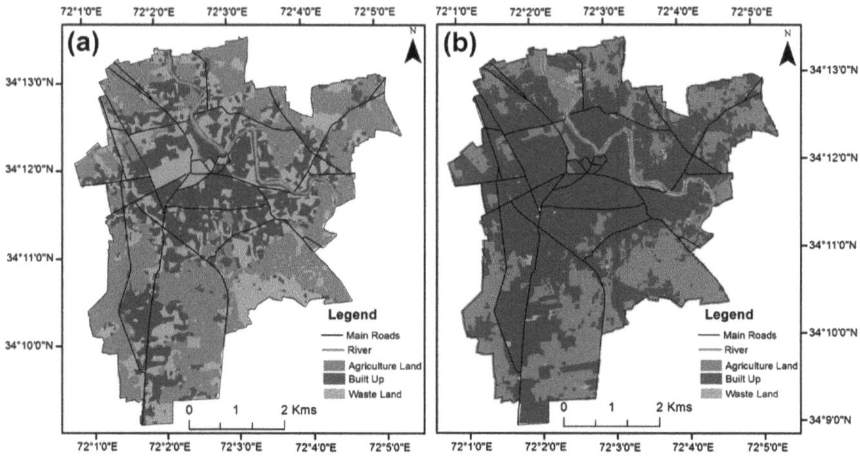

Figure 11.9 Mardan City: (a) Land use 1990; (b) land use 2010.

Table 11.2 **Mardan City: Land Use in 1990 and 2010**

	1990		2010		Change from 1990 to 2010	
Land Use	**Area (ha)**	**Percent Share**	**Area (ha)**	**Percent Share**	**Area (ha)**	**Increase (%)**
Built-up	953	30	1994.5	63	1041.5	52
Agricultural land	1339.1	42.4	1108.8	35.1	−230.3	−17
Wasteland	827.4	26.2	16.4	0.5	−815.4	−98
River	35	1.1	35	1.1	–	–
Total	**3154.7**	**100**	**3154.7**	**100**		

Source: LANDSAT (1990) and SPOT (2010)

area and a reduction in the food production land for urban dwellers. This is a serious threat to the urban food security of Mardan and is expected to further intensify in the near future.

To understand land encroachment in different land use categories, a change detection matrix is shown in Table 11.2 which also reveals that trends of land utilization and conversion during the past two decades (1990-2010). The spatial analysis indicates that farmland within the city limits has been continuously transformed into other irreversible land uses. This is one of the long-term, irreversible negative impacts on the area's carbon footprint and is a silent disaster for the citizens of Mardan. One of the contributing factors for the increase in built up area in Mardan after 1990 was the migration of Afghan refugees to Mardan, which was selected as a refugee campsite. Most of the migrants later settled around the Mardan area (Nasir, 2004).

The analysis reveals that the built-up area was the largest consumer of farmland and wasteland during the period from 1990 to 2010. A total of about 613 ha wasteland was

Table 11.3 Loss of Agricultural Land in Mardan, 1990–2010

Land Conversion	Area Lost in ha (1990 to 2010)	Percent of Total Changed Land
Waste land to built-up area	613	54.5
Farmland to built-up area	371	33
Waste land to farmland	141	12.5
Total	**1125**	**100**

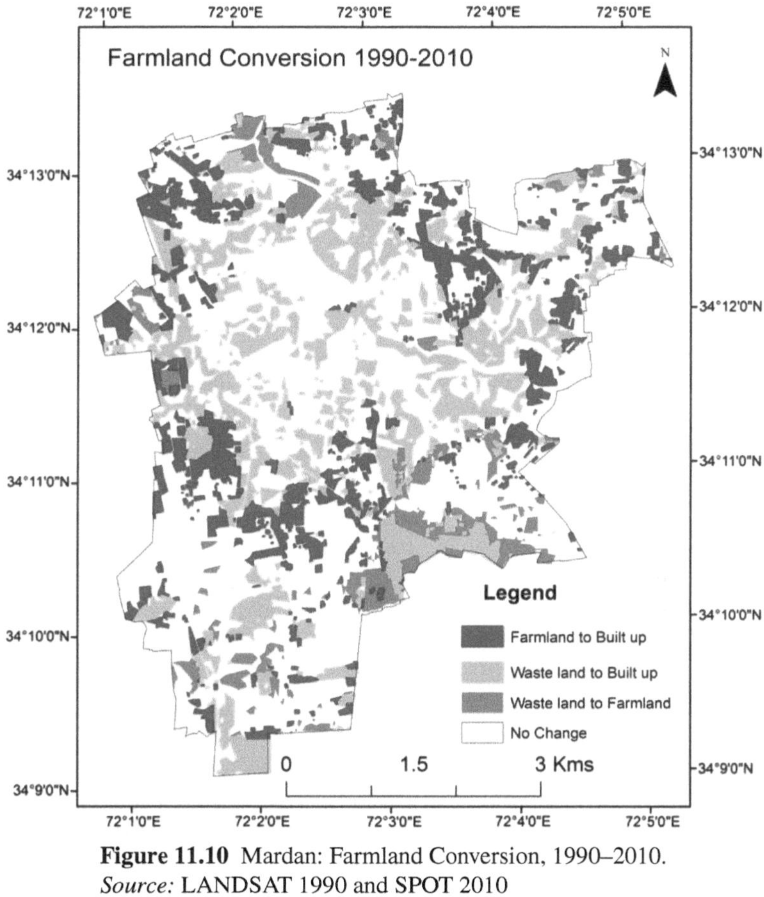

Figure 11.10 Mardan: Farmland Conversion, 1990–2010.
Source: LANDSAT 1990 and SPOT 2010

lost to built-up uses over the period of 1990 to 2010 (Table 11.3). The spatiotemporal land use change detection indicates that 52% of new development was reported over wasteland. Similarly, more than 33% of farmland was converted into built-up area during the same period. However, a large share of approximately 141 ha of wasteland was cultivated during the period from 1990 to 2010. As a whole, there is consistent farmland conversion into built-up environments, as shown in Figure 11.10.

11.4 Discussion

Asia's contribution to global food production (crops and livestock) has grown from 40.9% during the period from 1990–1995, to 46.3% in the period from 2005–2009. Although the food production increased by 1.7% in 2009, the production per capita is still below the world average (ADB report 2013). Considering that Asia is urbanizing rapidly and its population is growing at an alarming rate, food security will be a challenge in the near future. In addition, technological development and superiority is more important in deciding agricultural productivity than climatic and local conditions (Brown et al. 2008).

As evidenced from the cases of India and Pakistan, urban expansion is slowly reducing the agricultural production in the urban areas where people prefer to take other higher paying employment options. In the developing countries of Asia, the city governments lack expertise and weakly implement development plans. The boosting of various special economic zone (SEZ) projects on the periphery of bigger cities to attract investment focuses on the agriculture tracts and wastelands.

In Vietnam, ongoing economic development and structural change is leading to considerable change in land use, with much of the agricultural land, especially paddy fields, transforming into built-up land. Policy will play an important role in regulating this transformation (Dijk et al., 2013). The Mekong delta and the Red delta area will be exposed to climate change-induced flooding which will severely impact the rice fields. Even though Vietnam is the second largest exporter of rice, its households are net buyers of mainly processed food and, with the rise in food prices, food security is a real challenge for the future. In addition, climate change is slowing down the agriculture production by 0.2 % every year in Vietnam. In order to tackle this issue, Vietnam's government, through legislation (Decision 3119/QD-BNN-KHCH), is trying to increase agricultural production by 20% and reduce emissions and poverty by 20% by 2020. The Red River delta and the Mekong delta are main urban and industrial centers, with the four major cities of Hanoi, Hai Phong, Ho Chi Minh City, and Can Tho in close proximity. Both of these regions are experiencing a built-up area expansion that is impacting the paddy cultivation, as can be seen in Figure 12.6. Most of this expansion is happening in the peri-urban areas of the four cities as an effect of industrial growth (Dijk et al., 2013).

In India, the high food density in the urban centers is due to population size rather than the high rate of urbanization (such as in the case of developing countries like Japan), which will be more intriguing considering that the rate of urbanization is also increasing (Matsuake, 2009). In addition, a general trend in India is to export agricultural produce for generating finances rather than feeding local populations. The 2013 national food security act draws out schemes for bettering the public distribution system (PDS), but does not address the issues of reducing urban land use change and its impact on the food security in urban areas. Further improving storage of food after production is an important goal for India, especially in urban areas. As per OCP's final report (2013) India may face a 50MT food shortage by 2020. This would require imports worth approximately 80 billion dollars to supplement the shortage. Globally, 50% of the food produced is wasted due lack of proper storage. One major strategy to handle this issue is to better the infrastructure in urban areas. This strategy has also been supported by the 2013 national food security act.

Although there is hardly any agricultural land in Mumbai, there are several pockets in the peripheral areas of Mumbai where agriculture is a major economic activity. These areas have been marked under the no development zone (NDZ). Manori Gorai Uttan is one such area, with large tracts of approximately 792.90 ha of land (18.33%) of the entire area operating under traditional agriculture practices. Paddy and vegetables are grown in the area and are supplied to the local market. The largest amount of land under agriculture and plantation is in Manori and Uttan, which is around 32.50% and 24.24%, respectively (MMRDA Report, 2012).

In the case of Pakistan, approximately 42% of the land is owned by 9% of the total population. Figure 11.10 shows that the rate of conversion of wasteland to built-up land is higher in comparison to wasteland's conversion to agricultural land, which indicates a specific shift in land use. The agricultural land conversion to built-up land is higher than wasteland conversion to agricultural land, and, as a result, overall agricultural area is decreasing. In Mardan, crop yield in the province is only 23% of its potential capacity, and due its location away from the seaports, the cost of import and export of food products and goods are high. In addition, the area faces issues such as the division of land into smaller units, which reduces profit; limits water for irrigation; results in weak knowledge and technical know-how, weak credit facilities; low prices of agricultural output; and no support of agriculture-based industries (Khan, 2012). The area produces good quality wheat maize, sugarcane, barley, and fruits. Table 11.4 lists the commonalities and differences in the case of Vietnam, India, and Pakistan in relation to food security and urban expansion.

11.5 The Way Forward

From the previous discussions, the three main areas that need to be addressed as a priority to tackle the issue of food security emerge. These priority areas are rapid urbanization, population growth, conversion of agricultural land into built up land, and their collective implication on urban food security.

11.5.1 Stronger Policy and Land Regulations

Urban growth and housing is directly related to policies, land regulations, and land availability (Glaeser et al., 2006). The change in land use, cover, and developments are the result of reforms in policies to keep pace with the need of infrastructure development. The MMR was established on the basis of Gadgill committee recommendations for wider regional planning to depopulate Mumbai. This has lead to a decrease in Mumbai's population, but has increased the density of the peripheral areas and lead to a decline of agricultural land on the city's outskirts. The Industry Act of 1992 further relocates the hazardous and polluting industries from the core to the periphery, which has had an impact on the agricultural land. The recent land acquisition, rehabilitation, and resettlement (LARR) made amendments to the earlier bill by limiting acquisition of irrigated multi-cropped land through a notification (LARR Amendment Bill, 2015).

Households, especially those on the urban fringe that have merged seamlessly into the expanding city, often adapt their farming systems to new urban opportunities, such as the proximity to markets with better opportunities to collect market information and to sell directly to urban customers (Arce et al. 2007). Innovative urban farming techniques that require less land and have higher yields, are required to sustain the urban food market

Table 11.4 **Commonalities and Differences in the Case Studies**

Vietnam	India (Maharashtra, Mumbai)	Pakistan (Mardan)
Food security is impacted by climate change due to change in rainfall patterns and temperature, leading to a change in demand and land use.	Maharashtra is a drought-prone state which impacts food availability and prices.	Agricultural productivity is decreasing due to increasing land desertification, increasing temperatures, and loss of soil fertility.
Urban expansion is leading to a decrease in agricultural land. New built-up area is growing along the peri-urban areas of bigger cities to support industries (Red River delta and Mekong delta).	The urban population growth along the periphery of Mumbai is consuming agricultural land due to outward migration and the increasing rate of real estate.	Conversion rate of agricultural land to built-up area is higher than conversion of wasteland to agriculture land.
Limited land use rights limit large-scale production in Vietnam. Though the government in 2012 introduced new reforms, they have failed to link the regional and sectoral growth plans. Further awareness and understanding of community on the new reform is minimal.	Shift in government policy emphasizing trade liberalizing in agriculture. Weak institutional monitoring and policy implementation by the local government. The food security bill in India is focused on energy security and does not address the availability, accessibility, and stability of food resources. Preference of non-food crop over food crops.	Agriculture is encouraged as an export. As a result, local needs gets ignored. Less funding facility, conflict, and dominance of the state government over local government also play an important role in Mardan's food security.
Private sector involvement is limited. The government supports credit facilities for the farmers.	Involvement of private sector is limited. Inadequate credit facility for the farmers.	Limited involvement of private sector in agriculture. Inadequate credit facility for the farmers.
The rice cultivation is surplus in the Mekong delta and Red River delta regions. Vietnam is dependent on these two regions for internal consumption and export, which are highly exposed to natural disasters which impacts the supply chain of the regions.	Most of the food is supplied from the rural areas of Maharashtra and other parts of the country, leading to an increase in the cost and wastage due to lack of proper storage facilities.	Distance from the seaports and other logistics increases the food prices.

locally. Urban and peri-urban agriculture is more dependent on microunits of intensive livestock rearing and horticultural production, sometimes even without the need of cultivated land through utilizing rooftop, hydroponic, and container production (Prain et al., in press). Common produce includes green vegetables, dairy products, poultry, pigs, mushrooms, ornamental plants, herbs, and fish. In Mumbai, six different types of urban farming have been identified: railway gardens, balcony gardens, terrace gardens, farms, goat keepers, and Tabelas[1]. Of these types, open terrace and balcony gardens, which use limited resources and inputs, are most economical (Vazhacharickal et al., 2011). The Indian railways, under the "Grow more food" schemes of 1975, rented unutilized land near railway tracks and stations to railway class IV employees and nonrailway employees for cultivating vegetables and to protect railway land against encroachment from slum dwellers (Indian Railways, 2008). Vazhacharickal et al. (2013) stated that in MMR, 176 ha of land have been rented for urban agriculture along the railway tracks and the produce yielded is sold in local markets.

The urban areas often have very high water and soil contamination that impacts urban agriculture. Similar to Mumbai, rivers in MMR have high levels of contamination as a result of urbanization and industrialization in the case of the railway gardens. Urban agricultural production systems suffer from high weed growth, pests, and diseases to plants, which often cause skin diseases to the workers (cross referenced from Vazhacharickal, 2011).

11.5.2 Public Investment and Research in the Agricultural Sector

One of the major challenges for Asia's urban and rural agriculture is that there is a lack of public investment in agriculture and agricultural research. More investment could result in better health and nutrition levels, due to an increased access to food, lowered food prices, and the promotion of programs that would give access to nutritious food and health services (Mogues et al., 2012). Research in this field can boost production, public investment, and better support for communities taking up agriculture as an occupation in urban areas; which will prove useful to promote agriculture in cities. The Beijing declaration on the Asia-Pacific Economic Cooperation (APEC, 2014) food security act of 2014 also calls for the development of agricultural policies to support investment in food production, research, and development. New technologies, better management and models of food supply chains, and cold chain technologies will reduce food loss and decrease the logistical costs of food and increase the affordability rate in urban areas.

11.5.3 Strengthen Supply and Chain for Urban Food Security

Food must travel through a complex network in order to supply cities around the world. In order to better the efficacy of this system, the source of food can be made closer to the urban areas through innovative urban agriculture. The daily consumption of an individual generally contains both perishable items, such as vegetable and dairy products; and products that can be stored for longer periods, such as rice, cereals, and others. Bettering the supply chain management gives competitive advantage to individual firms and farmers. The Beijing declaration on APEC food security calls for Asian nations to form community-based networks and enhance the role of producers (i.e., farmers, fisherman, etc.) to establish a reliable and stable food supply chain. Considering that these chains are exposed to impacts of disasters, the identification and securing of critical nodes in the supply chain is necessary. For example, the Otagai project was set up after the 2011 to strengthen the supply chains and promote industrial cluster partnerships across various nations. In countries like India, strengthening the PDS can lead to bettering the supply chain management and, in case of disasters, increase its ability to act effectively by supporting relief work.

[1] Tabelas are cattle farms in India and mainly produce dairy product.

11.6 Conclusion

Urban food security in Asia is a major challenge considering the region's rapid urbanization and population growth rate. This challenge is more severe in developing countries, as seen in the case studies of Vietnam, India, and Pakistan, where the urban expansion impacts the peri-urban agriculture. Climate change and the spiraling increase in natural disasters exert additional pressure on food production and availability. Considering that production itself is not a guarantee of urban food security, other factors like higher export of agriculture products, increase in built-up areas to support urban population influx, and lower profit margins in agriculture and weak policies play a decisive role. Increasing investment and researching a better supply chain are two areas that should be focused on to tackle future challenges of urban food security.

Acknowledgements

The Ranit Chatterjee thankfully acknowledges the Japanese Government (Monbukagakusho) Scholarship; the Atta-ur-Rahman acknowledges higher education fellowship of the government of Pakistan; and the Tho Tran acknowledges the support from the government of Vietnam for conducting research in Graduate School of Global Environmental Studies in Kyoto University.

References

Abul, Q. (2001). Conversion of Agricultural Land to Non-agricultural Uses in Bangladesh: Extent and Determinants. *Bangladesh Development Studies, Vol. 34.*

Acharya, A. K., & Nangia, P. (2004). Population growth and changing land-use pattern in Mumbai metropolitan region of India. *Caminhos de Geografia, 11*(11), 168–185.

ADB Report. (2013). *Food Security in Asia and the Pacific.* Accessed at http://www.adb.org/sites/default/files/publication/30349/food-security-asia-pacific.pdf. on April 30, 2015.

Afshar, R. (2003). *Dynamics of Poverty, Development, and Population Mobility: The Bangladesh Case* Ad Hoc Expert Group Meeting on Migration and Development, organized by the Economic and Social Commission for Asia and the Pacific, Bangkok, August 27–29.

Arce, B., Prain, G., Valle, R., & Gonzalez, N. (2007). Vegetable Production systems as livelihood strategies in Lima-Peru: opportunities and risks for households and local governments. *Acta Hort (ISHS), 762*, 291–302.

Beijing Declaration on Asia-Pacific Economic Cooperation (APEC) Food Security (2014). Accessed online at http://www.maff.go.jp/j/press/kokusai/boueki/pdf/140926-02.pdf on April 30, 2015.

Bender, W. H. (1997). How much food we will need in 21st Century? *Environment, 39*, 7–11.

Bernstein, D. J. (1994). *Land-use Consideration in Urban Management (UMP).* Washington, DC: World Bank.

Bhagat, R. B., & Jones, G. W. (2013). Population Change and Migration in Mumbai Metropolitan Area; Implications for planning and governance. *Asia Research Institute Working paper series no, 201.* Accessed online at http://www.ari.nus.edu.sg/wps/wps13_201.pdf. on October 23, 2015.

Brown, M. E., & Funk, C. C. (2008). Food Security Under Climate Change, NASA Publications. Paper 131. http://digitalcommons.unl.edu/nasapub/131.

Burton, P., Lyons, K., Richards, C., Amati, M., Rose, N., Desfours, L., Pires, V., & Barclay, R. (2013). *Urban food security, urban resilience, and climate change*. Gold Coast: National Climate Change Adaptation Research Facility.

Dev, S. M., & Mungekar, B. L. (1996). Maharashtra's Agricultural Development: A Blueprint. *Economic and Political Weekly*, *XXXI*(13), A38–A48.

Department of Agriculture & Cooperation Ministry of Agriculture New Delhi. (2015). Agriculture Census 2010–11 (Phase-II). *Agriculture Census Division*. Accessed online at http://agcensus.nic.in/document/agcensus2010/allindia201011H.pdf. on October 21, 2015.

Dijk, V. M., Hilderink, H., Rooij, W., Rutten, M., Ashton, R., Kartikasari, K., & Lan, V. C. (2013). *Land-use change, food security and climate change in Vietnam; A global-to-local modelling approach*. LEI report 2013-020 ISBN/EAN: 978-90-8615-610-8 122. Accesed online at https://www.wageningenur.nl/upload_mm/5/c/2/e244a034-5b8b-47e9-8add-237d20f8fef4_2012-063%20vDijk_WEB_DEF.pdf. on October 20, 2015.

Engelman, R., & LeRoy, P. (1995). *Conserving land: Population and sustainable food production*. Washington, DC: Population Action International.

Glaeser, E. L., Gyourko, J., & Saks, R. E. (2006). Urban growth and housing supply. *Journal of Economic Geography*, *6*(1), 71–89.

Guerny, D. (1995). *Gender, migration, farming systems, & land tenure*. SDWP., November 1995. United Nations Population Information Network (Popin) UN Population Division, Department of Economic and Social Affairs, with support from the UN Population Fund (UNFPA).

Indian Institute for Human Settlements (IIHS). (2012). *Urban India 2011: Evidence Report from the Indian Institute for Human Settlements [Internet]*. Accessed May 30, 2012, from http://www.iihs.co.in/wp-content/themes/education/resources/Urban-Dynamics.pdf/.

Kim, M. (2012). *Peri-urbanization and its impacts on rural livelihoods in Mumbai's urban fringe*. 48th ISOCARP Congress 2012 Perm, Russia, 10–13 September 2012.

LARR Amendment Bill. (2015). *Amendments proposed on March 9, 2015*. http://www.prsindia.org/uploads/media/Land%20and%20R%20and%20R/Note%20on%20Amendments%20of%20LARR%20Bill,%202015.pdf. Accessed on October 21, 2015.

Lean, W., & Goodall, B. (1977). *Aspects of Land Economics*. 151 Warder Street- London: The Estate Gazette Limited.

Li, X., & Sun, L. (1997). *Driving Forces of Arable Land Conversion in China*. International Institute for Applied Systems Analysis (IIASA). INTERIM REPORT IR-97-076/Sept. Accessed online at http://citeseerx.ist.psu.edu/viewdoc/download?rep=rep1&type=pdf&doi=10.1.1.152.7572. on October 20, 2015.

Liu, J. (1991). *Rural economic development of China: Ten-year plan and the eighth five-year plan*. Beijing: Ministry of Agriculture Press (in Chinese).

Maharashtra Human Development Report (MHDR, 2012). Accessed online at http://www.in.undp.org/content/dam/india/docs/human-development/MHDR%20English-2012.pdf. on October 20, 2015.

MMRDA Report. (2012). *Development Plan Report—2012 for Manori Gorai Uttan Notified Area*. Accessed online at https://mmrda.maharashtra.gov.in/documents/10180/221387/DP%20REPORT%20for%20MGU%20NA%202013/8e63a407-6bb8-4b92-91eb-74ae-66919c17?version=1.0. on October 20, 2015

Mogues, T., Bingxin, Y. B., Fan, S., & McBride, L. (2012). *The Impacts of Public Investment in and for Agriculture Synthesis of the Existing Evidence*. IFPRI Discussion Paper 01217. National Food Policy Research Institute. October 2012.

MSME Development Institute Report. (2014). *The Industrial State Profile of Maharashtra—2013–2014*. Accessed on June 16, 2015, at http://msmedimumbai.gov.in/html/ISPMaharashtraFINAL.pdf.

Nasir, Z. M. (2004). *A rapid assessment of bonded labour in the carpet industry of Pakistan.* Geneva, Switzerland: Working Paper, International Labour Office.

Prain G., & Zeeuw H.D. (in press). Enhancing technical, organizational, and institutional innovation in urban agriculture. *Urban Agriculture Magazine*, 19. Accessed on April 29, 2015, at http://www.futureagricultures.org/farmerfirst/files/Add_Prain_de_Zeeuw.pdf

Rahman, A., & Shaw, R. (2015). Urban risk and reduction approaches in Pakistan. In A. Rahman, A. N. Khan, & R. Shaw (Eds.), *Disaster risk reduction approaches in Pakistan* (pp. 295–314). Tokyo: Springer.

Samiullah, S. (2013). *Expansion of built-up area and its impact on urban agriculture in Peshawar, Pakistan.* Ph.D thesis submitted to the Department of Geography. University of Peshawar-Pakistan.

Taubenbock, H., Wegmann, M., Roth, A., Mehl, H., & Dech, S. (2009). Urbanization in India spatio-temporal analysis using remote sensing data. *Computers, Environment and Urban Systems, 33*, 179–188.

Thanh, H. X., Anh, T. T., Quang, L. T., Giang, D. T., & Phuong, D. T. T. (2013). *Food security in the context of Vietnam's rural-urban linkages and climate change, Country Report, December 2013.* IIED.

Thompson, A. W., & Prokopy, L. S. (2009). In *Tracking urban sprawl: Using spatial data to inform farmland preservation policy* (26) (pp. 194–204). Land Use Policy 2009.

UNAPCEAM Policy Brief. (2012). *Urban agriculture: Improving food security and environmental health of cities.* issue no-1 Jan-Apr 2012 Accessed on-line at http://un-csam.org/publication/PB2012_1.pdf. on October 20, 2015.

United Nations. (2005). *World Urbanization Prospects, 2005 Revision.* United Nations: DESA, Population Division. Accessible from http://www.un.org/esa/population/publications/WUP2005/2005WUPHighlights_Final_Report.pdf. on October 20, 2015.

Vazhacharickal, P. J., & Buerkert, A. (2011). Sustainable cities: An overview of the urban and peri-urban agricultural production in Mumbai Metropolitan Region (MMR). *Leituras de Economia Política, Campinas, 19*, 69–87.

Vazhacharickal, P. J., Predotova, M., Chandrasekharam, D., Bhowmik, S., & Buerkert, A. (2013). Urban and peri-urban agricultural production along railway tracks: A case study from the Mumbai Metropolitan Region. *Journal of Agriculture and Rural Development in the Tropics and Subtropics, 114*(2), 145–157.

VNG. (2011). *Vietnamese Master Land Use Planning up to 2020 and Land Use Plan for the Period of 2011–2015.* Vietnamese government (In Vietnamese). Available from Vietnamese Institute for Legislation; accessed on September 2, 2015, from http://vnclp.gov.vn/UserControls/News/pFormPrint.aspx?UrlListProcess=/ct/cms/DuLieuChuyenDang/Lists/ThongTinKyHop/View_Detail.aspx&ListId=df23fb7d-b6b2-43ac-820a-21f835d1fa2d&SiteId=61e0af9c-c60b-48ee-af52-27908a86acd8&ItemID=242&SiteRootID=ae93a5bf-4d4f-412c-ba6e-3dfdba90bf10.

World Bank. (1992). *China's Environmental Strategy Paper.* Washington, DC: World Bank.

Wu, Q., Li, H. Q., Wang, R. S., Paulussen, J., He, Y., & Wang, M. (2006). Monitoring and predicting land use change in Beijing using remote sensing and GIS. *Landscape and Urban Planning, 78*(4), 322–333.

Yar, P. (2014). *Urban expansion and its impacts on agricultural land of Mardan city.* M.Phil thesis submitted to the Department of Geography. Pakistan: University of Peshawar.

Zezza, A., & Tasciotti, L. (2010). Urban agriculture, poverty, and food security: Empirical evidence from a sample of developing countries. *Food Policy*, 265–273.

Identifying Priorities of Asian Small- and Medium-Scale Enterprises for Building Disaster Resilience

12

Ranit Chatterjee[1], *Nafesa Ismail*[1], *Rajib Shaw*[2]
[1]Graduate School of Global Environmental Studies, Kyoto University, Kyoto, Japan;
[2]Professor, Graduate School of Global Environmental Studies, Kyoto University, Kyoto, Japan

12.1 Introduction

Small- and Medium scale Enterprises (SMEs) are vital contributors to any national economy across the world. In Asia, 90% of all enterprises are SMEs (ADRC, 2012) and across the regional economies of the Association of Southeast Asian Nations (ASEAN), the Asia-Pacific Economic Cooperation (APEC), and the South Asian Association for Regional Cooperation (SAARC), SMEs are major investors and employment generators. The number of SMEs varies. In Japan and Thailand, more than 99% of all enterprises are SMEs. The recently concluded Third World Conference on Disaster Risk Reduction in Sendai recognizes the impact of small-scale disasters and slow-onset disasters on SMEs and recommends their active engagement in disaster risk reduction (DRR). A report by the United Nations Economic and Social Commission for Asia and the Pacific (UNESCAP) and United Nations International Strategy for Disaster Reduction (UNISDR) (UNESCAP/UNISDR, 2012) suggests that present trends show a surge in physical and economic exposure to disasters in the Asian region. Considering that 75% or more of the total capital investment is made by the private sector in most national economies, the contribution of SMEs becomes a sizeable part of that.

In Asia, a majority of the urban areas are located along the costal areas or deltas and thus are exposed to a higher risk of climate-induced extreme weather events (IPCC, 2007). The preference of SMEs to be situated in urban areas is influenced by the locational advantage of logistics, a dependency on the urban market, and the presence of banking and financial institutions. Urban SMEs have a faster growth rate in comparison to their rural counterparts due to infrastructure support and strategic business conditions in urban areas (ADB, 2012). Further, the growth of urban markets, driven by an increase in demand and better transportation options, is influencing the migration of SMEs to the urban areas. Dodman et al. (2009) informs that informal settlements in urban areas that are frequently affected by floods and landslides are an increasing trend in SMEs.

A study by APEC (2011) on the business continuity plans of the private sector highlights a higher utilization of risk finance mechanisms in large-scale companies in comparison to SMEs. The Pakistan floods in 2010 highlights that 10% of the SMEs

Urban Disasters and Resilience in Asia.

closed down and 90% of the SMEs that reopened within six months incurred losses and were dependent on external support for recovery (Asgary et al., 2012). A similar trend is noticed in the case of the East Japan Earthquake and Tsunami (EJET) of 2011, where 656 SMEs stopped operations, mainly due to fund shortage and supply chain disruption (ADRC, 2012).

Until recently, disaster management was in the public domain, with no clear role for the private sector in the building of urban resilience. SMEs, in comparison to large-scale enterprises, are highly vulnerable to natural hazards due to their limited involvement in risk-management approaches and resources. Legal and regulatory frameworks can effectively encourage SMEs to investments in disaster risk reduction and reduce undependability for improving socioeconomic development (UNDP, 2013). Influenced by this fact, the international and regional frameworks consider the private sector, especially SMEs, as a major stakeholder in disaster resilience. Business continuity planning (BCP) is one of many ways of involving SMEs in disaster management, but various case studies suggest that other measures are equally important for the overall risk reduction of SMEs.

This chapter considers SMEs as important drivers of socioeconomic growth in Asia, and their engagement in disaster management is crucial for sustaining such growth. The first section focuses on various international frameworks, plus regional and national legislations and policies for engaging SMEs in disaster management activities. This chapter also presents three cases studies from Japan, Thailand, and Malaysia, which are studied on the basis of available secondary literature to draw out parallels. The final section discusses the lessons learned from the case studies to identify the priority areas of actions for accelerating the SMEs role in urban resilience.

12.2 Legislation for SMEs and Disaster Risk Reduction (DRR)

Legislation and policies at the international, national, and local government level are important tools to build capacity and remove gaps in local preparedness and to contribute effectively to resilience building. Different legislation at various levels has directed efforts to reduce the risk of SMEs to disasters. The recently-concluded Sendai Framework for Disaster Risk Reduction (SFDRR) recognizing the higher impact of small and frequent disaster on SMEs and have recommended a proactive role of the private sector in disaster risk governance through risk transfer, capacity building, and awareness generation, and sharing of good practices and information with a wider group of stakeholders (SFDRR, 2015).

In the Asia specially for the East, Southeast and South Asian region, since SMEs play a major role in all national economies, the Asian Ministerial Conferences for Disaster Risk Reduction (AMCDRR) has from time to time suggested important action points. The 3rd AMCDRR in Kuala Lumpur, Malaysia, in 2008, the 4th AMCDRR in Incheon, South Korea, in 2010, and the 5th AMCDRR in 2013 in Yogyakarta, Indonesia, all emphasized the role of the private sector in climate change and disaster risk reduction

(DRR). The 6th AMCDRR in 2014 stressed business continuity management (BCM)/ business continuity planning (BCP) case studies, describing resilient business practices for SMEs. Furthermore, long-term targets in the form of local government assistance for SMEs with BCM/BCP and coordination between large and small businesses were set in order to develop practical and actionable toolkits for Asian SMEs (AMCDRR, 2014).

The role of regional organizations in Asia—namely, APEC, ASEAN, and SAARC, for DRR in the Asian forum were highlighted in the 5th AMCDRR in 2012. Table 12.1 shows the institutional provisions in APEC, SAARC, and ASEAN for engaging SMEs in disaster management–related work.

SAARC envisions strengthening the private sector's role in DRR work by prioritizing work related to business continuity planning; reducing exposure; encouraging corporate social responsibility; creating public-private partnerships; and protecting the unorganized sector through microinsurance. Furthermore, the post-2015 framework of SAARC acknowledges the higher vulnerabilities of SMEs in cases of disaster, and notes that their reluctance to invest in insurance plays an important role in the business survivability. This is due to the linked supply chain, which also affects large corporations (SAARC Report, 2014).

At the national level, countries have developed specific SME policies that integrate the overall development of the sector with DRR. In the Philippines, through the Small Business Charter (RA6977, amended by RA8289), the SME Development Committee can offer support through a mutual aid system, which can be effectively used as a risk-control mechanism through insurances and tax exemptions in disaster scenarios. A similar safety net system is available in Indonesia but has a limited network and has no disaster-specific schemes. In Malaysia, in past flood events, the Credit Guarantee Corporation provided a guarantee for 80% of the financing for special disaster funds through the support of the banking institutions. Further, this corporation intervened to relax the repayment period of disaster-affected SME and urged insurance agencies to accelerate payment of claims (SMRJ Report, 2008). In Japan, METI supported SMEs affected by disasters via policy loans and safety net loans. After the East Japan Earthquake of 2011, the government made special consideration in the annual budget of 2013 for a disaster recovery project for joint facilities of SME associations, rehabilitation support cooperatives, and the development of temporary factories and stores (METI/JSBRI, 2011; METI/JSBRI, 2012; METI/JSBRI, 2013). For the Southeast Asian countries, although the hazard risks are high, no specific legislation or policies have been developed regarding SMEs and DRR. India is an observer to the Asia-Pacific Economic Council, which at its 37th High-Level Policy Dialogue in 2014 in Taipei shared the following commitments:

- Establish BCPs to facilitate trade and investment, focusing on information communication and technology (ICT), logistics, and auto parts industries.
- Implement a BCP Guidebook for better global supply chains. The guide needs to be prepared in various regional languages for effective understanding and implementation.
- Enhance public-private partnerships and cross-cutting cooperation.

The government of Sri Lanka in the post-tsunami phase created a three-pronged strategy for restoring livelihood within one year: a cash-for-work program, financial

Table 12.1 Institutional Mechanisms of APEC, SAARC, and ASEAN for SMEs

APEC	SAARC	ASEAN
Institutional Setup for SMEs		
Small and Medium Enterprises Working Group (SMEWG) under the aegis of APEC		ASEAN Economic Ministers (AEM), Minister of Economy, Trade and Industry (METI), Japan, AEM-METI Economic and Industrial Cooperation Committee (AMEICC), The Economic Research Institute for ASEAN and East Asia (ERIA) envision establishing a disaster-resilient society in East Asia.
Actions Taken		
SMEWG prepared a strategic plan for 2013–16 that envisions boosting disaster resilience to support secured development and reliable supply chains for SMEs and MEs in the Asia pacific region.	The SAARC Disaster Knowledge Network (SADKN) was set up for SAARC member countries for the exchange, sharing, and acquiring of disaster-related knowledge and information through the SADKN web portal.	
Action Points		
Capacity building and awareness of SMEs and MEs facing natural disasters Access to and sharing of information to ensure business continuity Propose projects to support the natural disaster resilience of SMEs and MEs	Sharing of voluntary targets for resource allocations	Satellite network systems for disaster preparedness and disaster relief ERIA to research the disaster preparedness of ASEAN's supply chain networks, institutional prevention mechanisms, effective development of infrastructure and the overall connectivity of the region To focus on an ASEAN country-specific disaster management system for disaster risk assessment and a business continuity plan

Source: APEC (2012); SAARC nd., ASEAN, 2012

assistance through banks for business recovery, and the National Development Trust Fund Scheme, which lead to the creation of 50,000 additional jobs (UNDP, 2013).

12.3 Case Studies

Earthquakes and floods are the two most potential natural hazards to impact the SMEs in Asia (ADRC, 2011). In this section, the case studies of East Japan Earthquake of 2011 and Thai floods of 2011 are considered as shown in Figure 12.1. The SMEs in both the economies are in a dominating positioning, making up for more than 99% of all enterprises. The EJET and Thai floods are termed as two of the most costly Asian disaster of the last decade, causing economic damage to tune of US$300 billion and US$46.5 billion, respectively (UNDP, 2013; APEC, 2012). What is important is here is the fact in both cases, the impact was not restricted to the region but had a far-reaching impact on a global scale. This section takes a look at the various good practices taken up by the two countries in the backdrop of the disasters. This learning is utilized to identify critical areas for action that can add to the resilience-building of SMEs.

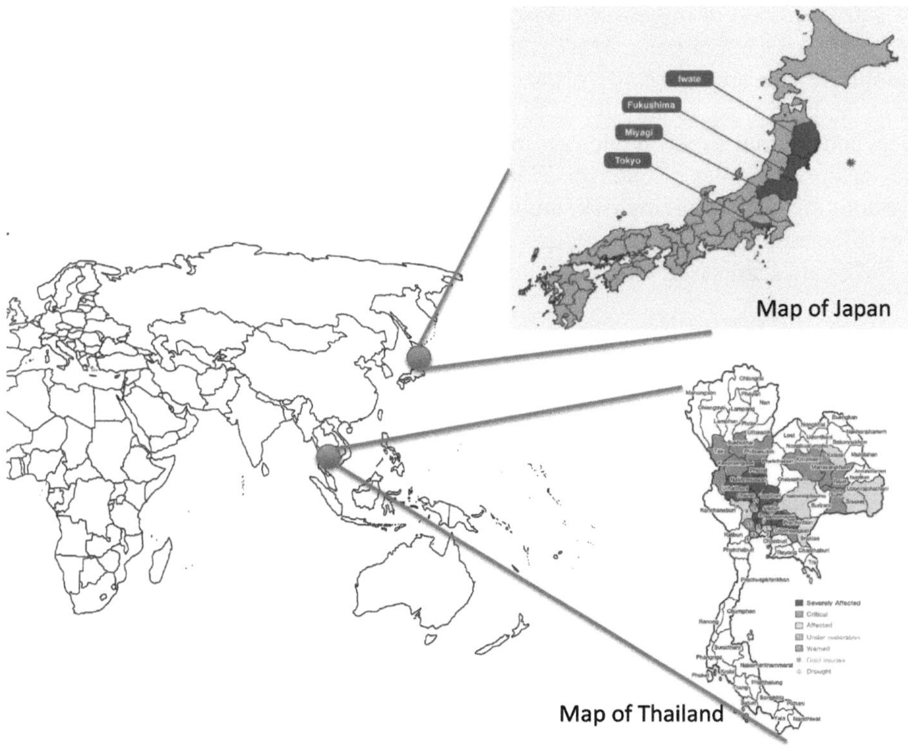

Figure 12.1 Key map with locations of disaster-impacted areas. (*Source:* Japan Meteorological Department and www.thaiflood.com)

12.3.1 Case Study of EJET

SMEs make up 99.7% of all enterprises in Japan, and employ 70% of the working population (Small and Medium Enterprise Agency, 2013). The EJET affected the economy across the country, with a loss of approximately US$309 billion. It is found that 13,127 SMEs had direct damage in disaster-affected prefectures and 2825 in other prefectures. At the same time, 6423 SMEs had indirect damage in disaster-affected prefectures and 123,683 in other prefectures (METI and JSBRI, 2011). The EJET caused 656 private companies to go bankrupt within one year, of which only 87 were located in the disaster-affected area; the rest were affected by indirect losses (Ranghieri and Ishiwatari, 2014). A study shows that among SMEs in the Iwate prefecture affected by EJET 2011, a majority of the respondents had no BCP (Figure 12.2).

The earthquake and tsunami highlighted the issues of lack of expertise and the disruption of supply chain for procurement of raw material and distribution of finished products. These caused the businesses in the shopping district to close; and a further shortage of power and water supply lead to a stoppage in production units.

The Japan Finance Corporation (JFC) issued equipment and working loans (Great East Japan Earthquake Special Recovery Loan system) through various centers set up on the same day of the earthquake at 0.5% less than the prevailing interest rate for a grace period of three years (Fukanuma et al., 2013). This effort was supported by the Shoko Chukin bank. Moreover, the government gave special unemployment benefits to the employees of disaster-affected SMEs, and also set up temporary stores and factories, rebuilt the local shopping area, supported the local core industries, and offered expert advice on reconstruction (METI and JSBRI, 2011). From the various examples of enterprises in post-EJET, it comes out clearly that innovative solutions, internalizing risk, mutual sharing, a strong supply chain, and the support of community, government, and financial institutions are crucial for the revival of SMEs after a disaster situation.

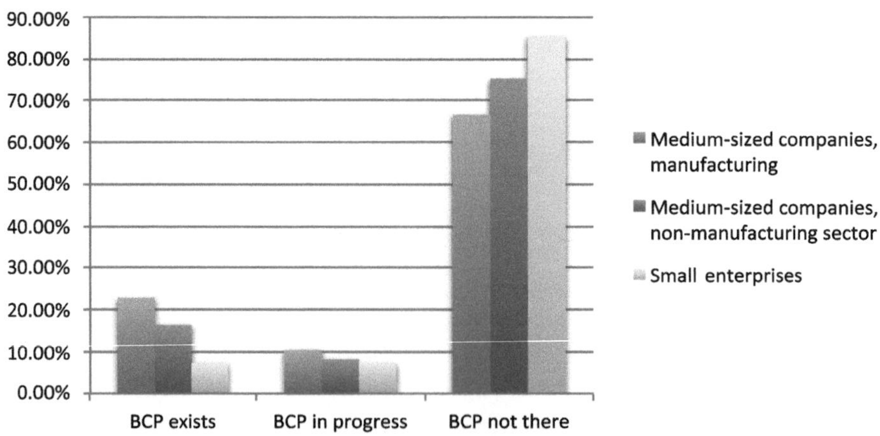

Figure 12.2 Existence of BCP in SMEs in Iwate Prefecture affected by EJET 2011. (*Source:* Tokio Marine & Nichido Risk Consulting Co., Ltd. and Tokyo Shoko Research, 2011)

Post-EJET, the government of Japan brought in structural changes by revising the SME Basic Act in 2013. Financial support was provided for SMEs to expand and reach out to the global market. Ranghieri and Ishiwatari (2014) shared that only 12% of the medium-sized companies had BCP done before EJET, and the reason for the ineffectiveness of the BCP was due to an underestimation of the impact of the disaster and a lack of training and drills by the companies following the preparation of BCP. More recently, in 2015, during the Sendai World Conference on DRR, the Ministry of Infrastructure and Environment, the Netherlands, and the Japan Bosai Platform (JBP), a business network platform representing 119 small to large corporations based in Japan, supported by UNISDR have agreed to develop links with suppliers of resilience building tools and services for urban resilience.

12.3.2 Case Study of Thailand Flood 2011

There are 2.7 million SMEs employing 11.8 million people (84%), which accounts for 98.5% of Thailand's total businesses and contributes 37% to the country's gross domestic product (GDP). The manufacturing industry in Thailand makes up 38.5% of the GDP and is an important player in the export sector. The Thailand government aims to make SMEs the primary economic driving force by 2016.

In 2011, 23% above the average normal annual rainfall caused terrible floods in Thailand, submerging seven major industrial states in up to 10 feet of water. These floods affected the country's GDP and resulted in high unemployment. According to the Office of SMEs Promotion (OSMEP) in Thailand, the floods affected 557,000 SMEs and left about 2.3 million employees jobless. The economic loss was to the tune of US$46.5 billion (APEC, 2012) and caused disruption in local and global supply chains, especially the manufacturing of cars and computers (World Bank GFDRR, 2012). Ranked 12th among the world automobile industries, Thailand has a considerable presence of automobile parts supply factories and production units. The factories are concentrated in the central and peripheral part of Bangkok, along the Chao Phraya River, so they were submerged during the flood disaster. Overseas automobile manufacturing companies such as Honda, Toyota, Mitsubishi, and Ford had to stop production, which resulted in an interruption of the global supply, which affected markets in North America and other parts of the world. Losses were also comprised of stock and inventory damages, production not obtained, sales not made, higher operational costs, lower revenue, and unexpected expenditure such as humanitarian assistance to affected employees and communities.

BCP is largely practiced by large companies in Thailand. Many SMEs failed to resume business after the disaster event due to a lack of management awareness, the absence of expertise, and limited financial resources to implement and carry out BCP (Ingirige et al., 2014; GARDRR, 2014). In the survey "Recommended Methods to Promote BCP Among SMEs in Thailand," the Asian Disaster Preparedness Centre (ADPC) found that 72.15% of the SMEs suggested that the government should play its part in informing and circulating the *BCP Guideline Manual* for SMEs in their respective sector. The survey also reported that 62.5% of the SMEs hope for more coverage, such as low premium rates from insurance companies for businesses with a standardized BCP.

According to World Bank (2012), the insurance penetration rate among SMEs was at only 4.67%, as the outreach is still very limited.

The Thai government had allocated a THB25 billion (US$900 million) fund to aid the flood-affected industries, aside from assuring both local and international investors in the long recovery period. Due to the devastating flood, the Thailand government created an insurance fund called the Catastrophe Insurance Policy to help companies guard themselves against future floods events. The insurance covers damages caused by natural disasters such as floods, earthquakes, and windstorms. For SMEs, the insurance intends to cover 30% of the total property (not more than US$1.5 million) with a premium rate of 1%. On the other hand, large industries enjoy coverage for up to 30% of property (with total property of sums over US$1.5 million) at a premium rate of 1.25%. However, in order to acquire the Catastrophe Insurance Policy, SMEs first need to purchase a fire insurance policy or Industrial All Risk (IAR) policy issued in Thailand (Willis International, 2012).

Due to the flood disaster, many were displaced and thousands of factories closed down, which affected the labor market demand. The Ministry of Labor helped by providing salary subsidies of USD60 per employee to encourage businesses to keep going in order to prevent unemployment. On top of that, the government also offered temporary assistance in form of relocation of machinery and raw materials during the flood event, import tax exemption for replacement of damaged machinery, and raw materials for those SMEs who were not insured, as well as credit measures by providing loans through the Government Savings Bank and other commercial banks. Table 12.2 lists the various measures taken by OSEMP for supporting flood-affected SMEs in Thailand.

Perwaiz (2015) reported that a project called Daruma was established between Japanese SMEs to encourage Thai businesses to resume operations after the disaster. This was carried out through product value creation, improved productivity, increasing exports to international trade, and expansion into the European markets. The Thai SMEs also stated that having a platform to share BCP experiences and expertise from large companies of the same sectors could also help them in terms of knowledge sharing and maintaining good relations.

Table 12.2 Measures Taken by OSMEP for SMEs After Thai Floods

Financial Measures	Revitalization Measures	New Entrepreneur Creation
Liquidity support Access to loans and interest rate subsidies Ease on loan condition and partial subsidy for interest	Provide SMEs ambulance Starting of SMEs Office Park and SMEs Factory Park Offer new business opportunities for the serious cases	Creation of new opportunities and supporting the unemployed population into entrepreneurship

Source: OSMEP (2012)

The Otagai Project was developed to protect Thai and Japanese SMEs from natural disasters. Through Otagai Business Continuity, the cooperation of those in the same business clusters means assisting and becoming each other's suppliers during a disaster, which boosts business as well as trust among customers. Additionally, strategic communication among major stakeholders (different government levels, communities, private sectors) is a vital component to accelerating the recovery process; policy initiatives, financial access, and technical support are also very important. These kinds of disasters draw attention not only to BCP, but also to the importance of performing standard risk assessments, which includes market risk, operational risk, and economic risk (Perwaiz, 2015).

12.4 Discussion

DRR goes hand in hand with development, especially for the urban centers. It is established that the private sector bears a major share of the losses, as shown in the cases of typhoons, floods, tsunamis, and earthquakes in Asia that have occurred in the recent past. Asia is urbanizing rapidly and has eight of the ten fastest growing urban centers (UNHABITAT, 2011). Considering the growing threat to coastal urban centers from hydrometeorological disasters and climate change, it is imperative that the private sector is brought in to support the city government in urban resilience. Corporate social responsibility, public/private partnerships, and industrial polices create an entry point for the private sector to engage effectively in disaster management work. But the results of a study by Miyaguchi and Shaw (2007) suggest that the involvement of the corporate sector is limited to the backdrop of various disasters, and the main engagement of the private houses is in providing training and technical support. Further, from the case studies of Japan and Thailand, it is seen that most of the SMEs are themselves affected by the disaster. As a result, building their own resilience and managing day-to-day stress will be essential for effectively addressing and adding to urban resilience. In the Asian context, considering the present frameworks and legislations focusing on SMEs, it is felt that there is a gap in translating them into practical actions and creating ownership. Chatterjee and Shaw (2015) suggest through case studies from India that the existence of legislative provisions does not ensure mainstreaming of disaster management in private sector at all levels. Voluntary commitments for engaging in disaster management-related work by the private sector is a deterrent in creating a long-term engagement of the government and private sector in managing urban risk. This needs to be supported by the regional regulatory bodies through standardization and certification for the private sector. The case studies show a strong dependency of the SMEs on the government in the disaster recovery phase.

12.4.1 Reducing Risk of Locational Hazards and Infrastructure Damage for Reducing Direct Loss

The Thai SMEs occupied flood plains due to rapid industrialization in 1970, leading to a decrease in agricultural land along the periphery of city. Similarly, in the case of EJET, the tsunami and landslides affected land close to coast and along the slope. Considering that most of the private sectors prefer to choose locations closer to the coast and delta for ease of transportation, reconsidering urban land use and updating building regulations are important measures for reducing the risk of disaster to these industries. Hayhoe et al., (2010) shares that there are not many studies on the interdependency of the infrastructure. Since SMEs depend heavily on urban infrastructures to support their growth, it is essential that such cross-cutting studies be taken up for the cities to identify linkages.

12.4.2 Securing Supply Chains to Reduce Indirect Loss

The case studies of EJET and the floods in Thailand suggest that companies that are not directly affected by a disaster can be equally vulnerable to the impact of the disaster. The reasons supporting this are the shared supply chain and resource networks, and dependency on each other for raw materials and loss of market (UNDP Report, 2013). The Zurich Financial Services Group and Business Continuity Institute (2011) shares that 51% of supply chains were affected by adverse weather in 2010, and 49% of businesses lost output from such disruption, which lead to an increase in cost by 38% and a decrease in profit by 32%. At both the government level and at the individual SME level, prioritizing the need to invest to secure infrastructure against rare events depends a lot on the availability of funds. Internalization of the supply chain risk and developing strategies as a part of BCP are effective in recovering from disaster, as seen in the later part in the chapter.

12.4.3 Securing and Restarting the Local Market Is Crucial for the Survival of SMEs

One of the focus areas for reviving the industries was to open the market in Japan. The Shiogama Port Recovery Market in Shiogama City, Miyagi Prefecture, the Hamakaze shopping area in Hisanohama in Fukushima Prefecture's Iwaki City, and Taro-chan House in Miyako City, Iwate Prefecture were started to support the sale of finished products of SMEs. Further, the demand for SME product grew due to the Home Eco Point Program at the end of July and a shift to digital terrestrial broadcasting, and a rise in reconstruction demand and replacement demand for energy-saving products (METI, 2012). The SME's innovation of various disaster management-related products created a new demand in the market and helped the SMEs to cope with the crisis. The government plan to market SMEs outside Japan provided a wider market for the products and reduced dependency on a single region for the selling of products. Around 40% of the SMEs reported a change in their procurement policies to make them robust against future shocks, with a higher rate of global procurement and distributed suppliers (METI, 2012). The Thai case also shows a similar trend where the OSMEP, in order to boost the SME sector, supported new entrepreneurs through funds and also set up SME office parks at different locations to support the market and aid production.

12.4.4 Securing Livelihood and Business Through Financial Support Systems

SMEs are one of the highest employment generators in Asia; as result, their well-being reflects directly on the well-being of the community. A UN report (2011) recommends that central and local governments can play a supporting role in recovery strategies for SMEs through financial, technical, and political support but they need to consider the central institutions for channelization due to their greater capacity and decision-making power than disaster-affected local governments. The Thai flood is a good example of similar approach. On the contrary, the EJET case shows that SMEs were supported through loans and interest cuts, and that the local government and community were actively engaged in the reconstruction process. A different approach is seen in post-2005 Mumbai floods, where financial relief was given to affected households and various committees were set up to remove encroachment but did not involve the community and other stakeholders in active planning and management (UNHABITAT, 2007). The creation of contingency funding through the various national-level regulatory bodies is crucial for building resilience of the SMEs. Further, national financial institutions, with the support of the government, can create special funds to help disaster-hit SMEs cope with losses.

12.4.5 Disaster Risk Insurance

The Thai floods resulted in approximately US$12 billion of insured loss, which lead to the collapse of the Thai insurance sector and the unavailability of flood catastrophe insurance. The government started the Catastrophe Insurance Policy for earthquakes, floods, and windstorms by pooling together the funds of 67 market players. The sum insured for SME property is not exceeding US$1.5 million, the coverage is limited to 30% of the property sum insured, at a premium rate of 1.0% of limit. This is an effective way of risk transfer, especially in cases like India where insurance penetration[1] in life, nonlife, and industry is 3.10%, 0.80%, and 3.90%, respectively (Willis International, 2012; IRDA, 2014). In the 2005 Mumbai floods, out of a total damage of around US$5000 million, only 15% were insured, the majority of which were large companies; this suggests a low penetration rate of the insurance sector among SMEs (Munich Re, 2006). The case of the Thai floods also shows a similar trend, with only 4.67% of the total SMEs covered by insurance. The low penetration rate of insurance in India is due to low income levels and the perception of insurance as a saving schemes rather than a risk-transfer mechanism (Gurenko, 2007). The premiums for earthquake insurance are very high, and as a result, most property insurance did not include earthquake insurance (Atmanand, 2003). In the case of EJET 2011, the EJET Recovery Emergency Guarantee Program supported SME Units in raising funds under the Credit Insurance System which followed the guidelines laid down by Act no 40 of 2011 of the Government of Japan. The Nippon Export and Investment Insurance (NEXI) announced various measures to assist SMEs affected by the earthquake and tsunami by a postponement of insurance policy procedures. SMES were given deferment, reduction, and waivers of insured parties'

[1] Insurance penetration is measured as ratio of premium (in U.S. dollars) to GDP (in U.S. dollars).

obligations, and also a reduction or elimination of the financial burden on insured parties (METI, 2012). In Japan, public-private national insurance systems for natural perils exist with obligatory or voluntary contributions where one can choose single-hazard comprehensive insurance (Bayer and Mechler, 2009). In India, there are innovative insurance schemes like Afat Vimo (disaster insurance), which is a microinsurance initiative started by All India Disaster Mitigation Institute (AIDMI) with an objective of supporting small businessmen, small vendors, home-based workers, and landless laborers for 19 types of disasters, covering damage to property, tools, equipment, stock, death, and loss of earning capacity. Such models are effective when supported by major insurance houses and governments (Bayer and Mechler, 2009). The need is to create awareness and increase penetration of the insurance sector among SMEs.

12.4.6 BCP for SMEs

A study conducted by the Asian Disaster Reduction Centre (ADRC) in 2012 for the Asia Pacific region suggests that only 13% of the SMEs have a BCP and 21% of the responders consider a BCP important for risk reduction. After the 2005 Mumbai floods, a one-day holiday was announced to support individual recovery, but this had a reverse impact on the business continuity of the banking sector and other industries (Graham and Kaye, 2006). This highlights two points: the need to integrate BCPs with the city's disaster management plan and to understand each other's strategies for the effectiveness of such planning process, and secondly, a need to reexamine BCPs to reevaluate risks. The EJET case study shows that out of the few SMEs who had prepared a BCP, about 80% of them benefited by the business recovery process, as shown in Figure 12.3.

Post-EJET 2011, there has been a rise in preparation of BCPs among SMEs. Yet the various constraints in the preparation of BCP are a lack of information and coordination, expertise, funding, the fact that it is not a requirement by client, or uninterested

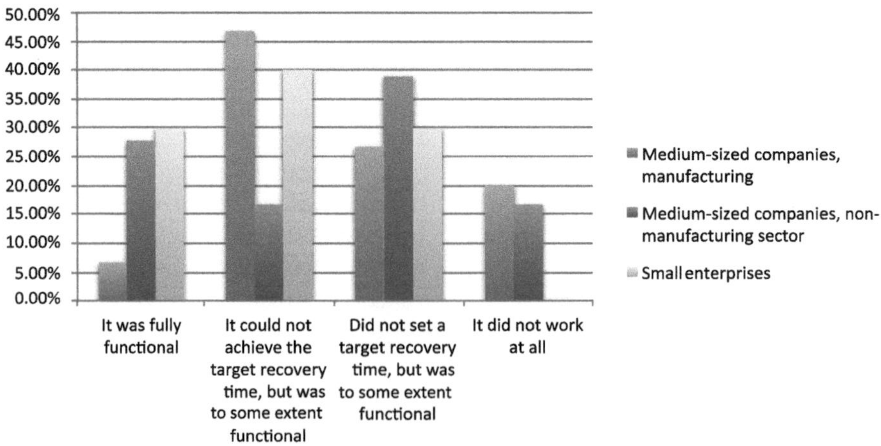

Figure 12.3 Effectiveness of BCP in Backdrop of EJET 2011.
(*Source:* Tokio Marine & Nichido Risk Consulting Co., Ltd. and Tokyo Shoko Research, 2011)

management (METI, 2012, ADRC, 2012). A UNIDO (2008) study further suggests that SMEs view enterprise social responsibility (ESR) differently from the large corporate houses, who implement CSR as a risk-reduction mechanism.

12.4.7 Creation of a Common Platform for Interaction

Regular interaction and resource pooling are common during the response phase, but having said that, most of these are large industries and they do not have capacities to control offsite emergencies. The Japan Platform, founded in August 2000, conducted international aid with a tripartite cooperation system where nongovernmental organizations (NGOs), the business community (Keidanren, the biggest business association in Japan), and the government of Japan (Ministry of Foreign Affairs) work in close cooperation, based on equal partnership, making the most of the respective sectors' characteristics and resources. In the same vein, the Bosai Japan platform is also an important platform for bringing the private sector and government to work together in disaster response and risk reduction. A similar engagement and utilization of South Asian Disaster Knowledge Network (SDKN)-like, Web-based resources are effective in having a coordinated approach towards reducing urban risks.

12.5 The Way Forward

Supplementing Asia's efforts to strengthen its urban resilience, SMEs and other private sector have an important role to play. As is seen across Asia, although SMEs contribute highly to the national GDP, there is no concrete step to reduce their hazard risk. SMEs are more engaged in concretizing their survivability in day-to-day life than investing in DRR for rare disaster events like EJET or the Thailand floods. Due to limited access to the policy and decision-making process and the absence of clear guidelines, SME's involvement in risk reduction is sporadic and needs widespread awarenessgeneration linking DRR and business continuity. The role of various financial institutions is crucial for supporting SMEs in pre- and post-disaster scenarios, and it needs to support innovation and local market. The insurance sector has to come up with new schemes and policies for region-specific disasters, as was seen in the case of Japan and Thailand. The importance of BCP and its benefits needs to be shared with SMEs, and city governments can support BCP preparation by sharing expert advice. The SME's role in DRR is an emerging area of study, hence there are many gray areas that need to explored through focused research, which can be transformed into actions to build synergies among the government and private sector in the future.

Acknowledgments

The first and second authors acknowledge the support of the MEXT (Government of Japan) scholarship for their studies at Kyoto University.

References

Asia-Pacific Economic Cooperation (APEC). (2012). *SME Monitor*. Issue 1, APEC SME Crisis Management Centre. Taiwan Institute of Economic Research.

Asian Disaster Reduction Center (ADRC). (2012). *BCP Status of the SMEs in the Asia-Pacific Region 2012*. Accessed June 9, 2015, at http://www.adrc.asia/publications/bcp/survey_2012.pdf.

Asian Ministerial Conference on Disaster Risk Reduction (AMCDRR). (2014). *Bangkok Declaration on Disaster Risk Reduction in Asia and the Pacific 2014, Bangkok, June 22–26*. Accessed at http://6thamcdrrthailand.net/6thamcdrr/Portals/0/Final%20Bangkok%20Declaration%20-6%20AMCDRR%20-final%2026%20June-0800%20hours.pdf.

ASEAN report. (2012). *Joint Media Statement on the ASEAN Roadshow to Japan*. 28 April 2012, Tokyo. Accessed online at http://www.meti.go.jp/press/2012/05/20120502003/20120502003-3.pdf on 23 October 2015.

Atmanand. (2003). Insurance and disaster management: The Indian context. *Disaster Prevention and Management: An International Journal, 12*(4), 286–304.

Bayer, J. L., & Mechler, R. (2009). Insurance Against Losses from Natural Disasters in Developing Countries, UN/DESA Working Paper No. 85, ST/ESA/2009/DWP/85.

Chatterjee, M. (2010). *Routledge handbook of climate change and society* 2010, Case study: floods in Mumbai, 162–177.

Chatterjee, M. (2011). Flood Loss Redistribution in a Third World Megacity: The Case of Mumbai. *Coping with global environmental change, disasters and security: Threats, challenges, vulnerability and risks*, 603–612.

Chatterjee, R., & Shaw, R. (2015). Public-Private partnership: Emerging role of the private sector in strengthening India's disaster resilience. In T. Izumi, & R. Shaw (Eds.), *Disaster management and private sectors challenges and potentials* (pp. 265–279). Tokyo: Springer.

Department of Fisheries Malaysia. (2004). *Annual Fisheries Statistics 2000–2004*. Kuala Lumpur, Malaysia: Percetakan Nasional Malaysia Berhad.

Department of Statistics Malaysia. (2011). *Annual Statistical Yearbook*. Kuala Lumpur, Malaysia: Percetakan Nasional Malaysia Berhad.

Department of Statistics. (2014). *Annual Statistical Yearbook*. Kuala Lumpur, Malaysia: Percetakan Nasional Malaysia Berhad.

Dodman, D., Hardoy, J., & Satterthwaite, D. (2009). *Urban Development and Intensive and Extensive Risk* Internationals, Institute for Environment and Development report as Contribution to the 2009 Global Assessment Report on Disaster Risk Reduction.

Fukanuma, H., Ishihara, Y., Matsui, Y., & Ota, T. (2013). Nihon Seisaku Kinyu Kouko niyoru Chusho Kigyou muke Sinsaikanren Yushi no Keizaikouka Sokutei nikansuru Ichi Kousatsu (Research note on lending programs for small business enterprises in Great East Japan Earthquake by Development Bank of Japan). *Nihon Seisaku Kinyu Koko Ronshu, 20*, 1–23.

Graham, J., & Kaye, D. (2006). *A Risk Management Approach to Business Continuity: Aligning Business Continuity and Corporate Governance*, 177–178.

Greyhound Report. (2014). *The Indian SME Survey: Analysing Indian SME Perceptions Around Union Budget 2014–2015*. Accessed online at http://firstbiz.firstpost.com/sme-report/pdf/Analysing-Indian-SME-perceptions-around-Union-Budget-2014-15_Final-new.pdf on 20th October 2015.

Gurenko, E. N. (2007). *Climate Change and Insurance: Disaster Risk Financing in Developing Countries*. Earthscan, 596–688.

Hayhoe, K., Robson, M., Rogula, J., Auffhammer, M., Miller, N., VanDorn, J., & Wuebbles, D. (2010). An Integrated Framework for Quantifying and Valuing Climate Change Impacts on Urban Energy and Infrastructure. *A Chicago Case Study, International Association for Great Lakes Research* July, http://www.sciencedirect.com/science/article/pii/S0380133010000547.

Insurance Regulatory and Development Authority of India (IRDA). (2014). *Annual Report—2013–2014.*

Intergovernmental Panel on Climate Change (IPCC). (2007). *Contribution of Working Groups I, II, and III to the Fourth Assessment Report of the Intergovernmental Panel on Climate Change*, Cambridge University Press, Cambridge, United Kingdom and New York, NY, USA.

Ingirige, B., Amaratunga, D., Kumaraswamy, M., Liyanage, C., Perwaiz, A., Towashiraporn, P., Wedawatta, G. (2014). *Private Investment and Disaster Risk Management.* Background Paper Prepared for the 2015 Global Assessment Report on Disaster Risk Reduction. Accessed at http://www.preventionweb.net/english/hyogo/gar/2015/en/bgdocs/Ingirige%20et%20al.,%202014.pdf.

Ministry of Economy, Trade, and Industry (METI) and Japan Small Business Research Institute (JSBRI). (2011). *White paper on Small and Medium-Sized Business in Japan.*

Ministry of Economy, Trade and Industry (METI) and Japan Small Business Research Institute (JSBRI). (2012). *White Paper on Small and Medium-Sized Enterprises in Japan.*

Ministry of Economy. (2013). Trade and Industry (METI) and Japan Small Business Research Institute (JSBRI). *White Paper on Small and Medium-Sized Enterprises in Japan.* Accessed online at http://www.chusho.meti.go.jp/pamflet/hakusyo/H25/download/2013hakusho_eng.pdf on 20th October 2015.

Ministry of International Trade and Industry (MITI). (2005). *Third Industrial Master Plan 2006–2020.* Accessed online at http://www.miti.gov.my/index.php/pages/view/1690 on 20th October 2015.

Miyaguchi, T., & Shaw, R. (2007). Corporate community interface in disaster management—A preliminary study of Mumbai, India. *Risk Management, 9*, 209–222.

Munich re Report. (2006). *Annual Report.* Accessed online at http://www.munichre.com/site/corporate/get/documents/mr/assetpool.shared/Documents/0_Corporate%20Website/_Publications/302-05252_en.pdf on 21 October 2015.

Office of Small and Medium Enterprises Promotion (OSMEP). (2012). Thai SMEs Sailing Through 2011 Great Flood Crisis: Helping Hands and Obstacles. *Presentation made in APEC SME WG Meeting SME Crisis Management, Jerudong, Brunei Darussalam* April 25–26, 2012.

Penang Economic Monthly. (2005). Impact of the Tsunami on Penang's Economy. *Socio-economic and Environmental Research Institute, 7*(1), 3–17.

Perwaiz, A. (2015). Thailand floods and impact on private sector. In T. Izumi, & R. Shaw (Eds.), In *Disaster management and private sectors challenges and potentials* (pp. 231–245). Tokyo: Springer.

Ranghieri, F., & Ishiwatari, M. (2014). *Learning from Megadisasters : Lessons from the Great East Japan Earthquake.* Washington, DC: World Bank. © World Bank https://openknowledge.worldbank.org/handle/10986/18864. License: CC BY 3.0 IGO.

SAARC Report. (2014). *Post-2015 DRR Framework for SAARC Region (HFA2):* SAARC Disaster Management Centre *SDMC, March.*

SAARC Report (nd.). SAARC Disaster Knowledge Network Portal Guidelines Report, SAARC Disaster Management Centre. Accessed Online at http://www.preventionweb.net/files/11672_SADKNPortalGuidelinesv3.81.pdf on 23 October 2015.

Sendai Framework for Disaster Risk Reduction (SFDRR). (2015). Third United Nations World Conference on Disaster Risk Reduction. *Adoption of the Final Outcomes of the Conference, Sendai, Japan*. Accessed online at http://www.preventionweb.net/files/43291_sendaiframeworkfordrren.pdf on 23 October 2015.

Tokio Marine & Nichido Risk Consulting Co, 2015 Tokio Marine & Nichido Risk Consulting Co., Ltd. and Tokyo Shoko Research Survey. (2011). Accessed July 10, 2015, at http://www.tokiorisk.co.jp/cgi-bin/topics/page.cgi?no=667.

UNHABITAT. (2007). *Enhancing Urban Safety and Security: Global Report on Human Settlements*, p. 324.

United Nations Development Programme (UNDP) Report. (2013). *Japan's Case Study*. New York: Bureau for Crisis Prevention and Recovery.

United Nations Office for Disaster Risk Reduction (UNISDR). (2012). *How to Make Cities More Resilient: A Handbook for Local Government*. Accessed from http://www.uclg.org/sites/default/files/toolkit_on_how_to_make_cities_resilient_0.pdf.

United Nations Economic and Social Commission for Asia and the Pacific (UNESCAP) and United Nations International Strategy for Disaster Reduction (UNISDR). (2012). *Reducing vulnerability and exposure to disasters: The Asia-Pacific disaster report 2012*. Accessed June 3, 2015, from http://www.unisdr.org/files/29288_apdr2012finallowres.pdf.

United Nations Industrial Development Organization (UNIDO). (2008). *CSR Perceptions and Activities of Small and Medium Enterprises (SMEs) in seven geographical clusters*.

Willis International. (2012). Thailand's New Government Insurance Responds to Flood Loss. *International Alert, 58*. Accessed from http://www.willis.com/documents/publications/Services/International/2012/IntlAlert_Thailand_0712.pdf.

World Bank. (2012). *Thailand Flood 2011: Rapid Assessment for Resilient Recovery and Reconstruction Planning*. Accessed from https://www.gfdrr.org/sites/gfdrr/files/publication/Thai_Flood_2011_2.pdf.

Zurich Financial Services Group and Business Continuity Institute. (2011). *Supply Chain Resilience 2011 Study*. Accessed at http://www.zurichna.com/internet/zna/SiteCollectionDocuments/en/corporatebusiness/riskengineering/Supply_Chain_Resilience_2011.pdf.

Urban Disasters and Risk Communication Through Youth Organizations in the Philippines

13

Glenn Fernandez[1], *Rajib Shaw*[2]
[1]Disaster Risk Management Systems, Asian Disaster Preparedness Center, Bangkok, Thailand; [2]Professor, Graduate School of Global Environmental Studies, Kyoto University, Kyoto, Japan

13.1 Introduction: Importance of Participatory Risk Communication

The Sendai Framework for Disaster Risk Reduction (DRR) 2015–2030 was adopted at the end of the Third United Nations World Conference on Disaster Risk Reduction. The Sendai Framework lists the roles and potential contributions of different stakeholder groups in DRR (UNISDR, 2015). One of the 13 guiding principles of the Sendai Framework states that DRR requires an all-of-society engagement and partnership as well as empowerment and inclusive, accessible, and nondiscriminatory participation, especially of people disproportionately affected by disasters.

Like the Sendai Framework, its predecessor, the Hyogo Framework for Action 2005–2015 (HFA), promoted participatory disaster risk reduction (DRR) and stated that "both communities and local authorities should be empowered to manage and reduce disaster risk by having access to the necessary information, resources, and authority to implement actions for disaster risk reduction" (UNISDR, 2005). The involvement of local governments and communities in the design and implementation of DRR initiatives is a generally recognized good practice (UNISDR, 2012). Unfortunately, although there has been some progress in establishing national DRR policies and legislation, this has not accelerated widespread local DRR practices (GNDR, 2011). For example, the gap between HFA and its implementation at the "front line" or local level had not been bridged (GNDR, 2011). The failure to strengthen local governments and to make progress in community participation means that the HFA rhetoric had not been translated into DRR action (Matsuoka & Shaw, 2012). All members of the community, regardless of their age, gender, ethnicity, religion, or socio-economic position, should be involved in thinking, planning, and deciding about DRR (GNDR, 2011). By involving people in the decisions that affect their lives through the provision of access to necessary information and resources, participatory DRR has the potential to make initiatives more sustainable, integrative, and empowering (Cadag & Gaillard, 2012).

Urban Disasters and Resilience in Asia.

In the case of the Philippines, the necessity of participatory DRR is made more prominent by the need to develop and strengthen local disaster prevention, mitigation, response, and reconstruction capacities to complement national capacities (Fernandez & Shaw, 2013). The Philippines is one of the world's most disaster-prone countries (IFRC, 2009) and has been consistently among the top five countries with the highest number of reported disaster events in the last eight years (CRED, 2014).

Risk communication is an important component of DRR. It is the provision of risk information to concerned and interested stakeholders to enhance risk awareness and understanding, which hopefully will lead to disaster preparedness action and eventually to protection and safety (Rød et al., 2012). Recently, risk communication is considered more as an open and transparent interactive exchange rather than a uni-directional or one-way transfer of information and knowledge between those responsible for managing risks and those who may be affected by those risks (Höppner et al., 2010). The purposes and functions of risk communication are shown in Box 13.1.

Risk communication has to use various communication channels and tools in order to be effective, taking into consideration the specific needs of the target audience, including language, disabilities, culture, degree of vulnerability, and geographic dispersal (Höppner et al., 2010; Cole & Murphy, 2014). It is no longer advisable to rely entirely on television, radio, and newspapers as the only sources of risk information. Social media has become one of the primary sources of information, particularly during disasters and emergencies. According to Tabell (2015), "With the rising popularity of smart phones, which have also become very inexpensive nowadays, people turn to social media to get or share public safety information, or stay connected with family and friends, or even request assistance from emergency response agencies."

The use of social media, particularly social networking sites, has become so extensive and widespread in the Philippines that the country has been tagged as "The Social Networking Capital of the World" (Vulcan Post, 2014). Social media played an important role in disaster response during the recent calamities in the Philippines, like the Bohol Earthquake and Typhoon Haiyan, both of which happened in 2013.

Box 13.1 Purposes and Functions of Risk Communication

- Raise awareness
- Encourage protective behavior
- Inform to build up knowledge on hazards and risks
- Inform to promote acceptance of risks and management measures
- Inform on how to behave during events
- Warn of and trigger action to impending and current events
- Reassure the audience (to reduce anxiety or "manage" outrage)
- Improve relationships (build trust, cooperation, networks)
- Enable mutual dialogue and understanding
- Involve actors in decision making

Source: Höppner et al. (2010)

In the case of Typhoon Haiyan, which was one of the strongest storms ever recorded on Earth, traditional communication channels became nonoperational (Takahashi & Tandoc, 2014). Social media was effectively used in connecting with loved ones; in notifying the authorities of the local situation and bridging the information gap between the government and the affected communities; in organizing relief efforts; in increasing public awareness; and in empowering and stress-debriefing for survivors (Third Team Media, 2014).

Risk communication is essential in pursuing Sendai Framework's Priority 1: understanding disaster risk. To achieve Priority 1, nonsensitive hazard exposure, vulnerability, risk, disaster, and loss disaggregated information must be made freely available and accessible. The Sendai Framework also encourages stakeholders to "strengthen the utilization of media, including social media, traditional media, big data, and mobile phone networks to support national measures for successful disaster risk communication, as appropriate and in accordance with national laws" (UNISDR, 2015). Similar to other DRR activities, risk communication should be participatory as well as interactive and continuous in order to have a positive impact and be effective.

13.2 The Role of Youth in Risk Communication in Urban Areas

Until recently, most DRR research traditionally assumed that youth are passive victims with no or very little role to play in communicating risks, participating in DRR-related decision-making processes, or preparing for, preventing, and responding to disasters (Anderson, 2005; Kirschke & van Vliet, 2005; Chatterjee et al., 2015). For example, Mitchell et al. (2008) observed that none of the more recent theoretical models or guidelines for good communication practice singled out the unique needs and potential role of youth as resources or potential informants within risk communication systems. Two explanations can be offered for young people not appearing in the models. First, risk communication has historically been associated with information flows from the top downwards, with scientific institutions at the top and the public at the bottom. These approaches assume an ignorant public whose knowledge "deficit" (compared to that of the experts) requires that they be provided with simple information. The second reason relates to paternalism and the commonly held belief that parents make decisions about the level of risk their children face. Parents and guardians are assumed to be able to pass risk information on to their children and/or act to protect their safety. This leads to disaster risk management practices that assume that parents make responsible and appropriate decisions and choices about the risks their children face. This paternalistic view of children does not recognize either the rights or the value and utility of young people as agents who are able to assimilate and manage information and to convey rational risk management choices among their families and wider networks.

But youth definitely can make contributions to risk communication. In the Solferino Youth Declaration, released in 2009 during the 150th anniversary of the birth of the idea that eventually led to the founding of the International Red Cross and Red Crescent Movement, youth volunteers issued an appeal to world leaders to recognize youth as agents of change; to encourage the unique abilities and skills that young people alone can bring to the table, such as intercultural communication and innovative use of technology; to include youth in decision-making and planning processes at all levels; to push for youth to have a stronger role in program development and implementation to empower their communities; and to increase focus on formal and nonformal peer education as a primary method of disaster prevention (IFRC, 2009). According to the International Federation of Red Cross and Red Crescent Societies (IFRC) Strategy 2020, focusing on young people is a crucial investment (IFRC, 2010a). Young people can have multiple roles as contributors, including being innovators; early adopters of communication, social media, and other technologies; inter-cultural ambassadors; peer-to-peer facilitators; community mobilizers; agents of behavior change; and advocates for vulnerable people (IFRC, 2010b). Young people bring much-needed skills when working alongside older people, which can lead to an intergenerational transfer of experience that is vital to both progress and stability in society.

In the case of the Philippines, youths (i.e., persons aged 15 to 24) constitute one-fifth of the total population, at 24 million (CIA, 2014). This is an age group that has still been largely untapped to contribute in DRR. But with the increasing pressure to promote inclusive participation in the Sendai Framework, the participation of young people, as well as other previously neglected stakeholders like women, children, the elderly, and persons with disabilities, is hoped to be increased in the coming years. Anecdotal examples have shown that young people indeed have the capacity to take part in DRR undertakings. However, it is acknowledged that access to the necessary information, resources, and authority is needed to empower youth to participate in DRR in their communities. Back et al. (2009) has noted that youth participation in DRR makes economic sense because in the long run, when young people learn and practice DRR "the benefits stream is integrated into the rest of their adult lives, yielding a higher benefit than when adults acquire the same skills, and embedding the changed behavior early enough for it to be passed on to subsequent generations."

Excluding youth from DRR processes threatens their safety from disasters and neglects a valuable resource for risk communication, education, advocacy, and action-oriented risk reduction activities (Anderson, 2005). The evidence confirming the value of young people's participation in DRR is getting stronger (Walden et al., 2009). For example, the experiences of an international nongovernmental organization (NGO) working with youth in the Philippines on community risk mapping and mitigation activities have shown that youth have a much greater capacity to participate in DRR than many people assume. Through community mapping, young people are able to assist in identifying hazardous areas, safe areas, and evacuation routes in their locality (Fernandez, 2012; Fernandez & Shaw, 2015a). Young people used their DRR knowledge to successfully persuade school officials and community planners to relocate their school, previously situated in a high-risk landslide zone, to a safer area (Mitchell et al., 2009). The growing popularity of volunteer geospatial initiatives (Ushahidi, Open Street Map, Wikimapia, etc.) is now considered as a way to engage communities, especially young people, in collecting data that will help in planning and managing disaster risk at the local level (GFDRR, 2014).

Young people can offer innovative ideas about managing risks, provided they are encouraged to learn about disasters, hazards, vulnerability, and coping capacities (Walden et al., 2009). They also have the ability to share and apply what they learned within their households, families, and the wider community (Shaw et al., 2009). Young people should be regarded as potential partners in conducting DRR activities, such as risk communication. Youth organizations can be one of the main actors in communicating disaster risks to communities (Mulyasari & Shaw, 2013). Before a disaster, youths can assist in disseminating disaster awareness information and in conducting preparedness drills. After a disaster, they can be mobilized to help in gathering information about the impact of the disaster and to report this to the community.

The Sendai Framework urges that youth "should be given the space and modalities to contribute to disaster risk reduction" (UNISDR, 2015). It is very important to increase our knowledge of young people's capacities in order to better understand the roles that they can play in reducing personal and community vulnerability and to effectively design and implement programs that encourage their active involvement. Although young people are vulnerable to disasters and need to rely on adults for various forms of protection and support, they nevertheless possess significant qualities that could serve as a significant resource for households and communities in preparing for, responding to, and recovering from disasters. Their knowledge, creativity, energy, enthusiasm, and social networks could be used during all phases of the disaster risk management cycle to help themselves and others (see Table 13.1).

Table 13.1 Young People's Contributions Throughout the Disaster Risk Management Cycle

Preparedness	Response	Recovery
Canal declogging and dredging projects to avert the perilous effects of flash floods Fund-raising for acquiring necessary life-saving gear and vehicles Conducting first-aid, cardiopulmonary resuscitation (CPR), and water safety training, as well as fire and earthquake drills Supporting the campaign for voluntary blood donation Risk communication through community mapping (identifying hazardous areas, safe areas, and evacuation routes)	Providing emergency medical services to disaster victims Helping in relief operations (distributing food, clothes, and medicines to families in evacuation centers) Joining search, rescue, and retrieval operations Operating a 24-h disaster and emergency response organization	Planting various high-yielding fruit trees on a denuded mountain after a devastating flash flood Acknowledging the contribution of those who helped in emergency response and recovery efforts Reviewing and modifying preparedness plans Volunteering to build houses for poor disaster victims

Source: Fernandez and Shaw (2013)

In addition, young people can also serve as a bridge between their parents and digital DRR information because of their knowledge of technology (computers, mobile phones, and the Internet), which may be considerably more sophisticated than that of older generations (Mitchell et al., 2009; Riddell et al., 2011). Youth can also employ social media and mobile devices to mobilize their peers and volunteers, disseminate important information, and assist in tracing missing community members when a disaster occurs (INEE, 2013). In communities with a high incidence of poverty (i.e., where parents are illiterate, do not have the time to attend trainings and meetings, or do not have access to information), youth already play a major role as interpreters and conveyors of messages to their households and communities (UNISDR, 2007). Young people play a significant role as articulate informants in DRR communication, from disaster mitigation and preparedness information dissemination to the use of early warning systems (Walden et al., 2009).

According to UN-Habitat (2013), by 2030 about 60 percent of all urban dwellers will be under the age of 18. A huge proportion of young people now live in urban areas and there is even a phenomenon referred to as the "urban youth bulge." Young people living in cities can capitalize on the "urban advantage" of having more access to the abundant goods, services, and amenities such as access to ICT infrastructure not found in rural areas (UN-HABITAT, 2013). For example, Mearns et al. (2014) mentions that majority of Twitter users are young and urban dwellers. Also, according to Mulyasari and Shaw (2013), one of the strengths of youth organizations in urban areas is the ability to raise funds from the government, NGOs, the private sector, and the community. Given this situation, youth living in cities will have more and better opportunities to participate in DRR activities like risk communication compared to their rural counterparts. However, the involvement of youth in urban areas in DRR will benefit not just their city but also peri-urban and rural areas that rely on urban areas as their market for goods and services.

The impact of urban disasters will go beyond city limits (Sanderson et al., 2012). The potential number of people and assets that will be affected by urban disasters will be huge and will require a large number of organizations, including youth organizations, putting efforts into disaster prevention, mitigation, and preparedness. According to Cumiskey et al. (2015), "Creating awareness of DRR and increasing communication related to DRR among young people is vital as it helps young people understand their capacity for contributing to DRR." When applied in the urban context, creating opportunities for youth engagement will not only directly benefit youth, but will also have positive impact on cities and surrounding areas (Arrighi et al., 2015).

13.3 Examples of Risk Communication by Youth Organizations in Urban Areas in the Philippines

There are many youth organizations in the Philippines. Presented in the next sections are six examples of youth organizations operating in urban areas and how they conduct risk communication. The Internet has been instrumental in the increased youth participation in risk communication and other DRR tasks. The Internet and the virtual networks it supports have reduced the costs of participation (time, effort, money) by increasing the availability of information (Boulianne, 2009). The Internet and social media can reach large numbers of young people fast and at a relatively low cost, enhancing the voice of young people significantly (Back et al., 2009). One advantage of activities involving social media is that they do not require physical presence. Young people can be engaged in DRR remotely and over a wide geographical area with an Internet connection, which usually are urban areas.

13.3.1 Youth Councils

Filipino youth have been highly encouraged to participate in the affairs of their community, and the Philippines is the only nation in the world that has a mechanism of involving the youth sector in local governance, through the *Sangguniang Kabataan* (SK, Youth Council) (Fernandez & Shaw, 2015b). In urban youth councils, possible risk communication roles for the youth include putting up warning signs; organizing emergency drills; and assisting in information dissemination through the use of social media, like Facebook, Twitter, and YouTube. As an example, most of the youth councils in Makati City have their own Facebook account where they post announcements about their upcoming activities, including those related to DRR, as well as share pictures of implemented projects. When Typhoon Haiyan hit the central Philippines in November 2013, *Barangay* Olympia (Olympia Village) Youth Council called for volunteers to help prepare relief goods for the typhoon victims. The youth council was able to find volunteers quickly by using Facebook and Short Message Service (SMS) or text messaging.

13.3.2 Red Cross Youth

The Philippine Red Cross is the premier humanitarian organization in the Philippines. The Red Cross Youth (RCY) is its youth volunteering program (PRC, n.d.). Several RCY chapters in urban areas use social networking sites to promote their various activities. For example, the University of the Philippine Red Cross Youth in Quezon City posts announcements on Facebook (https://www.facebook.com/UPRCY) about its slippers donation campaign, disaster preparedness seminar, blood donation drive, donation drive for the 2015 Nepal Earthquake victims, fundraising campaign (by selling t-shirts), etc. Another chapter, the Centro Escolar University Red Cross Youth in Manila City, utilizes Facebook (https://www.facebook.com/CEURedCrossYouth) to call for volunteers and for first-aid training participants. Their posts usually contain pictures or posters, contact mobile phone numbers, and hashtags, which are "a type of label or metadata tag used in social network and microblogging services which makes it easier for users to find messages with a specific theme or content" (Wikipedia, n.d.).

13.3.3 Boy Scouts of the Philippines

The Boy Scouts of the Philippines (BSP) is one of the largest youth organizations in the Philippines and has around two million members (Santos, 2015). Its online presence includes having its own official website (http://www.scouts.org.ph/), its own hashtag (#ScoutsPH), its own Facebook account (https://www.facebook.com/scoutsphilippines), and a Facebook community page (https://www.facebook.com/laginghanda). Posts on Facebook show announcements of suspension of classes due to typhoons, survival tips, reports on fire prevention seminar and disaster preparedness trainings, etc.

13.3.4 Girl Scouts of the Philippines

The Girl Scouts of the Philippines (GSP) has a special service program called SAVER, which stands for Service Auxiliary Volunteers for Emergency and Relief (GSP, n.d.). SAVER is undertaken by a team composed of 12 to 45 Girl Scouts and non–Girl Scout volunteers organized to assist in fact finding, communication, referrals, and relief operations. Their pre-emergency, emergency, and post-emergency activities include providing a communication service (transmission and reception of messages/information).

13.3.5 Youth Bonded Together

Buklod ng Kabataan (Youth Bonded Together) is a group of children and youth in *Barangay* Banaba of San Mateo Municipality. It advocates DRR and climate change adaptation through theater performances (Tanner et al., 2009). This group communicates their perceptions of local hazards, such as flooding and river bank erosion, and the impacts of these hazards on lives, livelihoods, and properties through singing, dancing, and acting. Through their performances, different stakeholders are encouraged to take action.

13.3.6 Youth Peer Education Network

The Youth Peer Education Network (Y-PEER) is a youth-to-youth initiative pioneered by the UN Population Fund (UNFPA). Although the primary advocacy of Y-PEER is adolescent sexual and reproductive health, when Tropical Storm Washi (locally known as Tropical Storm Sendong) hit southern Philippines in December 2011, UNFPA quickly mobilized youth volunteers through Facebook and SMS to various evacuation centers to help identify pregnant women in need of assistance and to organize information sessions with young people (UNFPA, 2012a; UNFPA, 2012b). Volunteers engaged other young people in the evacuation centers through focus group discussions to talk about their experiences in these temporary shelters.

Table 13.2 summarizes the risk communication activities of the six youth organizations presented above. With their extensive contacts and networks, and up-to-date knowledge of social media and mobile devices, young people can be a critical asset in risk communication and follow-through actions like sharing information about people who need help, making it easier for authorities to respond, assisting in relief operations and managing evacuation centers, raising awareness about the disaster, and fundraising.

Table 13.2 Youth Risk Communication Activities in Urban Areas in the Philippines

Youth-led and Youth-serving Organizations	Risk Communication Activities
Sangguniang Kabataan (Youth Councils)	Putting up warning signs; organizing emergency drills; assisting in information dissemination through the use of social media, like Facebook, Twitter, and YouTube; calling for volunteers for relief goods repacking via Facebook and text messaging.
Red Cross Youth	Using social networking sites to announce donation collection drives, disaster preparedness seminars, blood donation drives, fundraising campaigns, calls for volunteers and for first-aid training participants, etc.
BSP	Posting announcements on social media of suspension of classes due to typhoons, survival tips, reports on fire prevention seminar and disaster preparedness trainings, etc.
GSP	Has a special service program called Service Auxiliary Volunteers for Emergency and Relief (SAVER) to assist in fact finding, communication, referrals, and relief operations.
Buklod ng Kabataan (Youth Bonded Together)	Communicating risks and responses through youth-led theater in the Barangay Banaba, San Mateo, Rizal.
Youth Peer Education Network (Y-PEER)	When Tropical Storm Washi hit southern Philippines, Y-PEER mobilized youth volunteers through Facebook and SMS to various evacuation centers to help identify pregnant women in need of assistance and to organize information sessions with young people.

Aside from the long-standing formal youth organizations, there are also spontaneous youth groups that form when the need arises and disband after emergency situations have passed (INEE, 2013). There are also risk communication contributions made by individuals. Citizen reporters can provide timely and detailed information on the local situation (Mulyasari, 2014). Plan International supported young people from Typhoon Haiyan-hit provinces to share their views and be actively involved in disaster recovery through its Youth Reporter Project, which trained 280 youth in media skills (Plan International, 2014). Utilizing handy video cameras, young people working in small groups with their friends shared their thoughts and experiences about the typhoon that changed their lives and their plans for the future.

These group and individual contributions show that youth can be tapped to play an important role in DRR (Oposa, 2013). Even through simple tasks like reposting information about a disaster on their Facebook or Twitter accounts, young people already contribute significantly to increasing awareness about disasters and the plight of affected communities (De la Cruz, 2014).

13.4 Future Directions: Helping Youth Contribute Further to Risk Communication

Discussions leading to the Sendai Framework for DRR, as well as the strategies of international civil society organizations (CSOs) like the International Federation of Red Cross and Red Crescent Societies, have highlighted the potential significant contributions of the youth. But it should be noted that what is important for local actors like young people is how they will transform international priorities, goals, and strategies into local actions and results. At the operational level at the "front lines," the mechanisms for youth participation in information-sharing, consultation, decision-making, initiating action, evaluating results, and providing feedback should be clear if we are to expect better results in terms of the quantity and quality of youth involvement in DRR in general and in risk communication in particular.

There are several concrete steps that can be taken to help young people get more involved in risk communication. One is to provide media skills training, like what Plan International did in their Youth Reporter Project. Another is to provide youth with improved access to risk information before and after an emergency so that they are aware of what different stakeholders are doing and what resources are available so that they can plan ahead and be empowered to plan their own activities. Youth can also be provided organizational and institutional support, as proposed by the Bangkok Declaration on Disaster Risk Reduction in Asia and the Pacific 2014 (AMCDRR, 2014a; AMCDRR, 2014b). For example, the youth councils in the Philippines are not yet fully integrated into the DRR committees in their village. Changes in government policies can mandate the more active engagement of youth with DRR committees, which can provide opportunities for intergenerational interaction and learning, and can cultivate productive adult-youth partnerships. Mentorship and adult supervision can provide youth with positive role models and a safe environment to explore their roles in the community. But it is also important to provide enough space for them to maintain their independence in pursuing their own initiatives.

DRR activities like risk communication need to be inclusive and participatory to be efficient and effective. It is necessary to ensure the engagement of stakeholders, like young people, who possess significant qualities that can serve as a significant resource for households and communities in preparing for, responding to, and recovering from disasters. Their knowledge, creativity, energy, enthusiasm, and social networks make youth a valuable resource for risk communication, as shown by the examples presented earlier in this chapter.

Acknowledgments

We thank the Ministry of Education, Culture, Sports, Science, and Technology (MEXT) of Japan for the generous scholarship grant that made it possible for the first author to obtain his MA and PhD degrees from the Graduate School of Global Environmental Studies, Kyoto University.

References

Asian Ministerial Conference on Disaster Risk Reduction (AMCDRR). (2014a). *Bangkok declaration on disaster risk reduction in Asia and the Pacific 2014*. http://6thamcdrrthailand.net/6thamcdrr/Portals/0/Final%20Bangkok%20Declaration%20-6%20AMCDRR%20-final%2026%20June-0800%20hours.pdf.

Asian Ministerial Conference on Disaster Risk Reduction (AMCDRR). (2014b). *Bangkok Declaration Annex I: Statement of voluntary commitments of children, youth and child-centered organizations stakeholder group*. http://6thamcdrr-thailand.net/6thamcdrr/Portals/0/Annex%2001-%20Children%20and%20Youth%20final%2026%20June%202014.pdf.

Anderson, W. (2005). Bringing children into focus on the social science disaster research agenda. *International Journal of Mass Emergencies and Disasters*, *23*(3), 159–175.

Arrighi, J., Kadihasanoglu, A., & Suarez, P. (2015). *Engaging youth and technology in improving the urban environment*. https://ugecviewpoints.wordpress.com/2015/05/28/engaging-youth-and-technology-in-improving-the-urban-environment/.

Back, E., Cameron, C., & Tanner, T. (2009). *Children and disaster risk reduction: Taking stock and moving forward*. http://www.savethechildren.org.uk/sites/default/files/docs/Child_Led_DRR_Taking_Stock_1.pdf.

Cadag, J. R., & Gaillard, J. C. (2012). Integrating knowledge and actions in disaster risk reduction: The contribution of participatory mapping. *Area*, *44*(1), 100–109. http://onlinelibrary.wiley.com/journal/10.1111/(ISSN)1475-4762.

Central Intelligence Agency (CIA). (2014). *Philippines*. https://www.cia.gov/library/publications/the-world-factbook/geos/rp.html.

Chatterjee, R., Shiwaku, K., Das Gupta, R., Nakano, G., & Shaw, R. (2015). Bangkok to Sendai and beyond: Implications for disaster risk reduction in Asia. *International Journal of Disaster Risk Science*, *6*, 177–188.

Cole, J., & Murphy, B. (2014). Rural hazard risk communication and public education: Strategic and tactical best practices. *International Journal of Disaster Risk Reduction*, *10*, 292–304.

Center for Research on the Epidemiology of Disasters (CRED). (2014). Annual disaster statistical review. http://www.cred.be/publications.

Cumiskey, L., Hoang, T., Suzuki, S., Pettigrew, C., & Herrgård, M. (2015). Youth participation at the Third UN World Conference on Disaster Risk Reduction. *International Journal of Disaster Risk Science*, *6*(2), 150–163.

De la Cruz, G. (2014). *NYC: More involvement for young Pinoys in disaster planning*. http://www.rappler.com/move-ph/65995-more-involvement-disaster-management-young-pinoys.

Fernandez, G. (2012). *Youth participation in disaster risk reduction through science clubs in the Philippines*. Unpublished MA thesis. Kyoto, Japan: Graduate School of Global Environmental Studies, Kyoto University.

Fernandez, G., & Shaw, R. (2013). Youth council participation in disaster risk reduction in Infanta and Makati, Philippines: A policy review. *International Journal of Disaster Risk Science*, *4*(3), 126–136.

Fernandez, G., & Shaw, R. (2015a). Youth participation in disaster risk reduction through science clubs in the Philippines. *Disasters Journal*, *39*(2), 279–294.

Fernandez, G., & Shaw, R. (2015b). *Youth participation and community-based disaster risk reduction*. Singapore: Research Publishing Services.

Global Facility for Disaster Reduction and Recovery (GFDRR). (2014). *Understanding risk: The evolution of disaster risk assessment since 2005. Background paper prepared for the 2015 Global Assessment Report on Disaster Risk Reduction*. Geneva, Switzerland: UNISDR. http://www.preventionweb.net/english/hyogo/gar/2015/en/bgdocs/GFDRR,%202014a.pdf.

Global Network for Disaster Reduction (GNDR). (2011). *Views from the frontline 2011*. http://www.globalnetwork-dr.org/views-from-the-frontline/voices-from-the-frontline-2011.html.

Girl Scouts of the Philippines (GSP). (n.d.). GSP–SAVER (Service Auxiliary Volunteers for Emergency and Relief). http://girlscouts.org.ph/programs/special-programs/gsp-saver/.

Höppner, C., Buchecker, M., & Bründl, M. (2010). *Risk communication and natural Hazards*. http://caphaz-net.org/outcomes-results/CapHaz-Net_WP5_Risk-Communication.pdf.

International Federation of Red Cross and Red Crescent Societies (IFRC). (2009). *Solferino youth declaration*. http://www.ifrc.org/Global/Publications/youth/170700-Youth_declaration-EN.pdf.

International Federation of Red Cross and Red Crescent Societies (IFRC). (2010a). *Strategy 2020*. http://www.ifrc.org/Global/Publications/general/strategy-2020.pdf.

International Federation of Red Cross and Red Crescent Societies (IFRC). (2010b). *Guidelines for national societies to organize local youth actions and celebrate the International Year of Youth*. http://www.ifrc.org/Global/Publications/youth/201500-IYY-guidelines-EN.pdf.

Inter-Agency Network for Education in Emergencies (INEE). (2013). *Engaging youth-led and youth-serving organizations in disaster relief efforts*. http://www.ineesite.org/uploads/files/resources/201212_INEEGuidanceDocEngagingYouthOrganizationsinDisasterReliefEfforts_1.pdf.

Kirschke, J., & van Vliet, W. (2005). "How can they look so happy?" Reconstructing the place of children after Hurricane Katrina: Images and reflections. *Children, Youth and Environments*, *15*(2), 378–391.

Matsuoka, Y., & Shaw, R. (2012). Hyogo Framework for Action as an assessment tool of risk reduction: Philippines national progress and Makati City. *Risk, Hazards, and Crisis in Public Policy*, *3*(4), 18–39.

Mearns, G., Simmonds, R., Richardson, R., Turner, M., Watson, P., & Missier, P. (2014). Tweet my street: A cross-disciplinary collaboration for the analysis of local Twitter data. *Future Internet*, *6*(2), 378–396.

Mitchell, T., Haynes, K., Hall, N., Choong, W., & Oven, K. (2008). The roles of children and youth in communicating disaster risk. *Children, Youth and Environments*, *18*(1), 254–279.

Mitchell, T., Tanner, T., & Haynes, K. (2009). *Children as agents of change for disaster risk reduction: Lessons from El Salvador and the Philippines*. Brighton, UK: Institute of Development Studies.

Mulyasari, F. (2014). *Enhancing climate-related disaster resilience through effective risk communication in Bandung, Indonesia*. Unpublished Ph.D thesis. Kyoto, Japan: Graduate School of Global Environmental Studies, Kyoto University.

Mulyasari, F., & Shaw, R. (2013). Risk communication through community-based society organizations as local response to disaster in Bandung, Indonesia. In A. Neef, & R. Shaw (Eds.), *Risks and conflicts: Local responses to natural disasters* (pp. 231–250). Bingley, UK: Emerald Group Publishing.

Oposa, A. (2013). *Aid in practice—Mobilizing youth when disaster strikes*. http://blogs.adb.org/blog/aid-practice-mobilizing-youth-when-disaster-strikes.

Plan International. (2014). *Youth reporters: Life after the typhoon*. http://plan-international.org/about-plan/resources/videos/youth-reporters-life-after-the-typhoon/.

Philippine Red Cross (PRC) (n.d.). Red Cross Youth. http://www.redcross.org.ph/what-we-do/red-cross-youth.

Riddell, E., Rock, A., Clothier, P., & London, J. (2011). *Strengthening youth resilience to natural disaster with smartphone technology*. http://www.emergencyvolunteering.com.au/home/documents/Emergency%20App%20-%20Final%20Report.pdf.

Rød, S. K., Botan, C., & Holen, A. (2012). Risk communication and the willingness to follow evacuation instructions in a natural disaster. *Health, Risk, & Society*, *14*(1), 87–99.

Sanderson, D., Knox-Clarke, P., & Campbell, L. (2012). *Responding to urban disasters: Learning from previous relief and recovery operations*. London: ALNAP Lessons Paper. ALNAP/ODI.

Santos, R. (2015). *Fast Facts: The Boy Scouts of the Philippines*. http://www.rappler.com/newsbreak/iq/81614-fast-facts-boy-scouts-philippines-bsp.

Shaw, R., Takeuchi, Y., Shiwaku, K., Fernandez, G., Gwee, Q. R., & Yang, B. (2009). *1-2-3 of disaster education*. Kyoto, Japan: Kyoto University.

Tabell, A. (2015). *Social media: Critical for disaster managers*. http://www.rappler.com/move-ph/issues/disasters/84322-social-media-disaster-government.

Takahashi, B., & Tandoc, E. (2014). *Social media use in the face of disaster: An exploration of communication practices among stakeholders affected by Typhoon Haiyan in the Philippines*. http://www.colorado.edu/hazards/research/qr/qr251.html.

Tanner, T., Garcia, M., Lazcano, J., Molina, F., Molina, G., Rodriguez, G., Tribunalo, B., & Seballos, F. (2009). *Children's participation in community-based disaster risk reduction and adaptation to climate change*. http://pubs.iied.org/pdfs/G02813.pdf.

Third Team Media. (2014). *The role of social media in disaster response*. http://www.thirdteam.org/tips-tools/the-role-of-social-media-in-disaster-response.

United Nations Human Settlements Programme (UNHABITAT). (2013). *Youth and IT in sustainable urban development*. http://unhabitat.org/?wpdmact=process&did=MTg0LmhvdGxpbms=.

United Nations International Strategy for Disaster Reduction (UNISDR). (2005). *Hyogo Framework for Action 2005-2015: Building the resilience of nations and communities to disasters*. http://www.unisdr.org/2005/wcdr/intergover/official-doc/L-docs/Hyogo-framework-for-action-english.pdf.

United Nations International Strategy for Disaster Reduction (UNISDR). (2007). *Building disaster resilient communities: Good practices and lessons learned*. Geneva, Switzerland: UNISDR. http://www.unisdr.org/files/596_10307.pdf.

United Nations International Strategy for Disaster Reduction (UNISDR). (2012). *Towards a post-2015 framework for disaster risk reduction*. http://www.preventionweb.net/files/25129_towardsapost2015frameworkfordisaste.pdf.

United Nations International Strategy for Disaster Reduction (UNISDR). (2015). *Sendai Framework for Disaster Risk Reduction 2015-2030*. http://www.wcdrr.org/uploads/Sendai_Framework_for_Disaster_Risk_Reduction_2015-2030.pdf.

United Nations Population Fund (UNFPA). (2012a). *Technology to the rescue: Innovative tools for responding to disaster*. http://www.unfpa.org/news/technology-rescue-innovative-tools-responding-disaster.

United Nations Population Fund (UNFPA). (2012b). *World Humanitarian Day: Youth to the rescue*. http://www.unfpa.org/news/world-humanitarian-day-youth-rescue.

Vulcan Post. (2014). *Research confirms: The Philippines is still the social media capital of the world*. https://vulcanpost.com/12971/research-philippines-social-media/.

Walden, D., Hawrylyshyn, K., & Hall, N. (2009). *Children on the frontline: Children and young people in disaster risk reduction*. London, UK: Plan International and World Vision. http://www.preventionweb.net/files/10576_13121ChildrenontheFrontlineGPreport.pdf.

Wikipedia. No date. Hashtag. http://en.wikipedia.org/wiki/Hashtag.

Flood Risk Reduction Approaches in Dhaka, Bangladesh

14

Uttama Barua[1], Mohammad Shakil Akther[2], Ishrat Islam[2]
[1]Graduate Student, Department of Urban and Regional Planning, Bangladesh University of Engineering and Technology and Research Planner, Bangladesh Network Office for Urban Safety (BNUS), Bangladesh University of Engineering and Technology, Dhaka, Bangladesh; [2]Professor, Department of Urban and Regional Planning, Bangladesh University of Engineering and Technology, Dhaka, Bangladesh

14.1 Background

The Intergovernmental Panel on Climate Change (IPCC) has identified the interaction between climate change and vulnerability as one of the looming threats to settlement and society (IPCC, 2007). Low adaptive capacity in the context of development, overexploitation of resources of urban ecosystems, and increasing magnitude and frequency of climatic events make the cities of the developing world increasingly vulnerable (Reckien et al., 2011). In most of the cities in developing countries, the formal mechanism of disaster risk reduction (DRR) is either not working or completely absent. As a result, the destruction of the lives and livelihoods of city residents due to disaster is a common phenomenon.

Asia is the second-least-urbanized continent in the world, although half of the world's urban population lives there (UNHABITAT et al., 2010). Asia is also home to 11 of the 20 largest urban agglomerations in the world (Department of Economic and Social Affairs, United Nations, 2014b). Studying eight Asian cities, Satterthwaite (2008) identified 13 common characteristics, one of which is uncontrolled development of real estate at the cost of disadvantaged groups in the society. Although Dhaka was not one of the eight cities in the case study, it has many common traits with them. So in this study, we chose to focus on Dhaka as one representative Asian city.

Dhaka, the capital city of Bangladesh, is located in the center of the country. Topographically, the city sits on flat land, 2–13 m above mean sea level, with most of the urbanized areas at elevations of 6–8 m (Huq & Alam, 2003; Alam & Rabbani, 2007). It lies in the subtropical monsoon zone and receives approximately 2000 mm of rainfall annually, of which more than 80% occurs during the monsoon season (from June to September) (Dewan et al., 2007; Dewan & Yamaguchi, 2009a). So, the climate of the city is characterized by subtropical, high rainfall and humid conditions, with cool and short winters and long, humid, and hot summers.

Urban Disasters and Resilience in Asia.

Human settlement in Dhaka can be traced back as early as the 12th century (Ahmed, 1986) and has 400 years of recorded history as the capital of the region (Huq & Alam 2003; Alam & Rabbani, 2007). Historically, the city is surrounded by rivers and criss-crossed by numerous water bodies that drain water during the monsoon season (Khan, 2000; Dewan et al., 2007; Iqbal, 2013). Thus, in the past, the city was regarded as the "Venice of the East" or the "City of Channels" (Dani, 2009; Islam et al., 2010). James Rennel surveyed the rivers of east Bengal in 1765 and wrote that "the kingdom of Bengal, particularly the eastern part, is naturally the most convenient for trade within itself of any country in the world; for the rivers divided into such a number of branches that the people have the convenience of water carriage to and from any principal place" (Ahmed, 1986). The city is surrounded by the distributaries of the area's two major rivers, the Brahmaputra and the Meghna; i.e., Buriganga lies to the west and south, Turag to the west, Tongi canal to the north, and Balu to the east, Turag and Balu are connected by Tongi canal, and Turag, Buriganga, and Balu rivers are connected to the Dhaleshwari River (Figure 14.1). Thus, there are permanent wetlands in the area of the city that include rivers and a network of canals, and also temporary wetlands that include floodplains, marshes, and channels, which work as the natural drainage system of the city (Islam, 2014). The risk of flooding due to the low elevation of lands and the rivers and canals surrounding them have greatly constrained the direction of urban expansion of Dhaka (Chowdhury & Faruqui, 1989; Huq & Alam, 2003; Dewan & Yamaguchi, 2009b).

14.2 Hydrological Context and Growth of Dhaka

Due to the city's geographical location and climatic conditions, floods are a regular feature in Dhaka, and the severity of these events is increasing gradually (Alam & Rabbani 2007; Dewan et al., 2007; Islam et al., 2008; Haque et al., 2010; Parvin & Shaw, 2011; Islam, 2014). In the city, flooding results from water runoff from the rivers surrounding the city and heavy monsoon rainfall (Alam & Rabbani, 2007; Islam et al., 2008; Dewan & Yamaguchi, 2009a). Traditionally, the temporary water bodies of Dhaka have acted efficiently as water retention areas, storing and draining the excess water caused by excessive rainfall and thus helped to reduce flood risk and sustain the natural ecosystem (Khan, 2000; Islam et al., 2008; Sultana et al., 2009; Haque et al., 2010; Mahmud et al., 2011; Islam, 2014). However, the city has lost 5797 ha of the wetland each year from 1999 to 2010, which is an increase from 4887 ha annual loss during the previous decade of 1989-99 (Figure 14.2) (Shubho et al., 2014). Thus, the city is becoming more vulnerable to flood with the passage of time.

Regular and natural flooding actually can bring immense benefits for agriculture and food production, fisheries, and residents' livelihoods though increasing soil fertility, recharging aquifers, improving ecosystems, increasing fish production, and other advantages (Islam et al., 2008). However, the degree of this inundation sometimes becomes severe, which causes very serious problems to human health and livelihoods, especially for the poor. These disasters can have adverse effects on various sectors, including infrastructure, industry, trade and commerce, communication systems, utility services,

Figure 14.1 Location map of Dhaka.
Source: RAJUK (n.d.).

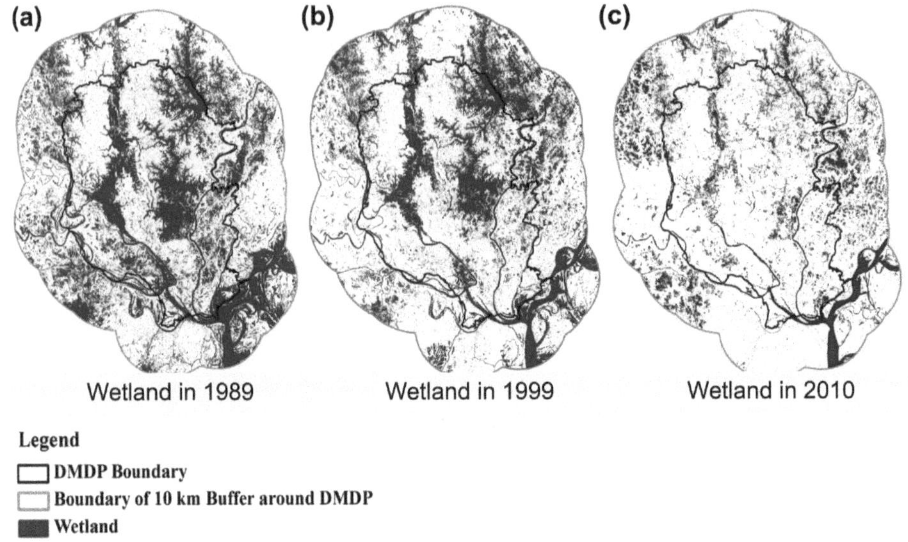

Figure 14.2 Diminishing wetlands of Dhaka (a) Wetland in 1989, (b) Wetland in 1999, (c) Wetland in 2010.
Source: Shubho et al. (2014)

sewage management, and the supply of electricity and gas (Mohit & Akther, 2000; Alam & Rabbani, 2007; Islam et al., 2008).

The fertile land of the delta influenced human settlement on this land, the growth of which reached its apex just after the liberation war in 1971 though which Bangladesh was established as an independent nation. (Dewan & Yamaguchi, 2009a; Kabir & Parolin, 2012). This growth is generating influential factors that increase the severity of flooding in the city. In addition, Dhaka has been subjected to a high rate of urbanization, where rapid migration is more prominent (DFID, 2004; Haque et al., 2010; Masum, 2010). These forces have resulted in rapid population growth, designating Dhaka as one of the world's largest megacities (Sultana et al., 2009; Haque et al., 2010). The city now accommodates about 10% of the nation's total population and about one-third of its urban population. According to World Urbanization Prospects 2014, Dhaka is now the 11th-largest megacity, which is projected to be 6th by the year 2030 (Department of Economic and Social Affairs, United Nations, 2014a).

Demographic and socioeconomic growth is influencing the expansion of the city, which is in turn influenced by the topography and land characteristics of the area (Sultana et al., 2009). During the initial period of development, settlements took place on the higher terraces, with elevations ranging from 10–12 m above mean sea level. The ever-increasing population growth and urban activities, shortage of high land, rise in land prices, encroachment by profit-minded private developers and individuals, lack of control of the government over development, and lack of maintenance and management by the government made it inevitable to expand Dhaka into its surrounding wetlands. This led to a gradual decline in water bodies, which endangered the environmental sustainability, and caused more frequent stormwater flooding (Sultana et al., 2009; Islam et al., 2010; Khan et al., 2012; Islam, 2014). In this way, many temporary wetlands have been partially or completely filled and compartmentalized (Sultana et al., 2009; Islam et al., 2010; Islam, 2014). This will result in the disappearance of all temporary wetlands in Dhaka before 2035 if this trend continues (Islam, 2014).

Dhaka does not have a proper and well-planned stormwater drainage system (Khan, 2000). The city depends on a natural drainage system through wetlands, infiltration resulting in reduced runoff, operation of a stormwater drainage system including pumps and regulators, and water level on peripheral rivers for draining water (Khan, 2000; Khan et al., 2012). Encroachment, land filling, siltation, blockage due to garbage disposal, and other factors are resulting in poor drainage of the existing natural drainage channels, and impervious areas are increasing day by day due to increases in built-up areas, resulting in reduced infiltration, increased runoff, and flooding and waterlogging (Khan, 2000; Khan et al., 2012). These factors lead to failure of the drainage systems of the city, resulting in frequent and longer durations of flooding. Climate change is worsening this situation through erratic and prolonged rainfall, with the increase in precipitation and river flow changes caused by sea level rise (SLR) (Dewan et al., 2007; Roy, 2009; Haque et al., 2010; Yahya et al., 2010; Rabbani et al., 2011).

14.3 History of Major Floods in Dhaka

The occurrence of floods in Dhaka can be traced back to as early as 1787–1788 (Hunter, 1877). There were severe floods in 1833–1834, 1870, 1954, 1955, 1962, 1966, 1970, 1974, 1980, 1987, 1988, 1998, 2004, 2007, and 2009 (Hunter, 1877; Rizvi, 1969; Huq & Alam, 2003; Alam & Rabbani, 2007; Islam et al., 2008; Yahya et al., 2010; Khan et al., 2012). In the flood of 1988, the depth of inundation ranged from 0.3 m to more than 4.5 m (Huq & Alam, 2003; Alam & Rabbani, 2007; Yahya et al., 2010). This flood affected 60% of the population directly (Mohit & Akther, 2000; Huq & Alam, 2003; Alam & Rabbani, 2007; Yahya et al., 2010). For residential buildings and institutions, the estimated damages were more than 4 billion Bangladeshi taka (BDT) or US$ 51 million and 400 million BDT or US$ 5 million, respectively (considering US$ 1 = 78.27 BDT) (Huq & Alam, 2003; Alam & Rabbani, 2007). The 1998 flood occured due to excessive rainfall and higher water level in surrounding rivers compared to normal time. (Mohit & Akther, 2000; Huq & Alam, 2003; Alam & Rabbani, 2007). The flood, which lasted for about 10 weeks, inundated 56% of the city, which included most of the eastern sector, and 23% of the western area of the city (Mohit & Akther, 2000; Huq & Alam, 2003; Yahya et al., 2010). The 2004 flood occurred due to heavy rainfall, which was at three times the normal level for that time of year (Alam & Rabbani 2007; Haque et al., 2010). The flood inundated 43% of the city, causing damage to housing and infrastructure worth 438 million BDT or US$ 5.6 million (considering US$ 1 = 78.27 BDT) (Dewan & Yamaguchi, 2009a). Both intercity and intracity bus links from eastern Dhaka were shut down due to the flooding (Alam & Rabbani, 2007). Disruption in the electricity supply occurred due to inundation of power grids (Rahman et al., 2005; Haque et al., 2010). The 2007 flood was a riverine flood that resulted in increased waterborne diseases in the eastern part of Dhaka (Islam et al., 2008; Haque et al., 2010).

14.4 Flood Management in Urban Policies and Plans

Dhaka is considered as one of the most vulnerable cities to flood in Bangladesh due to the swelling of the surrounding rivers and intensive rainfall that generates runoff, which is beyond the capacity of the existing drainage facilities (Yahya et al., 2010). To achieve physical resilience, flood management has been considered in various development plans of the city.

14.4.1 Plans During British Rule

The spatial growth of Dhaka over the last 50 years has been phenomenal. For the vast majority of its history, it was a sleepy rural settlement. Traditionally, the people of this delta adopted an approach of "living with flood." At the early days of the growth of Dhaka, the natural terrain of the city was duly considered and the development pattern was in harmony with nature. The Mughal made it their capital in 1608, and their development plans concentrated on building forts along the river to protect cities against external enemies (Mowla, 2013).

The physical renewal of Dhaka began in the second half of the 19th century (Ahmed, 1986). One of the projects completed during this period, which began in 1864, was the construction of a flood protection embankment (called Buckland Bund, after C. T. Buckland, the commissioner of Dhaka) on the eastern bank of the Buriganga. The scheme was completed in the 1880s. The objective for building this embankment was to protect the city from the overflows of the Buriganga and mitigate flood damage within the city (Ahmed, 1986; Mohit & Akther, 2000; Huq & Alam, 2003; Alam & Rabbani 2007; Islam et al., 2008; Parvin & Shaw, 2011). In 1905, Dhaka became the capital of the newly created province of East Bengal and Assam. The government appointed Sir Patrick Geddes in 1906 to make a development plan for Dhaka (Kabir & Parolin, 2012). After detailed analysis of the context, he submitted a plan in 1917 named "Dacca Town Planning Report 1917." He envisioned Dhaka as having parks and canals created by preserving Dhaka's natural canal network though horticultural development, water parks, and public amenities (Iqbal, 2013; Shafi, 2010). Even though Geddes did not provide detailed plan for the water bodies, he did emphasize their preservation. His plan was never implemented formally, though (Ameen, 1998; Kabir & Parolin, 2012; Iqbal, 2013). Geddes's focus on smaller water spaces such as canals was disregarded, leading to the disappearance of much of the wetlands during the construction of Dhaka during the late period of British rule (Iqbal, 2013). Thus, the trend of development encroaching water bodies was initiated during British rule, which led to reduction of water retention areas in the city.

14.4.2 Dacca Master Plan 1959

When the British left India in 1947, Dhaka became the capital of the province of East Pakistan. Due to the absence of any plan, the Planning Division of the Government of East Pakistan prepared a development plan spanning around 50 km^2. The plan was partially implemented, but the impacts of these development projects on the spatial structure of the city were quite significant (Ameen, 1998; Mowla, 2007). The East Pakistan government created the Dhaka Improvement Trust (DIT) in 1956 [renamed the Rajdhani Unnayan Kartripakkha (RAJUK) in 1987], which took greater interest and care in their road construction and city planning (Huq & Alam, 2003; Shafi, 2010). The DIT developed several residential areas, including the Gulshan Model Town in 1961, Banani in 1964, Uttara in 1965, and Baridhara in 1972 (though first conceived in 1962), most of which were agricultural lands and free from inundation (Huq & Alam, 2003; Dewan & Yamaguchi, 2009b; Kabir & Parolin, 2012).

The first comprehensive master plan, called the Dacca Master Plan 1959, was prepared by a consortium of the British firms Minoprio & Spencely, and P. W. Macfarlane in 1958, with the objective to establish planning principles rather than a detailed plan over a 20-year planning horizon (1958 to 1978) for an area of 220 m^2 (Parvin & Shaw, 2011; Kabir & Parolin, 2012; Iqbal, 2013; Islam, 2014; Shafi, 2010). The plan identified two main problems with the development of the city: a shortage of developable land above flood level and congestion in the old central area of Dhaka. As a consequence of these issues, the plan envisioned that the wetlands and flood plains of the city would be developed as housing estates in the future (Figure 14.3).

Although the plan acknowledged flooding as a major problem in Dhaka, it did not consider the role of water bodies, including rivers and canals, in flood management. It discussed the navigability of the Buriganga River and the internal canals of Dhaka for transportation. The plan also proposed to convert Buckland Bund into open space (i.e., a riverfront recreation area). At that time, Dhaka was crisscrossed by a network of canals that acted as a natural stormwater drainage system. The plan recommended the excavation of continuous canals along the natural lines of drainage to link to the existing canals where possible (Minoprio & Spencely & P.W. Macfarlane, 1959; Iqbal, 2013). But the plan also suggested that the eastern arm of the Dholai canal should be retained for navigation purposes, and the remaining part of the canal fills up to form an inner ring road and partly an open space for amenities (Minoprio & Spencely & P.W. Macfarlane, 1959; Iqbal, 2013).

After the creation of Bangladesh in 1971, the basic assumptions of Dacca Master Plan 1959 became obsolete because of the very rapid change in economy and demography. But the development authority of Dhaka took a long time to initiate a new plan for the city to meet the challenge of new realities. Due to the absence of any plan, several structural, nonstructural, and ad-hoc measures were undertaken to address all the city's problems, including flooding. In 1981, the Dhaka Metropolitan Area Integrated Urban Development Project (DMAIUDP) was prepared. It was the first urban development strategic plan in Bangladesh, with the United Nations Development Programme (UNDP) and Asian Development Bank (ADB) as financial assistants, and a planning commission with the Ministry of Planning as the executor. The aim of the project was to prepare a long-term development strategy for the future Dhaka city with flood protection as the main focus (Kabir & Parolin, 2012). The DMAIUDP was not approved or implemented, as it was not given importance at the highest levels of government (Islam et al., 2009).

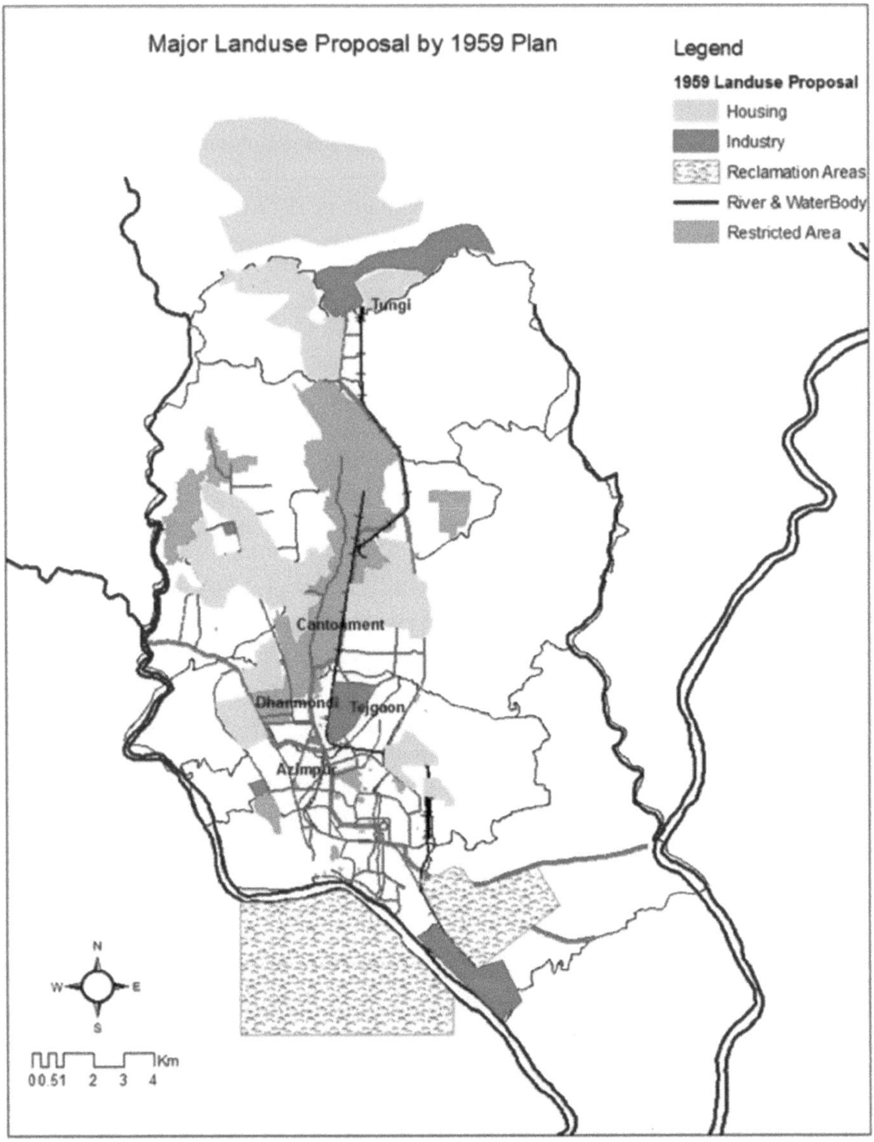

Figure 14.3 Major land-use proposals of the Dacca Master Plan 1959.
Source: Minoprio & Spencely and P. W. Macfarlane (1959).

Flood plain zoning was considered after the unprecedented flood of 1988, which was basically a nonstructural approach used to mitigate flood damages as a precautionary measure. A French consortium study assessed its necessity as a flood mitigation measure, but like previous plans, it also was never undertaken in any of the development plans or projects (Mohit & Akther, 2000).

14.4.3 Flood Action Plan (FAP), 1989

After the devastating flood of 1988, the Bangladeshi government established a committee with the primary objective of preparing a flood control plan for Dhaka (Figure 14.4). Eventually, as a part of the Flood Action Plan (FAP), two plans (FAP8A and FAP8B) were prepared for Dhaka (Islam, 2014; Kabir & Parolin, 2012; Haque et al.,

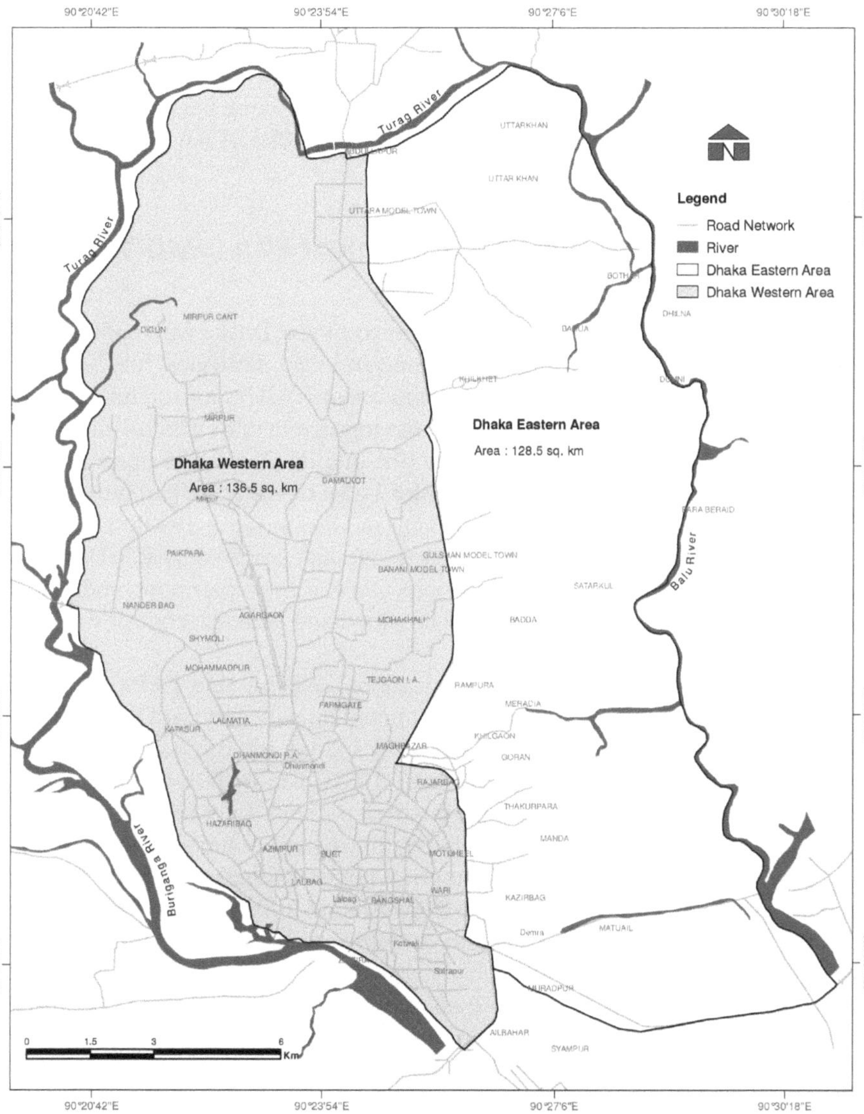

Figure 14.4 Area of FAP 8A and 8B.
Source: Das & Islam (2010).

2010; Bala et al., 2009; Islam et al., 2008; Alam & Rabbani, 2007; Huq & Alam, 2003; Mohit & Akther, 2000; Alam et al., 1996). The plans were aimed to provide a relatively flood-free and secure living environment through a long-term flood protection program and to improve environmental conditions for the promotion of a sustainable, long-term development in Dhaka (Alam et al., 1996; Islam, 2014).

FAP 8B, which covered 136.5 km^2 of the western side of the city, was included in the first phase of implementation, and FAP 8A, covering about 118.62 km^2 area of the eastern part of the city, was considered for the second phase (JICA, 1987, 1991, 1992). The major proposals of FAP 8A and 8B include (i) construction of an embankment and flood wall along the rivers surrounding Dhaka; (ii) improvement of canals within the cities; (iii) construction of a pumping station along the embankments; and (iv) acquiring land for retention pond (Huq & Alam, 2003; Alam & Rabbani, 2007; Islam, 2014).

14.4.4 Dhaka Metropolitan Development Plan (DMDP) (1995–2015)

The development of Dhaka is currently governed by the Dhaka Metropolitan Development Plan 1995–2015 (DMDP). The plan provided strategies for 20 years up through 2015 for the development of the area under RAJUK jurisdiction (1528 km^2 around Dhaka). This is a three-tiered plan, the top two tiers of which—the Structure Plan (the policy plan) and Urban Area Plan (the interim plan for 26 Special Planning Zone, SPZ)—were completed in 1995 (Figure 14.5). In 1997, the parliament adopted DMDP as the governing law for development in Dhaka (GoB, 1997). The plan was the first comprehensive development management plan prepared by RAJUK after the independence of Bangladesh. The prime objective was to prepare a hierarchical, multisectoral development plan for Dhaka where flood and drainage control was a major component.

The structure plan adopted the proposals of FAP 8A and FAP8B to prepare policies and strategies for the flood and drainage management of Dhaka. It identified Main Flood Flow and Subflood Flow Zones along the Buriganga, Turag Balu, Shitalakhya, and Dhalashwari rivers (Figure 14.6). The DMDP provided the first-ever policies on drainage and hydrology where policies were set for protection of flood flow zones, retention ponds, and flood control and drainage through development restrictions, and protection of retention ponds (Rahaman, 2008). The Urban Area Plan provided a midterm strategy from 1995 to 2005, mainly for urban areas. The proposals included the protection and preservation of retention ponds on western embankment (GoB, 1997).

The lowest tier, the Detailed Area Plan (DAP), was supposed to be completed by 2005. In 2004, the DAP project was started with the expectation that it would be completed by mid-2006 (Hammadi, 2008). After much deliberation and review, however, in June 2010, the government adopted the DAP as the final tier of the plan through a gazette notification (Ali & Byron, 2015). DAP prepared a land-use zoning plan along with an infrastructure development plan indicating locations of future land use as a part of development control. As the lowest tier of DMDP, DAP was supposed to conform to policies stated in the DMDP Structure and Urban Area Plan. But in DAP,

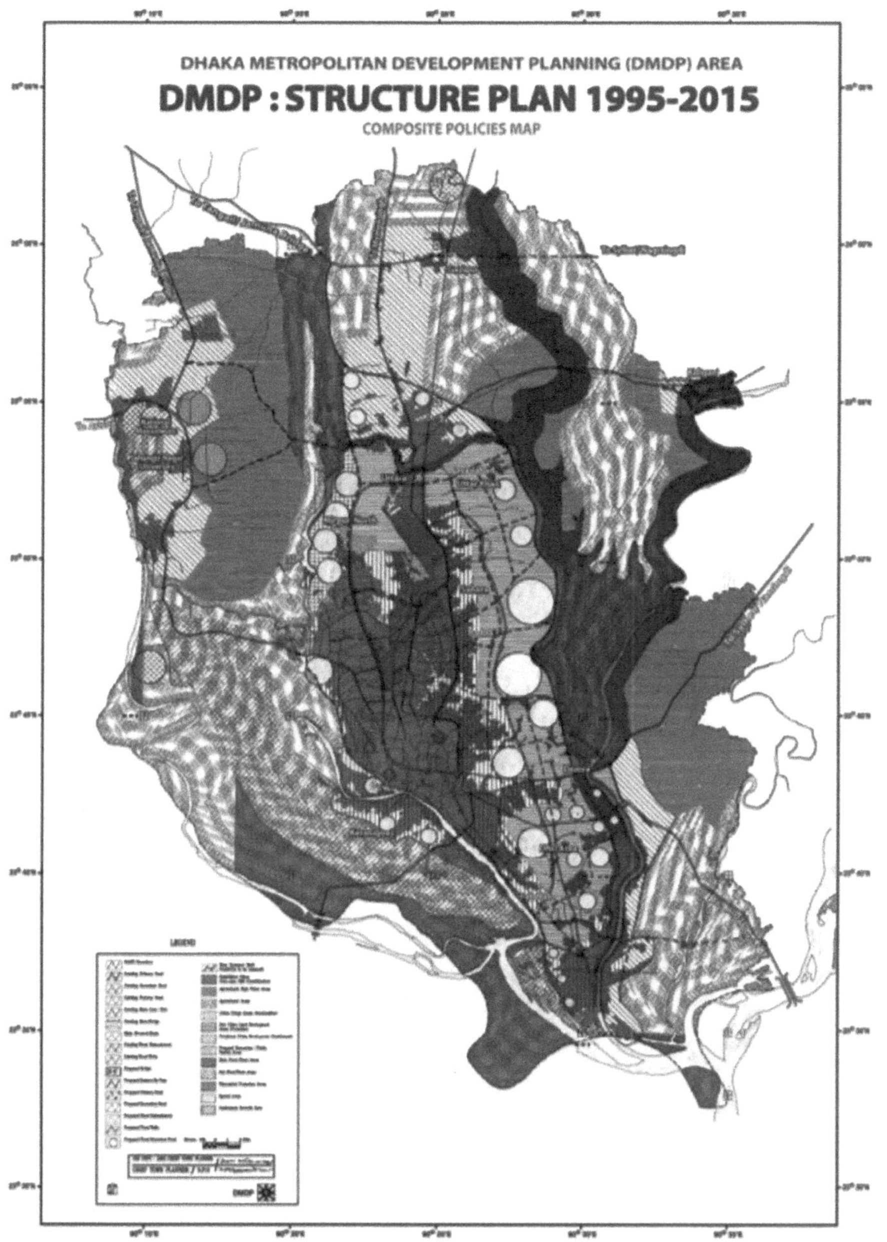

Figure 14.5 DMDP structure plan.
Source: GoB (1997).

the "subflood flow zone" (as proposed in the Structure Plan) was merged with the "Main Flood Flow Zone," as well as "urban residential zone," "rural homestead category," and "mixed-use category." DAP proposed only 1.09% of the DMDP area as retention ponds, whereas the JICA proposal designated 3.09% of the DMDP area for

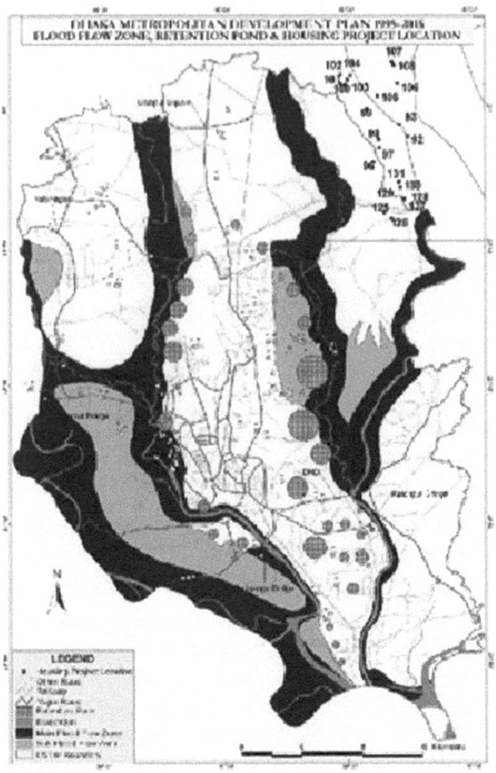

Figure 14.6 Flood Flow Zone and Water Retention Area in DMDP.
Source: Islam (2009).

this purpose (BIP, 2008). The proposed land-uses within flood flow zones are also contradictory to the Water Body Conservation Act of 2000, Bangladesh Environment Conservation Act of 1995, and the Ramsar Convention.

14.4.5 Flood Management Initiatives by Organizations

Following the devastating flood of 1988, the Dhaka Water Supply and Sewerage Authority (DWASA) undertook a Storm Water Drainage Improvement Plan in 1989 in order to solve the drainage and waterlogging problems in Dhaka. As part of this plan, different parts of natural canals were replaced by box culverts, including Dhanmondi Canal, Paribag Canal, Begunbari Canal, Mahakhali Canal, Segunbagicha Canal, and Dholai Canal. Unfortunately, most of these canals and low-lying areas are no longer visible due to encroachment. DWASA has begun to construct a Storm-Water Drainage Network consisting of canals, underground pipes, box culverts, and pumping stations. DWASA, Dhaka City Corporation (DCC), and Bangladesh Water Development Board (BWDB) have also constructed pumping stations and installed small pumps to drain stormwater (Huq & Alam, 2003; Parvin & Shaw, 2011).

Realizing the importance of wetland preservation, the government of Bangladesh enacted the Water Body Conservation Act in 2000, which defined natural water bodies as places that are demarcated in the master plan as rivers, canals (Khals), depression areas (Beels), lakes, streams or wetlands, or places that are declared as flood flow zones by the local government notification; these should also include lands that retain stormwater.

14.5 Implementation of Plans and Policies Toward Resiliency

Although Dhaka has adopted plans for development since 1959, the implementation of these plans has been very slow. In fact, organic, unplanned informal development has taken place which rapidly ignores plans and sensitivity of land use against flood. It was expected that DMDP would be an important tool for improving resilience against flood. The Structure Plan and Urban Area Plan of DMDP followed the principles of FAP 8A and 8B. FAP 8B proposed the preservation of 21 canals and retention ponds in three areas along the western embankment, but very little work has been done to maintain the canals, and the government has failed to acquire the required amount of land proposed as retention pond sites. The performances of the FAP proposals were observed from the floods of 1998, 2004, and 2007. A large-scale pump station functioned well in all three floods, though the performance of temporary pumps located inside the city were not satisfactory (Rahman et al., 2005; Islam et al., 2008; Bala et al., 2009). Dewan and Yamaguchi (2009b) claimed that due to the poor urban planning and reclamation and development of natural areas, such as wetlands and low-lying areas, the severity of flooding increased in 2004. Several researchers (Alam & Rabbani, 2007; Bala et al., 2009) mentioned that the flood-protective embankment of Dhaka worked against river flooding in the western part of the city. But it caused waterlogging or internal drainage congestion due to inadequate pumping facilities and the lack of proper planning and design of infrastructure for drainage, which becomes severe during heavy rainfall. Thus, the problem of internal waterlogging is getting worse rather then external flooding in the western part of Dhaka (Bala et al., 2009).

Similar proposals like that in the western part of Dhaka have been proposed for the eastern part of Dhaka in FAP 8A. DMDP also adopted those proposals in its Structure Plan, and the DAP also accommodated these proposals to some degree after facing great pressure from professional and environmental organizations. DAP was prepared 13 years after the acceptance of the Structure Plan and Urban Area Plan, and as a result, during this time many areas delineated as flood flow zones were filled up and settlements developed. It is an accepted fact that without keeping sufficient amount of retention pond areas, it would be very difficult to address urban flooding after the construction of the eastern embankment. On the other hand, without the embankment, flood scenarios would be even worse because both river and rain floods are a concern in that case. In such a context, the conversion of low-lying areas in eastern Dhaka is a crucial flood threat to the city. Islam (2014) showed the uncontrolled and widespread land-filling activities of private developers in the eastern part of Dhaka, even as the locations of retention pond areas are disappearing gradually.

Although the government accepted DAP, its sincerity regarding implementation of the plan has been questioned (Bhuiyan, 2014). In addition to the informal violation of DAP by the general public, the Ministerial Committee for reviewing the plan approved 30 projects on the designated flood flow and water retention areas of the plan. The committee also declared that all privately owned small water bodies (below the size of 0.134 hectares) must not be connected to the water flow network and privately owned open spaces (below the size of 0.134 hectares) must be declassified from their respective land use class based on the needs of the owner. It seems that the regulatory authority of Dhaka has become a major violator of DAP, and its decisions seriously jeopardize the core policies of the plan and reduce the resiliency of the city.

14.6 Conclusions

It is evident from this study that Dhaka is facing a severe threat of both river and rain flooding. The planning documents and flood management projects have considered this threat with due importance. But implementation of the policies is limited; Table 14.1 provides a summary of the key issues of different development plans/policies and the implementation status of the plan. The nexus of powerful developers, politicians, and government authorities has created the major obstacle toward implementation of the flood management policies laid out in DAP.

It is important to note that land speculation by developers and private entities has been identified as the major cause of destruction of flood retention areas (Islam, 2009). In addition to the integration of flood resiliency in planning documents, it is very important to adopt economic tools of land-use control, such as imposing an inheritance tax, capital gain tax, the Land Ceiling Act, and other legislation. Moreover, it is necessary to develop adequate infrastructure which is now attracting the developers. Institutional capacity and decision-making power are major aspects that can ensure implementation of the rules and regulations. In the context of Bangladesh, the political will and philosophy of the ruling government plays the major role in the area of development. It is important to realize that devastating floods in the recent past has caused very serious damage to infrastructure, as well as significant economic losses. The time span of the DMDP (1995–2015) completed in July 2015. RAJUK has already started working on a Structure Plan for the next 20 years (2016–2035), the draft of which was completed in December 2014. The first draft of Dhaka Structure Plan (2016–2035) is being prepared under Regional Development Planning (RDP) project funded by ADB. The plan is being prepared with a vision: "Making Dhaka a livable, functional, and resilient metropolis respecting local socio-cultural fabric and environmental sustainability." It is expected that this plan would address the flood management issue and adopt necessary implementation tools to ensure the resilient development of the city.

Table 14.1 **Key Issues Regarding Flood Management in Various Plans for Dhaka**

Year	Plans of Dhaka	Key Issues Regarding Flood Management	Implementation Status
1864	Construction of Buckland Bund	Protecting the city from flooding of the Buriganga and mitigating flood damage within the city	Construction was completed by 1880.
1917	Dacca Town Planning Report 1917	Emphasized the preservation of water bodies	The plan was never accepted by the government.
1959	Dacca Master Plan 1959	Acknowledged flooding as a major problem, but did not consider the role of water bodies, including rivers and canals, in flood management	The plan was accepted by the government and partially implemented.
		Envisioned that the wetlands and flood plains of the city will be developed as housing estates in the future	
1981	Dhaka Metropolitan Area Integrated Urban Development Project (DMAIUDP)	Long-term development strategy for the future of Dhaka city, with flood protection as the main focus	The plan was not accepted by the government.
1989	Flood Action Plan (FAP)	Construction of embankment and flood wall along the rivers surrounding Dhaka	Western embankments and flood walls were constructed.
		Improvement of canals within the cities	21 canals were rehabilitated.
		Construction of a pumping station along the embankments	Three permanent pump houses were constructed.
		Acquiring land for retention pond	The government failed to acquire the required amount of land for a retention pond.
1989	Water Drainage Improvement Plan	Preservation and improvement of the canal network	This was partially implemented as part of the development of a Storm Water Drainage System.
1995–2015	DMDP (1995–2015)	Identified Main Flood Flow and Subflood Flow Zones of the rivers surrounding Dhaka	In DAP, the "subflood flow zone" proposed in the Structure Plan was merged with the "Main Flood Flow Zone," as well as the "urban residential zone."
		Proposed protection and preservation of retention ponds on western embankment of the city	In the name of DAP review, the government is approving a project in the designated flood flow and subflood flow zone.

References

Ahmed, S. U. (1986). *Dacca: A study in urban history and development*. Glenn Dale: The Riverdale Company.

Alam, B. N., et al. (1996). Effects of flood mitigation measure: Lessons from Dhaka Flood Protection Project, Bangladesh. *Asian Profile*, *24*(6), 511–524.

Alam, M., & Rabbani, M. G. (2007). Vulnerabilities and responses to climate change for Dhaka. *Environment and Urbanization*, *19*(1), 81–97.

Ali, T., & Byron, R. K. (January 27, 2015). New 20-year DAP amid failure of the old one. *The Daily Star* Dhaka. 25.

Ameen, M. S. (1998). Growth potentials of a premature mega city: An event of grim portent. *Journal of Civil Engineering*, *26*(2) 151–72.

Bala, S., et al. (2009). *Performance of Flood Control Works Around Dhaka City During Major Floods in Bangladesh* 2nd International Conference on Water & Flood Management (ICWFM-2009), Dhaka, Institute of Water and Flood Management (IWFM), BUET.

Bangladesh Institute of Planners (BIP). (2008). Detail Area Plan (DAP) of Dhaka: Shattering the vision of DMDP. *The Daily Star*, 18 Dhaka, December 20.

Bhuiyan, K. (2014). *80pc Wetlands and Flood Flow Zones Filled Up*. Accessed January 26, 2015, from http://www.steinandpartners.com/eco/80pc-wetlands-and-flood-flow-zones-filled-up.

Chowdhury, A. M., & Faruqui, S. (1989). Physical growth of Dhaka city. In S. U. Ahmed (Ed.), *Dhaka: Past present future* (pp. 43–63). Dhaka, Bangladesh: The Asiatic Society of Bangladesh.

Dani, A. H. (2009). *Dhaka: A record of its changing fortune*. Dhaka, Bangladesh: Asiatic Society of Bangladesh.

Das, B., & Islam, I. (2010). Analyzing the proposals of FAP 8B project of Dhaka and present context of retention pond areas and canals. *Journal of Bangladesh Institute of Planners*, *3*, 89–102.

Department of Economic and Social Affairs, United Nations. (2014a). *World urbanization prospects: The 2014 revision*. New York: United Nations.

Department of Economic and Social Affairs, United Nations. (2014b). *World urbanization prospects: The 2014 revision highlights (ST/ESA/SER.A/352)*. New York: United Nations.

Department of International Development (DFID). (2014). *Rural and urban development case study—Bangladesh.* Oxford Policy Management.

Dewan, A. M., & Yamaguchi, Y. (2009a). Land use and land cover change in Greater Dhaka, Bangladesh: Using remote sensing to promote sustainable urbanization. *Applied Geography*, *29*(3), 390–401.

Dewan, A. M., & Yamaguchi, Y. (2009b). Using remote sensing and GIS to detect and monitor land-use and land cover change in Dhaka metropolitan of Bangladesh during 1960–2005. *Environmental Monitoring and Assessment*, *150*(1–4), 237–249.

Dewan, A. M., et al. (2007). Evaluating flood hazard for land-use planning in greater Dhaka of Bangladesh using remote sensing and GIS techniques. *Water Resources Management*, *21*(9), 1601–1612.

Government of Bangladesh (GoB). (1997). Dhaka Metropolitan Development Plan (1995–2015), Government of Bangladesh. *SRO 184/97*.

Hammadi, S. (2008). Hotchpotch map on faulty plan. *The New Age* Dhaka, Bangladesh.

Haque, A. N., et al. (2010). *Assessment of adaptation measures against flooding in the city of Dhaka, Bangladesh*. Rotterdam, Netherlands: IHS.

Hunter, W. W. (1877). *A statistical account of Bengal*. Vol. V (Districts of Dacca, Bakerganj, Faridpur, and Maimansingh). London: Trubner.

Huq, S., & Alam, M. (2003). Flood management and vulnerability of Dhaka city. In A. Kreimer, M. Arnold, & A. Carlin (Eds.), *Building safer cities: The future of disaster risk* (pp. 121–135). Washington, DC: World Bank Publications.

Intergovernmental Panel on Climate Change (IPCC) (Ed.). (2007). *Climate change 2007: Impacts, adaptation, and vulnerability. contribution of Working Group II to the fourth assessment report of the Intergovernmental Panel on Climate Change*. Cambridge, UK: Cambridge University Press.

Iqbal, I. (2013). First Master Plan for Dhaka City. *South Asia Chronicle, 3*.

Islam, A. K. S., et al. (2008). *Hydrological aspects of flood 2007*. Dhaka, Bangladesh: Institute of Water and Flood Management (IWFM), BUET.

Islam, I. (2009). *Wetlands of Dhaka metropolitan area: A study from social, economical, and institutional perspective*. Dhaka, Bangladesh: AH Development Publishers.

Islam, I. (2014). Wetland management in Dhaka: Institutional efforts and associated constraints. In G. A. Choudhury, K. Nakagami, J. Li, & K. Fukushi (Eds.), *Strategic adaptation towards water crisis*. Dhaka, Bangladesh: University Press Limited, 207–224.

Islam, I. M., et al. (2009). Urban consolidation approach for Dhaka city: Prospects and constraints. *Global Built Environment Review*, *7*(1), 50–68.

Islam, M. S., et al. (2010). Changes in wetlands in Dhaka city: Trends and physico-environmental consequences. *Journal of Life and Earth Science*, *5*, 37–42.

Japan International Cooperation Agency (JICA). (1987). *Study on storm water drainage improvement in Dhaka city*. Dhaka, Bangladesh: Japan International Cooperation Agency (JICA).

Japan International Cooperation Agency (JICA). (1991). *Master plan for greater Dhaka protection project (study in Dhaka metropolitan area), FAP 8A, Main Report and Supporting Reports I and II*. Dhaka, Bangladesh: Japan International Cooperation Agency (JICA).

Japan International Cooperation Agency (JICA). (1992). *Feasibility study on greater Dhaka protection project (study in Dhaka metropolitan area), FAP 8A, Main Report and Supporting Reports I and II*. Dhaka, Bangladesh: Japan International Cooperation Agency (JICA).

Kabir, A., & Parolin, B. (2012). Planning and development of Dhaka—A story of 400 years. In *15th International Planning History Society Conference*. Sao Paulo, Brazil: International Planning History Society, 1–20.

Khan, D. M., et al. (2012). Flood damage assessment for Dhaka city, Bangladesh. In F. Klijn, & T. Schweckendie (Eds.), *Comprehensive flood risk management: Research for policy and practice*. London: CRC Press, 138–147.

Khan, N. (2000). Temporal mapping and spatial analysis of land transformation due to urbanization and its impact on surface water system: A case from Dhaka metropolitan area, Bangladesh. *International Archive of Photogrammetry and Remote Sensing*, *33*, 598–605.

Mahmud, M. S., et al. (2011). Remote sensing and GIS-based spatio-temporal change analysis of wetland in Dhaka city, Bangladesh. *Journal of Water Resource and Protection*, *3*(11), 781.

Masum, F. (2010). Improving governance for fringe land development and management. In *46th ISOCARP Congress* Nairobi, Kenya.

Minoprio & Spencely, & P. W. Macfarlane (1959). *Master Plan for Dacca*. Dhaka, Bangladesh: Dhaka Improvement Trust.

Mohit, M. A., & Akther, M. S. (2000). Delineation of flood-damaged zones of Dhaka city based on the 1998 flood by using GIS. In M. A. Ali, S. M. Seraj & S. Ahmad (Eds.), *Engineering concerns of flood* (pp. 303–318). Dhaka, Bangladesh: Bangladesh University of Engineering and Technology.

Mowla, Q. A. (2007). A review of Dhaka's urban morphology. *Jahangirnagar Review, Part II: Social Sciences XXXI*, 15–29.

Mowla, Q. A. (2013). Dwindling urban water-bodies of Dhaka and the city fabric: A post-mortem. In S. Jahan & A. K. M. A. Kalam (Eds.), *Dhaka metropolitan areas and its planning: Problems, issues, and policies*. Dhaka, Bangladesh: Bangladesh Institute of Planners, 1–11.

Parvin, G. A., & Shaw, R. (2011). Climate disaster resilience of Dhaka City Corporation: An empirical assessment at zone level. *Risk, Hazards, & Crisis in Public Policy*, *2*(2), 1–30.

Rabbani, G., et al. (2011). Climate change implications for Dhaka city: A need for immediate measures to reduce vulnerability. In K. Otto-Zimmermann (Ed.), *Proceedings of the Global Forum 2010: Resilient Cities: Cities and adaptation to climate change* (pp. 531–541). Bonn, Netherlands: Springer.

Rahaman, G. (2008). Legal framework for development control and practice. In G. Rahaman (Ed.), *Town planning and the political culture of planning in Bangladesh*. Dhaka, Bangladesh: A. H. Publishing House, 277–336.

Rahman, R., et al. (2005). *Investigation of hydrological aspects of flood—2004 with special emphasis on Dhaka city*. Dhaka, Bangladesh: Institute of Water and Flood Management (IWFM), BUET.

Rajdhani Unnayan Kartripakkha (RAJUK). (n.d.). Dhaka City Map. Accessed on July 14, 2015, from http://www.rajukdhaka.gov.bd/rajuk/image/template/DHAKA%20CITY-ARMY%2048X%2072FINAL%20MAP.jpg.

Reckien, D., Wildenberg, M., & Deb, K. (2011). Understanding potential climate change impacts and adaptation options in Indian megacities. In K. Otto-Zimmermann (Ed.), *Resilient Cities* (pp. 15–34). Netherlands: Springer.

Rizvi, S. N. H. (1969). *East Pakistan district gazetteers—Dacca District*. Dhaka, Bangladesh: East Pakistan Government Press.

Roy, M. (2009). Planning for sustainable urbanisation in fast-growing cities: Mitigation and adaptation issues addressed in Dhaka, Bangladesh. *Habitat International*, *33*(3), 276–286.

Satterthwaite, D. (2008). Understanding Asian cities: A synthesis of the findings from eight city case studies. *Global Urban Development*, *4*(2), 1–24.

Shafi, S. A. (2010). Implementation of Dhaka City Master Plan. *The Daily Star*, *20* Dhaka, February 23.

Shubho, M. T. H., et al. (2014). An improved semiautomatic segmentation approach to land cover mapping for identification of land cover change and trend. *International Journal of Environmental Science and Technology*, *12*(8), 2593–2602.

Sultana, M. S., et al. (2009). Pre-and post-urban wetland area in Dhaka city, Bangladesh: A remote sensing and GIS analysis. *Journal of Water Resource and Protection*, *1*(6), 414–421.

UNHABITAT., et al. (2010). *State of the Asian Cities 2010/11*. Fukuoka, Japan: UN-HABITAT.

Yahya, S. M., et al. (2010). Climate change impacts on flood vulnerability for Dhaka city. *International Conference on Environmental Aspects of Bangladesh* Kitakyushu, Japan.

Postdisaster Urban Recovery: 20 Years of Recovery of Kobe

15

Michiko Banba[1], Rajib Shaw[2]
[1]Education and Research Center for Disaster Education, University of Hyogo, Japan;
[2]Professor, Graduate School of Global Environmental Studies, Kyoto University, Kyoto, Japan

15.1 Introduction

It has been 20 years since the Hanshin-Awaji Earthquake, which struck Hyogo prefecture, occurred at 5:46 a.m. on January 17 in 1995. Except for damage that occurred on Awaji Island, most of the destruction occurred in urban areas, including the cities of Kobe, Ashiya, Nishinomiya, and Takarazuka. For that reason, the Hanshin Awaji earthquake has been characterized as an urban disaster. Since this disaster, the heavily damaged infrastructure, commercial buildings, and residential houses have been rebuilt, and the city of Kobe has recovered. The slogan of the recovery from the earthquake was "Fukko (Reconstruction)," which meant to restore the damaged cities of Hyogo prefecture (including Kobe) so that they are not just the way they were, but even better, to be leading cities in the 21st century. The recovery of Kobe is the focus of this chapter.

The recovery of Kobe can be classified roughly into four categories: infrastructure recovery, housing recovery, industrial recovery, and recovery of livelihood. The features of this recovery includes a number of key factors, such as public consensus, planning, and community development. Large-scale construction projects were implemented to achieve this recovery. At the same time, the victims' livelihoods and mental states were negatively affected by the disaster, and local governments took measures to solve those problems.

15.2 Damages

15.2.1 Physical Damage

Hanshin-Awaji Earthquake was the first major earthquake to occur in close proximity to a large Japanese city since the end of World War II, and severe seismic activity was recorded in the densely populated Hyogo prefecture. The epicenter was in the Awaji Strait between Awaji Island and Kobe, and the hypocenter was an active fault located 16 km below the Earth's crust. The magnitude of the earthquake, which describes the extent of destruction at the focal point, was 7.3. On the Japan Meteorological Agency (JMA) scale of seismic intensity, which quantifies the severity of the tremors felt on the Earth's surface, the maximum intensity level was 7, which is the largest. The distribution of area with a seismic intensity level of 7 spread from Awaji Island, where the epicenter was located, Takarazuka City to the east (as shown in Figure 15.1).

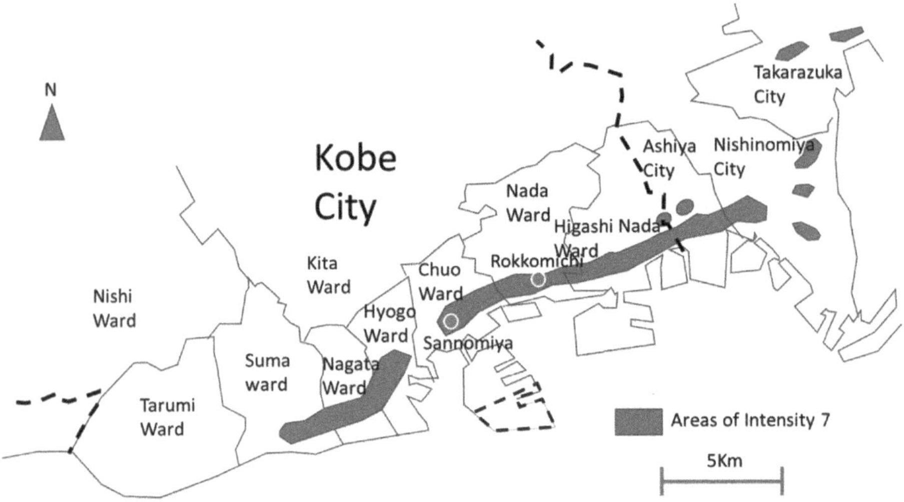

Figure 15.1 Distribution of seismic intensity 7 throughout Kobe.
Source: Kobe City (2011).

Table 15.1 Damage to Persons from the Hanshin-Awaji Earthquake

	Total	Hyogo Prefecture	Kobe City
Dead	6434	6402	4571
Missing	3	3	2
Injured	43,792	40,092	14,678

Note: The number of dead includes seven suicides.
Source: Figures come from the Hyogo Prefectural Government as of December 27, 2006. Figures of total and Kobe are from the city of Kobe as of January 1, 2007, and the Fire and Disaster Management Agency as of May 19, 2006, respectively.

The damages caused by the earthquake were tremendous (Table 15.1 and Table 15.2) and economic loss was approximately 10 trillion yen. In total, 6437 persons were killed or missing, and 4573 of them were caused in Kobe, respectively. More than 40,000 people were injured in Hyogo prefecture, approximately 35% of which lived in Kobe. Many houses were destroyed in Kobe, and Nagata and Higashinada ward in particular suffered serious damage. Many of destructed houses were wooden structure and built under the building standard before 1981 revision (Murosaki, 1998).

Lifelines and urban infrastructure, such as expressways, railways, subways, waterworks/sewage works, gas/power supplies, and communications were also seriously damaged. This damage severely disrupted escape and rescue activities and increased human suffering. It also affected people's lives during the recovery phase for almost a year after the event. Damage to critical facilities, such as hospitals, schools, or government offices, also affected rescue, relief, and response services after the disaster.

Table 15.2 **Housing Damages from the Hanshin-Awaji Earthquake**

Ward	Destroyed	Habitable	Total
Higashinada	16,174	48,845	65,019
Nada	10,050	33,280	43,330
Chuo	5964	38,271	44,235
Hyogo	7984	30,237	38,221
Nagata	23,301	36,186	59,487
Suma	10,761	52,032	62,793
Subtotal	**74,234**	**238,851**	**313,085**
Tarumi	3094	81,847	84,941
Kita	922	61,002	61,924
Nishi	1033	63,750	64,783
Citywide Total	**79,283**	**445,450**	**524,733**

Source: Kobe City (2011)

15.2.2 Population Changes After the Earthquake

Influences of the Hanshin-Awaji earthquake disaster can be seen in the migration of people from each ward (Figure 15.2). After the earthquake, the population of the most heavily affected areas decreased significantly as victims who lost their houses left those areas to move into temporary housing, relatives' houses, or private rentals.

The populations of Higashinada, Nada, and Chuo wards have returned to, or even exceeded, the level before the earthquake, while those of the Nagata, Hyogo, and Suma wards have not. Now, the center of gravity of Kobe is shifting toward the east side, closer to Osaka. Although those changes were not just due to the earthquake, the earthquake accelerated the population decline in these areas. The decrease of population in the western part of Kobe also affected the local commercial sector, with the shopping streets in Nagata ward are experiencing a decrease in sales.

15.3 Recovery Planning

15.3.1 Comprehensive Recovery Plan

The city of Kobe established the Headquarters for Reconstruction and organized the Committee for Recovery Planning in three weeks. By the end of March the Committee issued the "Kobe Recovery Guideline", which was followed by the "Kobe Recovery Plan". About three months after the earthquake, Kobe City Recovery Planning Council was organized with three subcommittees (Citizens' Life, City Vibrancy, and Safe City) to discuss about the recovery plan, and "Kobe Recovery Plan" was issued in the end of June (Kobe City, 2011). This plan does not include physical or spatial aspects, but rather visions, concepts or policies.

The plan consisted of basic concept for recovery, recovery plans classified by goal, resilient city development plans, urban area recovery plans, symbolic projects, and

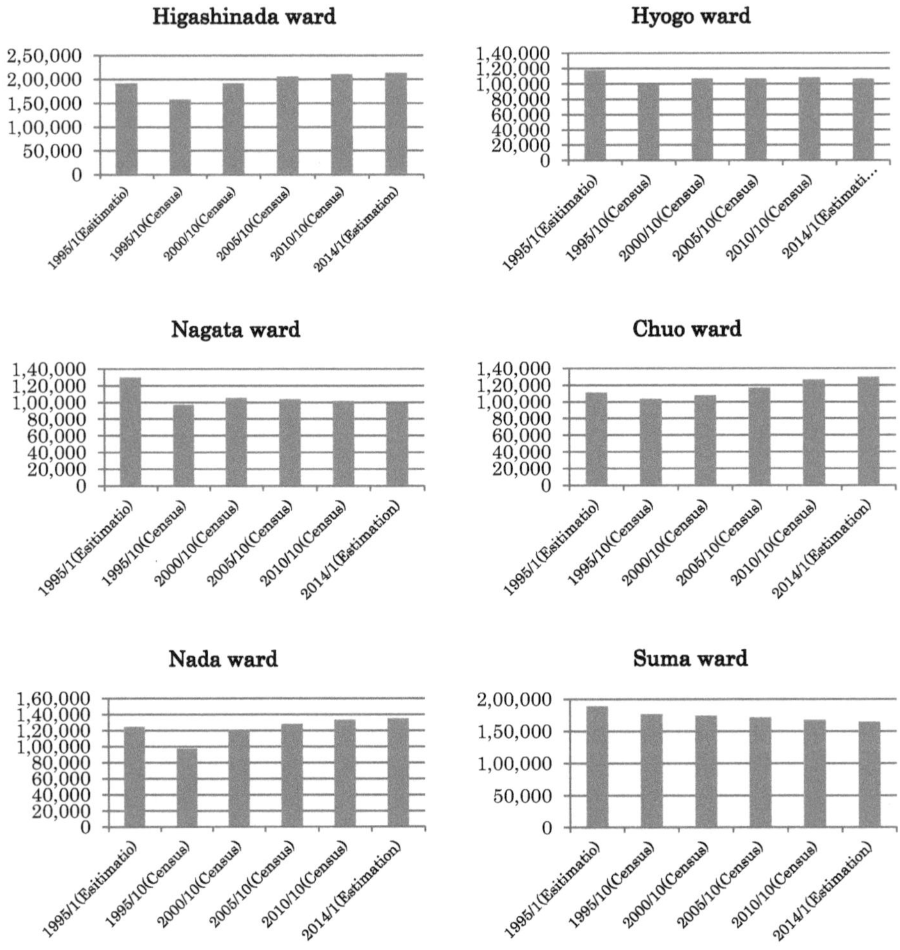

Figure 15.2 Population transition of wards in Kobe.
Source: Kobe City (2015).

implementation programs. Kobe Recovery Plan was developed in consideration of relationship with existing spatial plans or construction plans for the city and wards (Kobe City, 2011).

However, Kobe Recovery Plan was not developed as upper-level plan of city master plans, district plans or construction project plans, but as the regional plan from the long-term perspective (Kobe City, 2011).

The relationship among plans related to the recovery of Kobe is shown in Figure 15.3.

15.3.2 Recovery Plan Factors

In promoting the recovery of Kobe after the earthquake, six factors were involved: reconstruction of infrastructure, revitalization of economy, support for small business, housing recovery urban planning and recovery of livelihoods (Figure 15.4).

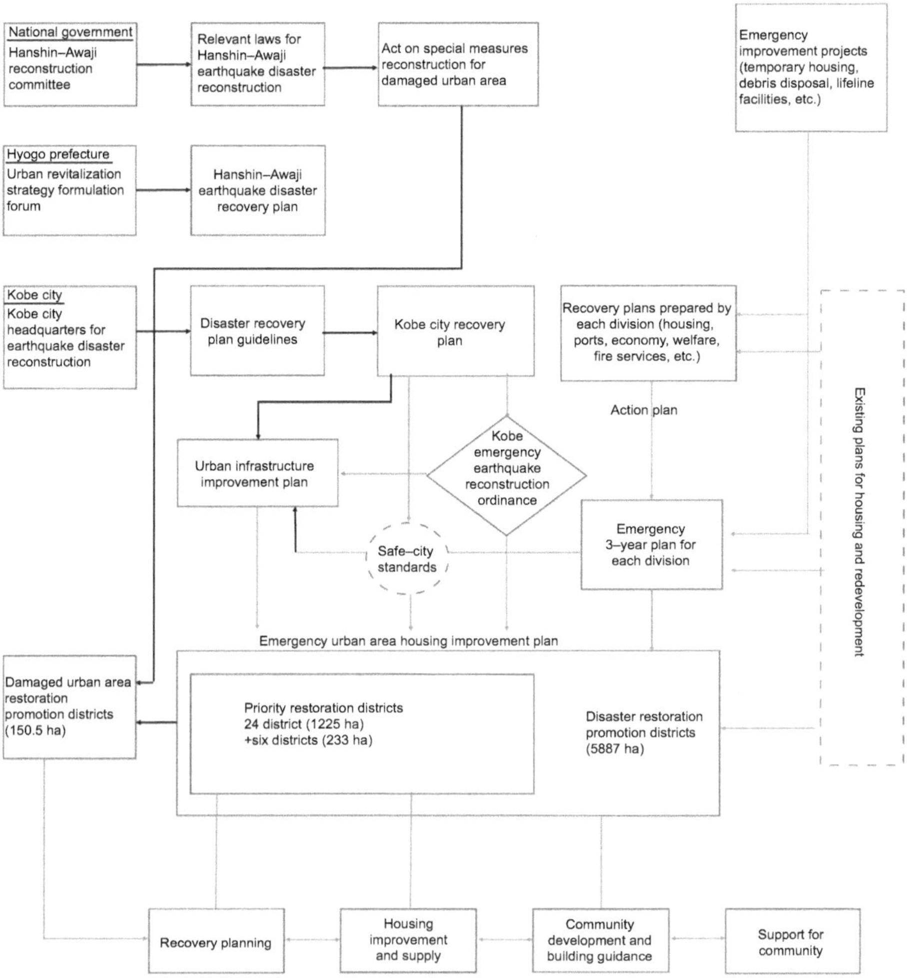

Figure 15.3 Relationship of related plans.
Source: Kobe City (2011).

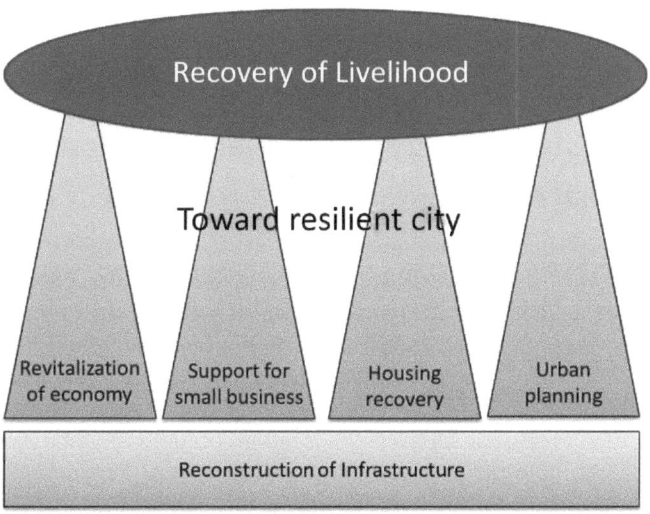

Figure 15.4 Factors related to the recovery of a resilient city.

Those factors work together to achieve a disaster-safe city. Restoration of infrastructure is essential for the restoration and recovery of other factors. Urban planning and housing reconstruction following infrastructure restoration provide victims with space to live in. Revitalization of Kobe's economy, as well as of small and medium-sized companies, are important for the livelihood of disaster victims, which in turn affect the recovery of the city's overall livelihood.

15.4 Two-Stage Planning Process for Reconstruction of the City

Apart from "Kobe Recovery Plan", plans to reconstruct the city physically were needed to be developed. Developments in the disaster stricken area were restricted by law for two months after the earthquake. It was required to develop a reconstruction plan to be approved during two months, which became constraint to complete the procedure if the city tried to obtain the consensus of residents. In order to proceed the planning process in the limited time, a two-stage city planning process was adopted to implement the reconstruction projects. In the first stage, Kobe City designated a 5-ha area to be reconstructed and determined the method of implementing projects and important urban facilities to make vulnerable areas earthquake resistant. The plan was passed on March 16. It is a framework for the detailed plan developed in the second stage. Table 15.3 shows the details of those projects, which involve land readjustment and redevelopment. Figure 15.5 shows the locations of those project areas.

Afterward, the Community Development Council was established with the participation of residents in the heavily disaster-stricken area with the assistance of consultants. The Council developed the vision for recovery of the city and proposed to the city of Kobe. Then, Kobe city developed a detailed plan based on the vision, with the locations of roads, parks, and urban facilities based on the proposal (the vision) prepared by the Council. It was the second stage of the planning process. Detail of each stage is explained in the following sections.

15.5 City Restoration by Urban Planning Measures

15.5.1 Regional Framework for City Restoration

Urban planning projects were implemented by the city of Kobe for restoration purposes. The Emergency Earthquake Reconstruction Ordinance was passed just a month after the earthquake. This law designated Disaster Restoration Promotion Districts and Priority Restoration Districts to promote supply of housing and improvement of urban areas (Figure 15.5). The area of those districts (between Suma and Higashinada wards) was approximately 5887 ha. Then, Priority Restoration Districts within the Disaster Restoration Promotion Districts (24 locations occupying a total area of 1225 ha) were created to promote projects to restore infrastructure, regenerate urban areas,

Table 15.3 City Planning Projects for Recovery

	Districts for Land Readjustment Projects					Districts for Land Redevelopment Projects	
Names	**Moriminami**	**Rokkomichi Station West**	**Matsumoto**	**Misuga**	**Shin-Nagata,Takatori**	**Rokkomichi South**	**Shin-Nagata Station South**
Area(ha)	16.7	19.7	8.9	10.1	69.2	5.9	20
Damage ratio	66	68	81	88	86	65	83

Source: Kobe City (2011).

and supply housing on March 17. It was said that political discretion influenced how the districts were drawn, as some areas had urban problems, such as inner-city problems (Inner city problem is caused by the degradation of the urban environment which results in the decrease of population worsens the city management) and lower quality of infrastructure.

15.5.2 Projects for City Restoration

15.5.2.1 City Planning Projects for Recovery

Regarding city planning projects related to disaster restoration in Kobe, seven districts (five for land readjustment projects and two for urban redevelopment) were designated for advance implementation in the context of a major project for public-led urban restoration. In the process of making the city planning decisions, Kobe adopted a two-stage planning process for land readjustment projects; the first stage concerned zoning (areas where projects were to be implemented and areas to be designated as disaster-hit urban renewal promotion districts), and the second stage concerned planning decision making for core city facilities (city streets and neighboring parks). Regarding urban redevelopment projects, some of the plans were revised in the process of implementation based on suggestions and proposals by Community Development Councils.

Other disaster restoration–related city planning projects were also implemented in Sannomiya District (5 sites, occupying 70.6 ha), and improvement projects for city roads (10 routes in 13 districts, including the Yamate Kansen Morikita District).

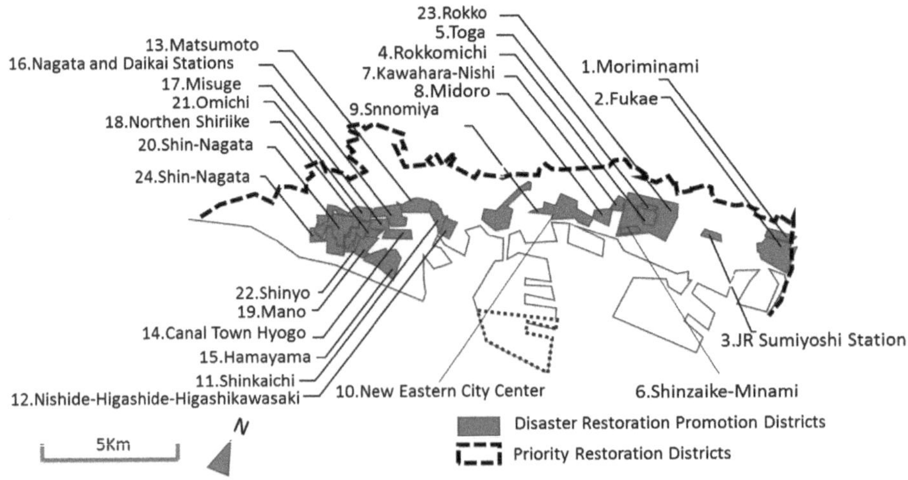

Figure 15.5 Designation of disaster restoration promotion districts and priority restoration districts.
Source: Kobe City (2011).

15.5.2.2 Housing Projects

Schemes for urban renewal included projects for residential area development based on the Residential Areas Improvement Act in four districts (5.48 ha in total): Toga (0.82 ha), Bancho-4 (1.27 ha), Bancho-5 (3.14 ha), and Higashikawasaki 7-chome (0.25 ha). Also, optional projects not covered by legal systems, which are Urban Residential Area Comprehensive Improvement Projects and Densely Built-up Residential Area Improvement Promotion Projects were implemented with housing programs, lot procurement, construction, and rent subsidies for public and semipublic housing, and improvement of related public facilities. Eight districts were designated for Urban Residential Area Comprehensive Improvement Projects as follows: Rokko (296.7 ha), areas around the New Eastern City Center (168.1 ha), areas around Kobe Station (58.0 ha), areas around Matsumoto (22.4 ha), the area south of Hyogo Station (35.6 ha), Misuga (29.1 ha), Shinyo (8.2 ha), and Shin-Nagata (224.0 ha). Nine districts (508.9 ha in total) were designated as Densely Built-Up Residential Area Improvement Promotion Project Districts: Fukae (49.1 ha), Harada-Iwaya (86.9 ha), Miyamoto-Azuma (98.9 ha), Nishide-Higashide- Higashikawasaki (22.6 ha), Hamayama (25.0 ha), Northern Shiriike (25.0 ha), Mano (39.0 ha), Southern Nagata (63.2 ha), and Eastern Tarumi (99.2 ha).

15.5.2.3 Other Projects

Focused on supporting private housing reconstruction, the urban renewal projects for nonpriority areas were implemented. Along with the promotion of joint housing projects using voluntary projects, such as the Improvement Projects to Provide Better Buildings, various subsidy and support systems were established. For example, the Road Improvement Type Group Reconstruction System was established to facilitate renewal plan preparation and subsidizing interest for housing fund loans and construction of private roads.

15.5.2.4 Example of a City Recovery Project

The Rokkomichi North district (16.1 ha) is located in the eastern part of Kobe. Before the earthquake, the district included both residential and neighborhood commercial establishments. After the earthquake damaged two-thirds of the district, it was reconstructed by utilizing a land readjustment project (Figure 15.6).

The land readjustment project was designed to improve the living environment and effective use of the land in urban areas by constructing necessary infrastructure and public facilities such as roads and parks. It required individual landowners to provide some portion of their lands for constructing roads and parks as the value of the land will be increased by the land readjustment project (Figure 15.7).

15.6 Community Development in Machidukuri

Under the Kobe Community Development Ordinance (enacted in December 1981), if any Community Development Council is certified as a civic organization by the

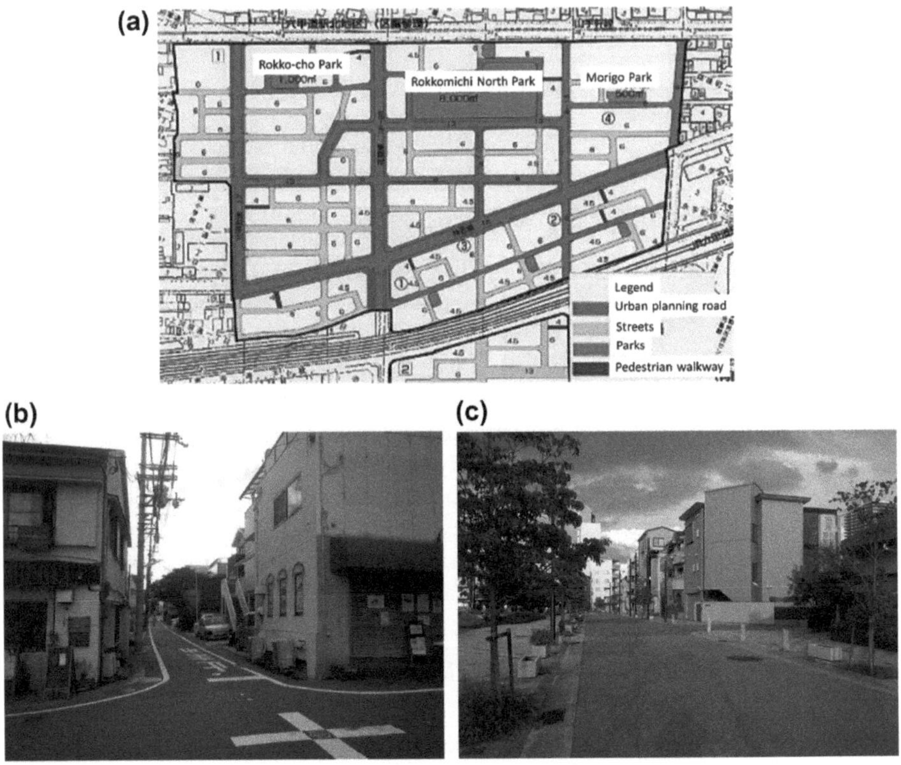

Figure 15.6 (a) Rokkomichi North project plan; (b) the undamaged area near Rokkomichi North district before the earthquake; (c) the reconstructed area in Rokkomichi North district after the earthquake.
Source: Kobe City, "Rokkomichi North Land Readjustment Project, http://www.city.kobe.lg.jp/information/project/urban/adjustment/jl00044.html (2010.9.1).

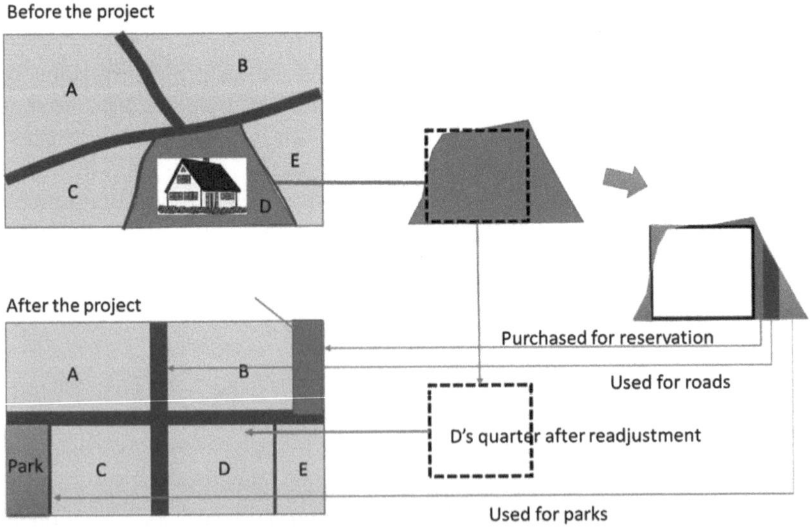

Figure 15.7 Land readjustment process.

municipality, it will be entitled to make a proposal for community development to the mayor. If they reach an agreement, they will enter into a community development agreement. The ordinance has supported partnerships between residents and administration for the community development.

On that basis, including Councils established before the earthquake there were more than one hundred Councils in Kobe after the earthquake. In districts where community development activities were in place even before the earthquake, civic recovery activities immediately started after the earthquake in the hope of creating a stable future for these areas. In case of emergency, their daily activities (not just knowledge, but culture) play a major role.

The process is as follows: First, residents who intend to solve problems related to living environment and improve them establish a Community Development Council. A draft plan developed through the discussion at the Council is made public to the community to ask for opinions of residents. Then, the draft plan is revised and asked for opinions, which is repeated until the proposal is finalized. The proposal is submitted to the mayor. Once the city of Kobe accepts the proposal, the residents work on community development through mutual understanding of their roles and partnerships.

15.6.1 Participation of Residents in the Recovery Process

The second stage is the image, vision and detail planning following the approved plan developed in the first stage by the city. This process is proceeded with the participation of residents centering Community Development Council. To support the participation of residents in the planning process, an on-site consultation office was set up near the project site so that citizens could have discussed with city officials. Then, if the Council asked, consultants and experts were dispatched to the Council to work with (Figure 15.8).

In the beginning of the second stage of two-stage planning process, Kobe City faced protests of residents against the plan developed by the city in the first stage, and could not obtain their cooperation (Nakayama, 2011). It was not easy to convince residents of disaster affected areas to establish the Community Development Council, but as the time passed and debris had demolished, the attitude of residents had became changed to be positive to participate in planning (Nakayama, 2011).

Residents of the planned project site participated in the Council voluntarily to discuss about the image and detail of the reconstruction plans. Experts supported their discussion providing professional knowledge about laws and regulations related to land, planning and developments. The proposal was developed through the discussions in the Council to be submitted as explained earlier.

Community Development Council was a vehicle for residents to negotiate about the future of the community. It provides residents to give their opinions in formulating the development images of the community and detail designs in the proposal.

15.7 Housing Recovery

Approximately 82,000 housing units were damaged in the Hanshin Awaji earthquake. Many of those damaged houses were wooden older houses built under the old building

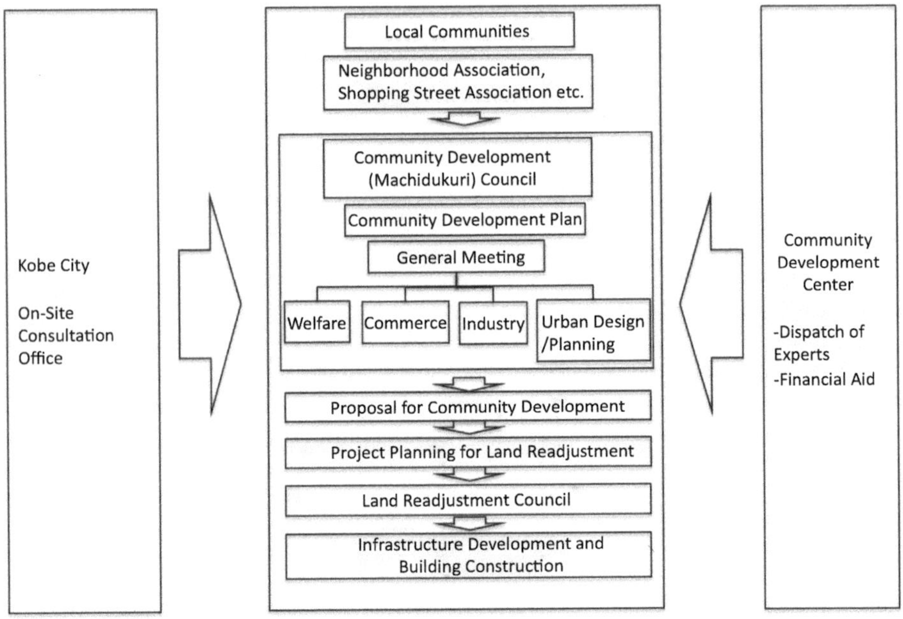

Figure 15.8 Framework for citizen participation in the recovery process.
Source: Revised figure of Kobe City (2011).

code and relatively low earthquake resistant. Moreover, in the densely built-up areas with old small wooden houses, the fire after the earthquake spread to burn down many houses.

After victims lost their houses (completely damaged or half collapsed), they restart living from shelters. Then, temporary housing are prepared for victims a few months after the earthquake at the earliest. It took a few years or more until victims are settled in permanent housing finally. There are three alternatives for permanent housing: rebuilding individual housing, entering public housing, or renting private housing units. The housing recovery process is illustrated in Figure 15.9.

To promote housing recovery and restoration, the Kobe City Emergency Three-Year Plan for Housing Reconstruction was formulated in July 1995. Within a framework for housing supply volume, a three-year target was set at establishing 82,000 housing units, which included at least 10,000 units then under construction, so the actual target number of newly constructed housing units was 72,000 units (Table 15.4). This overall volume of new housing was comparable to the numerical target of 79,000 units for Kobe's Five-Year Plan for Phase 6 (1991–1995).

15.7.1 Public Housing

Disaster restoration public housing for earthquake disaster victims is a type of housing supplied by local municipalities, with national grants, for renting to low-income residents whose houses were damaged by disasters such as earthquakes. There are three types of public housing; (a) constructed by Kobe City or Hyogo Prefecture,

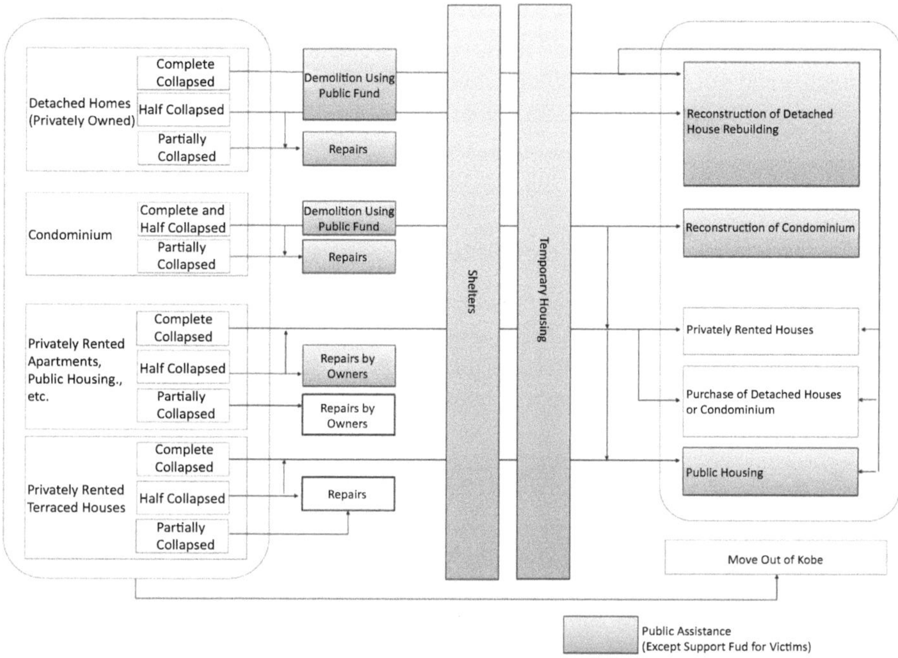

Figure 15.9 Housing reconstruction process.
Source: Revised figures of Kobe City (2011).

Table 15.4 Number of Housing Units to Be Rebuilt in the Three-Year Plan

Public housing	16,000
City of Kobe	10,500
Hyogo Prefecture	5500
Specially designated high-quality rental housing	6900
City of Kobe	5700
Hyogo Prefecture	1200
Redevelopment housing	4000
Public corporation housing	13,500
Urban Development Corporation	10,500
Kobe City Housing Supply Corporation	2000
Hyogo Prefectural Housing Corporation	1000
Private housing units	31,600
(with public grants)	(4600)
Total	**72,000**

Source: Kobe City (2011).

(b) subsidized the room-rent of privately rented housing by the government, or (c) constructed for people who sold lands for the projects. Public housing are all apartment houses and not single-family detached house. Public housing specially designed for elderly people were also built. In registration of tenants, elderly and disabled were yielded precedence. Also, if requested, residents (victims) in the same community were able to move in public housing together.

15.7.2 Housing Supply Supported by Government

As a support for middle income victims who cannot afford to reconstructing own houses, affordable housing were supplied. For example, government supported to construct subsidized rental apartments with high quality. Public corporations also constructed apartments or condominiums to supply housing for middle income households.

15.7.3 Assistance in the Reconstruction of Privately Owned Houses

In constructing their own houses, victims faced various kinds of problems, such as lack of funds, boundaries with next door, lack of knowledge about assistance programs and so on. In particular, lack of funds became serious problem. In addition, quite a few victims had housing loan, thus, if they rent money to reconstruct their houses, they would suffer from double loan. In order to support victims to reconstruct their houses, government provided low interest financing or paying money equivalent to interest of housing loans. To support victims whose lands are too small to reconstruct houses, government recommended to reconstruct a condominium together with other victims and subsidized it. Many of these programs were implemented utilizing recovery funds which were provided by the national government as special grants.

15.7.4 Evaluation of Housing Reconstruction Programs

The most important mission for the city of Kobe and Hyogo prefecture was to supply enough housing as soon as possible to victims who lost their housing to the disaster. There were many advanced policies and programs implemented for housing recovery by Kobe and Hyogo prefecture after the earthquake. Four points are discussed in this section.

First, approximately 40,000 public housing units were supplied eventually, which after 20 years have become a problem for both the city and Hyogo prefecture. The large stock of public housing is a serious issue in terms of management of housing supply, as Murosaki (2013) has pointed out, and Kobe and Hyogo prefecture need to deal with this in the long term.

Second, financial support to supply for housing loan interest was effective for victims who reconstructed their own houses as the interest rate was higher back in twenty years ago than now. This program is applied to victims of Great East Japan Earthquake, however, it is not so effective as the time of Hanshin Awaji Earthquake.

Third, a recovery fund plays a key role in the implementation of various housing recovery programs that were not institutionalized. In the recovery from Niigata Chuetsu earthquake in 2004, the recovery fund was utilized, but it was not with the Great East Japan earthquake, which occurred off the Pacific coast of Tōhoku in 2011. In order to make the best use of the recovery fund system, institutionalization is considered necessary.

Finally, local governments made efforts to place residents from the same regional community before the earthquake as a group in the same public housing complex, so that the community members could be together. This was a reflection of the dissatisfaction caused by placing victims into temporary housing without consideration of the communities they were part of.

15.8 Recovery of Livelihood of Victims

15.8.1 Policies and Programs

In order to focus more on alleviating the impact of disasters on victims' lives, the city of Kobe and Hyogo prefecture developed programs to help victims recover their livelihoods two years after the earthquake. In the plan developed by the city of Kobe, health care, jobs, and housing were important elements of the recovery of livelihoods. Meanwhile, Hyogo prefecture emphasized the importance of community development, mental care, and the elderly. A recovery fund was set up by the city and prefecture, with the assistance of the central government, to provide victims with grants to support their lives and financial assistance to rebuild their houses, as well as implement various other assistance programs.

15.8.2 Support for the Elderly and Disabled

As postdisaster public housing were occupied mainly by elderly people, it was significant to promote their health and independence of living. In order to cater to aging residents of public housing, Hyogo prefecture assigned caseworkers to look after the interests of elderly people and set up spaces to support their independence in the community. It also operated programs to support community development.

15.8.3 Support for Community Activities and Community Development

Quite a few victims were settled in public housings which are located in places apart from where they used to live. Through those projects, migration of population took place, and so the members of the communities changed. Therefore, it was necessary to redevelop or establish community ties. For example, the city of Hyogo placed regional welfare coordinators in each district to support community development by working with careworkers and regional organizations.

15.9 Industrial Recovery

The biggest impact on the economy of Kobe was caused by damages to Kobe harbor, which is one of the important harbors in Japan and fifth biggest port in the world in terms of the amount of freight handling. After the occurrence of earthquake, it decreased dramatically, and has continued to decrease to drop to the 50th in the world. Although the port of Kobe has been recovered physically, it has not recovered economically.

Small scale business in kobe also suffered. There were many small-scale chemical shoes manufacturing factories in the Nagata ward of Kobe. Due to heavy damage from the earthquake coupled with economic depression, chemical shoes manufacturing industry of Kobe was devastated. The retail sector also suffered because of the decrease of population in the affected area. Shopping streets were damaged physically by the earthquake and lost their customers as well.

In order to assist and revitalize the city's small businesses and shopping streets, local governments provided financial support and provided spaces where businesses could reopen. Nowadays, local governments provide financial assistance to either restore small businesses or start up new ones. For the revitalization of shopping streets, support for area development or planning events is also provided.

In order to revitalize the city, Kobe City and Hyogo Prefecture proposed the plans to develop enterprise zones or international economic hubs. Special districts are determined to promote specific industries or business by exemption of taxes. Five districts are determined to be enterprise zones and two districts are determined to be international economic zone to promote functions to support life and culture, information technologies, global activities, large-scale retailers, logistics, medical industry, environmental-related industry, innovative technologies and aerial industry.

15.10 Conclusions

Advanced efforts have been made by the city of Kobe and Hyogo Prefecture in the recovery process after the tragic Hanshin Awaji earthquake. Many programs to assist victims to reconstruct their houses and livelihoods and revitalize business sectors and economic activities were created, and recovery was achieved, with the goal of making cities more resilient. The effectiveness of the innovative programs to promote recovery have been reflected in the recovery that took place after the Niigata Chuetsu earthquake and the Great East Japan Earthquake, two disasters that have occurred since.

One of the features of these recovery programs was recovery funds that enabled Kobe City and Hyogo prefecture to help their residents. Although it is a powerful tool but the framework for the operation needs to be considered to institutionalize it. Another feature of the recovery of Kobe is that funds were granted to victims to help them to recover their livelihoods. Hyogo prefecture passed a Natural Disaster Victims Relief Law, which ensured the provision of grants, but note that it was not applied to victims of the Hanshin Awaji earthquake. It was used to help victims of the Niigata Chuetsu earthquake in 2004, however.

The recovery after the Hanshin Awaji earthquake was designed to make the damaged city of Kobe more resilient and sustainable than it was before the earthquake. Although Kobe has achieved this recovery, it is also burdened with social, economic, and financial issues whose resolution also may hold lessons for the future.

References

Kobe City. (2011). *Comprehensive Strategy for Recovery from the Great Hanshin–Awaji Earthquake.*

Kobe City. (July 30, 2015). Population of 2013. Retrieved from http://www.city.kobe.lg.jp/information/data/statistics/toukei/jinkou/suikeidata/25jinugoki_kij.pdf.pdf.

Kobe City. (September 01, 2010). Rokkomichi North Land Readjustment Project, http://www.city.kobe.lg.jp/information/project/urban/adjustment/j100044.html.

Murosaki, Y., & Editorial Dpt. of Science (1998). *What was the mega earthquake disaster:After the mega earthquake disaster*. Iwanami Publishing.

Murosaki, Y. (2013). Housing recovery and housing issues after the Hanshin Awaji earthquake. *Studies in Disaster Recovery and Revitalization*, *5*, 107–113.

Nakayama, H. (2011). Disaster Reconstruction Projects of Kobe. Gakugei Shppannsha.

Community Resilience Approach for Prioritizing Infrastructure Development in Urban Areas

16

Sameer Deshkar[1], *Vinayak Adane*[2]
[1]Assistant Professor, Department of Architecture and Planning, Visvesvaraya National Institute of Technology, Nagpur, India; [2]Professor, Department of Architecture and Planning, Visvesvaraya National Institute of Technology, Nagpur, India

16.1 Introduction

Cities make up only 2% of the land area of the Earth, but they concentrate more than half of the world's population and the majority of its physical capital in the form of buildings and infrastructures (Potter and Lloyd-Evans, 1998; Pelling, 2003). Sen (1981) described cities as places of refuge from famine where food stores, economic opportunity, and political accountability provide a buffer from environmental change. But today, cities are better described as hotspots of disaster risks. The major cities of the world are largely located in areas prone to one or another type of natural disaster. Along with this, increasing poverty and inequality, failures in governance, high population density, crowded living conditions, and the siting of residential areas close to hazardous industry or in places exposed to a natural hazard are some of the major sources for rising risks in urban regions.

The challenge confronting urban planners is not just that of spatially ordering society to avoid disasters, but to define implementable measures to minimize the effects of natural disturbances. The primary objectives are to reduce the loss of life, to reduce the level of social and economic disruption, to make it possible to resume functions and activities as rapidly as possible after an event, and to achieve total recovery as soon as feasible. Making the urban systems more resilient and hence less vulnerable to the disruptions of natural disasters is emerging to be an efficient and cost-effective idea.

16.2 Urban Infrastructures and Disaster Risk Reduction

Infrastructure systems are an integral part of urban development and are crucial in determining urban form, delivering services to communities, and supporting economic growth. Though the meanings and approaches of "infrastructure" planning vary widely in different domains such as engineering, economy, environment, and society, it is clear that good infrastructure provides basic services and improves quality of life. From the perspective of risk resilience, urban infrastructures are significant because

Urban Disasters and Resilience in Asia.

they enable the enhancement of the community's resilience by contributing to the mitigation and adaptation of adverse scenarios.

The very nature of urbanization makes people's lives more dependent on the infrastructural networks of the city to sustain their quality of life. "Cities are the summation and densest expressions of infrastructure, or more accurately, a set of infrastructures, working sometimes in harmony, sometimes with frustrating discord, to provide us with shelter, contact, energy, water, and means to meet other human needs" (Herman and Ausubel, 1988). As people rely on the continuous agency of infrastructure for almost all of the complicated functions of their lives, urbanites often have few or no real alternatives when the complex infrastructures that sometimes manage to achieve this are removed or disrupted. Infrastructures like telecommunications, electrical power systems, gas and oil storage and transportation, banking and finance, transportation, water supply systems, and emergency services (including medical, police, fire, and rescue) are considered to be critical, because their incapacity or destruction would have a debilitating impact on the defense or economic security, public health, and safety of the population. While disruptions of infrastructure systems are most felt among the inhabitants of urban areas, and especially the cities of the developing world, they ripple through the social and economic fabric to affect systems of interconnected cities, and the larger regions and hinterlands with which cities interact.

While much emphasis has been given to the development of physical infrastructures for urban growth, community resilience demands integration of all domains. The integrative planning approach has so far been adopted for managing the complex physical inter-linkages of urban infrastructures. The success of such integration is questionable in many developing countries, mainly because of the weak financial positions of the governing authorities and the lower standards of existing infrastructures, the revamping of which is a daunting task for the city governments. Hence, this study tries to find an alternative framework that can help the authorities to adopt an appropriate approach for infrastructure planning, focusing on generating community resilience.

16.3 Planning for Infrastructure-Based Risk Reduction: Some Issues

16.3.1 The Age of Infrastructures

Much of the infrastructure in use today was designed and constructed decades if not centuries ago. Many of these infrastructure systems are vulnerable to a variety of anthropogenic or natural disruptions even though their functioning is vital to the creation and maintenance of the quality of life in a region. The rapidly expanding residential zones and aged city cores of many cities have a basic infrastructure that was built without regard to appropriate construction standards. While the local authorities struggle to provide basic amenities to citizens, issues of environmental pollution and hazard management are not accorded priority until matters reach the proportions of a crisis. As commonly observed, catastrophic and chronic hazards are some of the root causes of decaying urban infrastructures, mostly in the cities of the developing world

(Pelling, 2003). There is technical literature on critical infrastructures, particularly urban water, but it remains outside the lens of most social science research.

16.3.2 Network Complexities

The life support networks cover a large number of sectors, including transportation networks, telecommunications and information systems, water systems, electricity distribution systems, sewage and sanitation, etc. However, these infrastructures are closely dependent on one another and sometimes are even coupled with information technology systems for communications and control. Communication and information technologies are integral components of an infrastructure system's design, construction, maintenance, operations, and control, and so a more pervasive coupling of these technologies is inevitable. But this inter-dependence is a major challenge for risk management owing to various complex issues in procuring and provisioning basic urban services. In the complex multitude of these interrelated infrastructures, the failure of one component can provoke, by the domino effect, the failure of other components up to the point of a network-wide failure (Benoît et al., 2003). A cascading failure in an engineered system occurs when a failure in one of the collection of interconnected parts that delivers a service triggers the failure of successive parts. In the case of infrastructure, a cascading failure can be conceptualized as a situation where an infrastructure disruption spreads beyond itself to cause an appreciable impact on other infrastructures, which in turn causes more deleterious effects on still other systems. The extent of the effects and their severity depends on how tightly the infrastructure components are coupled, how compelling the effects are, and whether counter measures are in place or not.

16.3.3 Physical Developments

It is difficult to draw conclusions on the degree of disaster risks that any city area is exposed to with regards to levels of infrastructures, since both the planned and unplanned developments are seen to be contributing to an increasing city-wide environmental risk. In most of the developing and especially in the poorer cities, urban growth takes place primarily through the informal housing sector, which is outside the formal planning control.

Squatting on peripheral land or marginal land in the city creates growing populations without access to basic services and infrastructures. In a double whammy of risk production, many urban squatter settlements are also located close to key components of urban environmental infrastructures, like sources of drinking water, drainage canals, or rivers. Settlement in these locations makes the maintenance of basic services difficult and at the same time introduces new pathways in the production of environmental risk, for example through the potential contamination of drinking water or reduced storm drainage capacity. However, it is important to note that squatters living in these places are merely the vehicles of risk production in the city; the real driving factor is a lack of affordable and appropriately located urban housing and infrastructures. The formation of squatters and their location can be attributed to various issues like poor governance, political nexus, systematic exclusion from the formal economy of the city, and the impossibly high entry barrier into legal land and housing markets. But the fact that poor people live in hazardous sites and are exposed to multiple environmental health risks via poor sanitation and water supply, little or no drainage or solid waste services, air and water pollution, and the recurrent threat of being evicted add to the overall vulnerabilities of urban locations.

16.3.4 Climate Change

Climate change is set to become an increasingly important strategic economic and political concern as it starts to affect the nation's economic growth rates as well as the lives and livelihoods of millions of people. Anticipated climate change impacts on cities are beginning to change how urban infrastructure is designed and located, populations are settled, and the some city services are organized. The predicted changes in climatic and environmental compositions are serious and rapid. If cities witness any of these conditions, the failure of infrastructures are the main cause for them to collapse, glimpses of which were seen during the 2005–2006 floods in Mumbai and Surat in India.

The projected implications of climate change such as drought, river and inland flooding, extreme rainfall events, cyclonic storms, storm surge and coastal flooding, mean and extreme sea-level rise, and environmental health risks are going to exacerbate immense pressure on the infrastructure provisioning in Indian cities as well as in rural regions. Studies focusing on the implications of climate change on human settlements highlight that the economic impact of climate change on infrastructure disruptions are similar or even greater in magnitude, and potentially regionally more concentrated, because the bulk of the infrastructure is typically located in urban areas.

16.4 Resilient Infrastructures and Infrastructure-Based Resilience

The significance of urban infrastructures finds mention in most of the literature pertaining to community resilience. The capacities of urban infrastructures are deterministic for disaster risk resilience and hence, for effective provisioning of services by local authorities, the conventional approaches of supply and augmentation must include demands for the resilience of the local population. There are two important approaches that address infrastructure and community resilience. One focuses on the resilience of infrastructures, which highlights the engineering capacities of the services and the existing infrastructures levels. Under this approach, the physical capacities of the infrastructures are evaluated against natural disasters, and the policy frameworks normally deliberate more on the mitigation options. It is argued that since the failure of infrastructures may lead communities to disasters, the resilience of infrastructures is a must for generating community resilience (Chang, 2009). However, even if infrastructure systems are designed to full capacities against adverse situations, issues emerging from poor governance, weak resource availability, loose development regulations, non-cohesive social structure, etc. may hamper the effectiveness of such systems.

Another approach considers infrastructures as a mode of generating resilience, which emphasizes the inter-linkages between infrastructure provisioning and community capacities for dealing with negative scenarios. This approach highlights more the socio-economic ideas of infrastructures and is based on the premise that, along with the robustness of infrastructure systems, the community's access to them is important in the advent of any disaster. The complementary nature of infrastructures in enhancing the capacities of the local population is the highlight of this approach. The policy frameworks based on this approach deal with the institutional arrangements and economic reforms for generating better access to infrastructures and services.

Attempts to measure resiliency focusing on the urban infrastructure consider the qualities of redundancy, robustness, resourcefulness (capacity to mobilize resources), and rapidity (speed of recovery) (Godshalk, 2003; Bruneau et al., 2003; O'Rourke, 2007). The livelihoods approach focusing on the quantity and quality of assets also includes access to services, infrastructure, communications, credit markets, and emergency relief (Twigg, 2007; World Bank, 2003). Measures of the built environment and infrastructure provide an overall assessment of the amount of public and private property that could be damaged by disasters and the likely economic losses. This also provides an understanding of the community response capacity as well as the status of critical infrastructures.

16.5 Indian Perspective on Urban Disaster Risks

As per the World Urbanisation Prospect (UN, 2014), the rise in India's urban population has been phenomenal and is expected to be a major contributor to the world's urban increment during 2025–2050. The contribution of the urban sector in the nation's economic growth has been significant, as the development of cities has been regarded as a key solution for socio-economic progression. However, to support this fast growing urban population, a large scale enhancement of the urban infrastructure has become necessary.

As India becomes more globalized and urban, it faces a steady rise in the occurrence of catastrophes. It is ranked one of the more vulnerable and risk-prone countries in the world (IFRC, 2004) with recurrent extreme events such as floods, droughts, cyclones, earthquakes, and landslides (GoI/NDMD, 2004). In spite of its economic growth, the estimates of disaster-related deaths in India have been nearly constant, from 24% during 1991–2000 (GoI/NDMD, 2002) to 22% during 2002–2011 (IFRC, 2012) of all disaster-related deaths in Asia. In year 2001 alone this share was as high as 73.1%. The intensity of the impact of natural disasters in recent years has exceeded even the worst in history, and locations that were considered less vulnerable have proven to be quite fragile against calamities.

With over 13% of the total urban population living below poverty line (GoI/RBI, 2015) and an estimated urban housing shortage of 18.78 million dwelling units

(GoI/MHUPA, 2012), the cities have become the most vulnerable locations. The increasing concentration of the population as well as economic assets in cities and megalopolis that are prone to disasters not only increases risk exposure but also results in huge fiscal requirements for relief and reconstruction activities in case of calamity. The gross estimates of the losses caused by direct natural disasters equate to up to 2% of India's GDP and up to 12% of the government's total revenues (World Bank, 2003).

The rate of provision of safer and developed lands and other life support systems lag far behind the pace of urban population growth. Due to poor land regulations and building by-laws, as well as inadequate and inequitably distributed infrastructures, urban planning has to deal with large, negative externalities caused by haphazard and unconstrained land use (Morris et al., 2009). A large portion of a typical Indian city's building stock is aged, dilapidated, and partially engineered, implying that they do not meet contemporary standards of building safety. Technical measures to strengthen and retrofit these buildings are well known, but have never been implemented at the city scale. This would require considerable institutional and financial innovation and a considered set of incentives and regulatory mechanisms to make it possible.

Also, external forces such as global warming and climate change are being held responsible for accentuating the vulnerabilities of urban locations to natural disasters (UNEP, 2002). A substantial increase in extreme precipitation is expected over a large area of the west coast and central India. The results of climate change, such as heavy rainfall, heat waves, floods, storms, fires, droughts, etc., are expected to bring about a severe depletion of resources and increase in disaster risks in the cities (IPCC, 2007). Many Indian cities could be forced to adapt by altering their exploitative relationship with the rural regions due to the climate change-induced disruptions and the reduction in net primary productivity by human systems. Climate change may force the increase of rural-urban migration by catalyzing the ongoing agrarian crisis in rural India (Revi, 2008), driven by increases in extreme events, greater monsoon variability, endemic drought and flooding, and resource conflict.

16.6 Status of Urban Infrastructures in India: An Overview

The cities and towns of India are visibly deficient in the quality of services they provide, even to the existing population (GoI/MUD, 2011). Considering that the Indian economy is now one of the fastest growing economies in the world, and standards are rising, current service levels are too low relative to the needs of urban households. They are also low relative to what will be required to sustain the economic productivity of cities and towns. Deteriorating infrastructures, weak municipal institutions, and poor delivery systems have constrained the urban economy and its ability to generate employment and incomes. With the rising population level, the corresponding demands for supporting infrastructure facilities are multiplying quickly. It is therefore extremely difficult to match the supply of basic urban services to meet the rising demands of a rapidly increasing urban population. The deficient urban basic infrastructure situation often leads to unhealthy and unhygienic living conditions and also adversely affects economic productivity and efficiency. As a result of this increasing gap between the demand and supply, cities today face a severe shortage of water supply, sewerage, developed lands and roads, housing, and other facilities. Much of the infrastructure in use today was designed and constructed decades if not centuries ago. Many of these infrastructure systems are vulnerable to a variety of anthropogenic or natural disruptions even though their functioning is vital to the creation and maintenance of quality of life in a region.

In India, although 82% of the urban population is reported to have access to safe drinking water, there are severe deficiencies with regard to the quantity of water available to urban residents. In a City Sanitation Study (2010) conducted by the Ministry of Urban Development, none of the 423 cities was found to be "healthy" and "clean." Close to 190 cities in the study were rated to be in a state of emergency with respect to public health and the environment. A large portion of a typical Indian city's building stock is aged, dilapidated, and partially engineered, implying that they do not meet contemporary standards of building safety. Technical measures to strengthen and retrofit these buildings are well known, but have never been implemented at a city scale. This would require considerable institutional and financial innovation and a considered set of incentives and regulatory mechanisms to make it possible.

16.7 The Evaluation Framework for Infrastructure-Based Community Resilience (IBCR)

Accessibility and availability of urban infrastructures services are considered to have a paramount significance in the resiliency of a place. The ability of an urban system and community to bounce back after a disaster primarily depends on the physical response and recovery mechanisms for which the urban infrastructures become a backbone. The time proximity to life support services and the availability of basic commodities for survival during adverse hours influence the community resiliency by supporting the overall preparedness of the local community (Deshkar et al., 2011). Access to safety and services are the basic components for the response to adverse situations, while the availability of services reveals gaps in the augmentation of basic services and the risks arising from it. Also, restricted access to amenities and basic services in a way induces vulnerability but at the same time forces communities to develop assets against adverse scenarios of non-availability. These community-level measures for infrastructure mitigation need to be considered when estimating socio-economic resilience in developing countries.

The study considers water-related adversities viz. water scarcity and flooding in urban areas. Conventionally, the blue (water supply) and gray (transportation, sewage, and sanitation) infrastructures are given priority in urban management. Improving their capacities are seen as imperatives for dealing with any negative scenario mainly related to water or for improving the urban quality of life. However, there is also increasing recognition of the green (open spaces) infrastructures in mitigating the additional pressures on these "hard" infrastructures. Therefore, the study takes into account the blue, gray, and green infrastructures together with key land use components for risk resiliency.

An Infrastructure Based Community Resilience Index (IBCRI) is proposed and defined as the ratio of the sum of the vectors of components of accessibility to life support services, availability of basic infrastructures, and community measures to the total vulnerability of a place (Deshkar et al., 2011).

16.7.1 Indicators of IBCR

The indicators of various components (Table 16.1) considered in this study represent the status of the basic infrastructures and life support services essential for the response and recovery of urban communities. The very basic parameter defining the resiliency of a community or a place is in terms of the time to respond and to bounce back after a calamity. This research has considered some of the land use components to which access "in time" is more crucial to negate the implications of a disaster and thereby contributes to gaining resiliency.

The component of accessibility in present research essentially addresses the average time distance to reach or avail a critical service in the advent of a calamity. The time distance could be influenced by several factors including the land use distribution, density pattern, road network, road surface characteristics, congestion levels, and mode(s) of transportation. The measure of time distance is considered in this study as a unit indicating the status of land use compositions and facility planning and thereby also directing the significance of physical developments in generating resiliency.

Table 16.1 Indicators of Infrastructure-Based Community Resilience

Component	Detail of Component	Indicator
AC: Accessibility	Rescue	Time from emergency services
	Transit place	Time to open space
	Health care	Time to hospital
	Food supply	Time to retail market
AM: Amenity	Water supply	Duration of municipal water supply per day
	Electric supply	Duration of metered electric supply per day
	Garbage disposal	Frequency of garbage collection
	Sewerage and drainage	Time for water runoff from property
CM: Community Measures	Mobility	Time for evacuation
	Communication network	Number of hours of network availability
	Water storage	Provision for number of days
	Alternative electric supply	Number of hours of supply availability
V: Vulnerability	Flood	Frequency of flood
	Water scarcity	Frequency of water scarcity
	Physical safety	Distance from hazardous land use
	Fire	Frequency of fire hazards

Source: Deshkar et al., 2011

Proper siting of public infrastructures and critical facilities as well as regulating land use and development are some of the basic tools for land use management that may be adopted to enhance the "time" factor for achieving resiliency. In selecting the indicators, aspects such as rescue, transit shelter, health care, and food supply, which are critical in responding to disasters, are considered; an indicator is selected that corresponds to each stage. The performance of critical facilities of a city (hospitals, fire stations, police stations) is mandatory for rapid response and rescue operations. Inadequate capacities of these services or longer time duration to avail these services may have a very adverse impact on the sustenance of the affected population and their resiliency.

Vulnerability due to lack of safe open spaces is still a major issue in cities, with too little effort put into mitigation. Here, open space serves a dual role: as a place for transit shelter as well as an element of green infrastructure. In the present study, the component of vulnerability considers natural vulnerabilities as well as those emerging from the urbanization processes together. Gas and petroleum storage facilities pose another threat to a city when damaged. Lack of supply is one issue; pollution by oil spills, fire in refineries or storage tanks, and fire ignited at service line connections to buildings are others. A key parameter to vulnerability is usually the age of construction and the proximity of such services to the populated localities.

As the selected components and variables are mutually exclusive and yet interdependent, their coefficients are estimated using a logistic regression analysis. The vector of a component is measured as the sum of the product of the regression coefficient of each variable and the actual value of the variable. The present study emphasizes a community's evaluation of urban infrastructures for measuring the resilience and therefore the method of choice experiment is adopted. It is fundamentally a stated preference method and is based on a simulated future scenario under which people make decisions for selected attributes. The values obtained from the regression analysis are analyzed statistically and spatially. When making interpretations, the distribution of selected components of resiliency is correlated with other parameters of spatial and socio-cultural aspects, which are observed to be influencing physical developments in the city of Nagpur. For converting the preferences into policy actions and prioritization of infrastructure provisions, a technique of quality function deployment viz. the Kano model is employed.

16.8 The Case of Nagpur, India

Nagpur, popularly known as the Orange City, has a history of 300 years and is situated almost at the geographical center of India (see Figure 16.1a). It has an administrative significance as being the second capital city of the Maharashtra state. The city of Nagpur is the third largest urban center in the Maharashtra state, next to Greater Mumbai and Pune, with a population of 2.05, which is 5% of the state's urban population. The city hosts around 40% of its population in slums, with squatters spread in pockets.

16.8.1 Natural Profile

Climatically, Nagpur is hot, dry, and tropical, which is typical of the Central Deccan Plateau, and it is characterized by an extremely hot summer and a cold winter. The mean annual temperature ranges from 25 ° to 27 °Celsius, rising to 48 °C in the summer and lowering to 6 °C to 8 °C in the winter. The average rainfall varies from 1000 to 1300 mm and occurs mainly in the monsoon period. Nagpur falls into three distinct

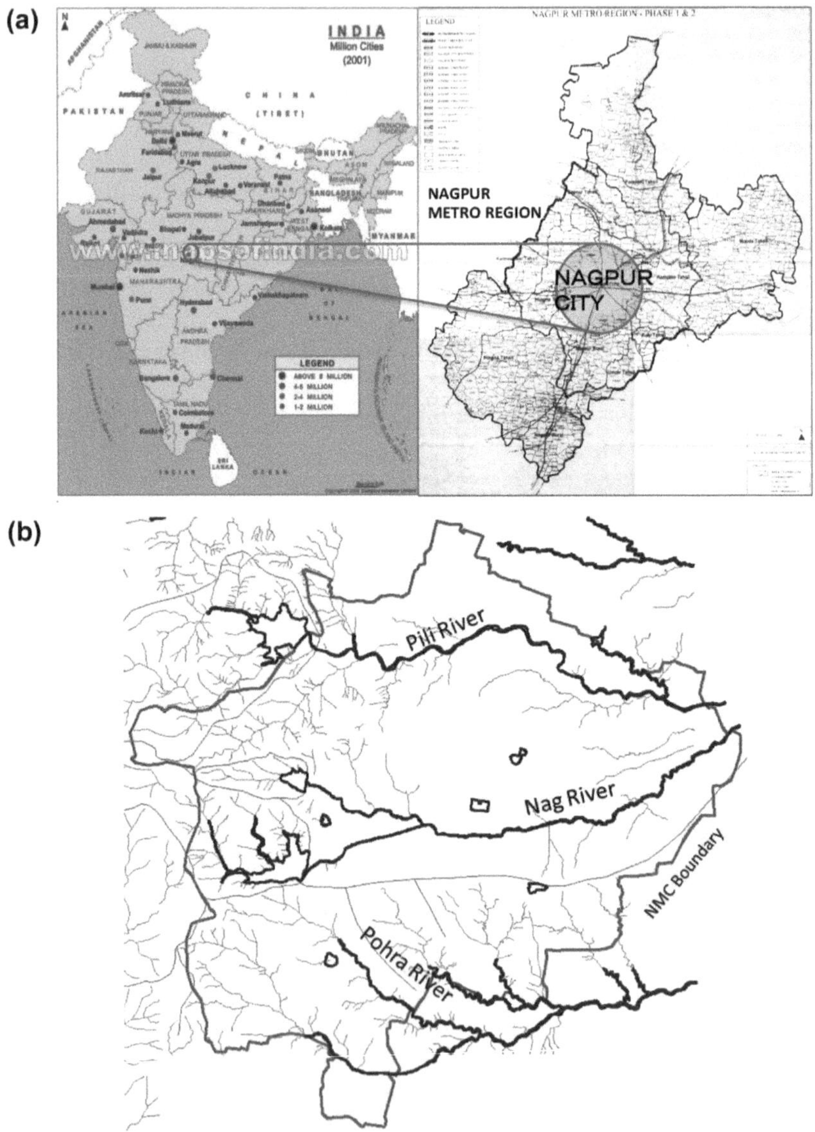

Figure 16.1 (a) Location of Nagpur and (b) natural drainage pattern of Nagpur (prepared by the author based on the Survey of India Topographic Sheet for Nagpur City).

drainage basins of the Pili River, Nag River, and Pohra River, which traverse the city in a west-to-east direction (see Figure 16.1b). The Pili and Nag rivers originate from the reservoirs situated on the western brim of the city. There are about eleven lakes situated in and around Nagpur.

16.8.2 Water Infrastructure Profile

A review of the infrastructure profile of the case study city revealed that services pertaining to water assume more concern from a vulnerability point of view. As far as the water supply is concerned, currently the city meets demand just sufficiently, but the low competent supply system creates a deficit situation during extreme summers and results in a grim scarcity, especially in the areas that are not served by piped water. About 85% of the city population receives the piped water supply while the rest depends on hand posts or tankers for meeting water requirement (NMC, City Sanitation Plan, 2006). There is also a discrepancy in the per capita distribution of water, which varies between 70 lpcd to 200 lpcd. Supply is also dissatisfactory in terms of the number of hours of water supply and water quality. The piped water facility of NMC supplies water to approximately 66% of the slum areas but the quality and the regularity is unsatisfactory. Although 72% of the slums are connected to the city's sewer system, most of the lines are defunct or chocked. Additionally, most of the existing sewer lines in the slums also act as storm water drains. The resulting flow of human waste through open drains is both a health and an environmental hazard. There are instances where water pipelines are in close proximity to sewer lines, increasing the probability of epidemics.

The city, with the tributaries of rivers flowing through its boundaries, is expected to have better drainage infrastructures. However, as evident from Figure 16.2, the haphazard developments and the rise of slums along these tributaries have reduced

Figure 16.2 Haphazard developments across Nag River, Nagpur.
Source: Google Earth, 2011

the capacity of the drains, and as a result, areas adjoining these river channels get flooded during the monsoon season. The unhygienic conditions due to direct dumping of waste into the river stream from the adjoining developments raises the risk of epidemics. Many other parts of the city have inefficient drains but they do not create a flood situation. However, there is water stagnation in low lying area or areas where the drains are not maintained and get clogged with debris from the roads. Under the extreme scenarios created by climate change, this could be a major concern because the existing capacities of the drains would prove to be far inadequate.

16.8.3 Disaster Risk Profile

The city of Nagpur has been very susceptible to water-related adversities in the form of flash floods and severe water scarcity. There is a probability of drought in every 5.5 years. In recent times, with sudden seasonal variations, the occurrence and impact of these calamities have been intensive in nature. A review shows that the city of Nagpur has a high vulnerability to major disasters such as earthquake, floods, and flash floods, and it has a medium level of vulnerability to calamities like fires and epidemics, especially water-borne diseases (Krishnan, 1999). At least 127 settlements in the city are sensitive to floods, with around 42 of them identified as most critical (NMC, Flood Control and Action Plan, 2000). These locations are mostly in the low lying areas in the river basins and lake catchments, indicating the lack of risk-based land use zoning and also the weak implementation of the development regulations. In some places, the improper disposal of solid waste is causing the drains to clog, leading to local floods; at many locations, particularly in the slums, there are no or inadequate storm water drainage facilities, which results in floods in monsoons. It is also interesting to note that a few localities witness floods due to faulty engineering methods and due to road construction. The incidences of fire also are drastically increasing, as are the losses reported. Water scarcity during the monsoon is mainly on account of the poor quality of the water, while during the summers, the inadequacy of water availability causes major concerns.

16.8.4 Status of Disaster Risk Management

Although the concerns for vulnerability reduction has resulted in a National Disaster Management Framework that provides guidelines at various sectoral and administrative levels, their integration with the local development plan preparation has hardly been documented. In the case of Nagpur, the much emphasized paradigm shift from response and recovery to preparedness in National Policies on Disaster Management has not been turned into any local action plans. The Nagpur Municipal Corporation, like many other local authorities in India, has been preparing seasonal contingency plans to deal with the adverse situations of floods during monsoons and water scarcity during summers. It has only lead to a compilation of locations prone to flood hazards, vulnerable populations, and required financial outlays for response and recovery. The mitigation actions include dewatering of water logging areas in heavy rains, removal and lifting of fallen trees on streets and clearing of streets, and making available all equipment to handle the trauma/disaster, mainly floods (NMC, City Sanitation Plan, 2006). However, no such data is available for water scarcity, which in recent times has become a very serious concern, especially in the newly developed outskirt areas with a ground water deficit. Nagpur falls in the seismic zone II with lower proneness to earthquakes. However, addressing the significant proportion of aging buildings (mainly in the city center), failing infrastructures, and vulnerable construction practices with un-engineered, unsafe practices, the city was included in the Urban Earthquake Vulnerability Reduction Programme (UEVRP), a joint initiative of Government of India and UNDP (2002–2007). But the programme initiatives in the city were meagre token efforts, addressing only public awareness generation and capacity building.

16.9 Methods for Data Collection, Analysis, and Interpretations

The components of accessibility, amenities, and the community measures comprise several measurables as well as non-measurable variables such as the physical configuration of urban areas, the age of development, land use patterns, service networks, distance proximity to critical services, duration of availability of services, commodity prices, socio-economic aspects, demographic features, levels of awareness, perception of risks, etc. The actual values of variables can be determined based on the recorded data such as strategic development plans of local authorities or primary observations. In Indian cities, however, there is generally a paucity and inconsistency in the recorded data available with concerned agencies, particularly related to infrastructures (GoI/MoUD, 2011). This is attributed to various factors such as a time lag between planning and implementation of development projects, data recording, etc. This study, therefore, considers the primary observations and data collected through interviews of the local population as the basis for determining the actual values of variables. The coefficients of these variables are calculated through a stated preference survey or choice experiment and then using generalized linear model in R software.

Since the survey required respondents to make choices carefully and the number of situations in which to make such choices was about 28, there was a possibility of the respondent answering casually in order to finish the survey quickly. Kroes et al. (1988) suggest that a respondent should be presented with no more than 16 choices and Pearmain et al. (1991) reinforce this by suggesting that more than this number of choices would result in respondent fatigue. Further, it is observed that the combination of text and pictorial explanation can enable a fairer choice selection by the respondent (Abley, 2000). In order to make the process of answering questions easier, a combination of text and visual approach was adopted in this research. The pictorial interpretations of disastrous situations were also adopted to help the respondent visualize the situation in consideration better and hence eliminate the gaps between the risk perceptions.

The vectors of accessibility, amenity, and community measures were normalized to evaluate the preparedness index of the local community. The interpretations of logistic regression coefficients are normally done by taking the odds ratio, which is the ratio of probability of occurrence of an event to the probability of its non-occurrence and expressed as the exponentiated value of the coefficient. The spatial distribution of coefficients is analyzed using Arc-GIS.

16.10 Decoding the Community Preferences

16.10.1 Statistical Analysis

The estimated odds ratio for selected variables of community resilience is represented in Table 16.2. It can be inferred that the change in predictor of time distance to retail market by 1 unit of time interval and keeping all other predictors constant will change the odds (probability of enhancement or reduction) of levels of accessibility by a factor of 0.784. An odds ratio of less than 1 indicates a negative relationship between the predictors and the response. It means that as the time distance increases, the odds of enhancing accessibility at a particular location reduce and vice versa. In the case of emergency services, since the odds ratio is almost near to 1, there is a near probability of 0.5, or a 50:50 chance that it will have an equivalent reciprocating effect on the odds of levels of accessibility. Thus, based on the odds ratios, it can be inferred that in the case study city of Nagpur, the residents have relatively more concern and thus a preferred choice for the proximity to retail market, followed by hospitals and open space.

Similarly, it is observed that people have a stronger preference for water supply and garbage collection. In the case of community measures, the water storage shows a very strong positive relationship with the odds of a community's preparedness. It indicates that as the water storage sufficiency increases by one day, the odds of enhancing community's preparedness almost triples (increases by a factor of 2.83). Similarly, an increase in the availability of an alternative source of electric power by one hour increases the odds of preparedness by 13%. It can be seen that the variables of flood, water scarcity, and hazardous land use indicate a negative relationship whereas a fire hazard shows a positive

Table 16.2 Odds Ratio of Logit Coefficients for Selected Variables of Infrastructure-Based Community Resilience

Components	Variables			
Accessibility	Retail Market (0.784)	Hospital (0.819)	Open Space (0.893)	Emergency Services (0.942)
Amenity	Water Supply (1.069)	Electric Supply (0.988)	Garbage (1.168)	Drainage (0.99)
Community Measures	Mobility (0.935)	Communication (0.605)	Water Storage (2.835)	Power Backup (1.127)
Vulnerability	Flood (0.844)	Water Scarcity (0.743)	Haz. Land use (0.634)	Fire (1.197)

Note: Number in parenthesis indicates odds ratio value.

relationship with the odds of vulnerability. In terms of values, it can be interpreted that an increase in the return period of natural calamities of flood and water scarcity by one unit of year, with other parameters remaining constant, reduces the odds of an increase in vulnerability by a factor of 0.845 and 0.74, respectively. Further, as the distance from hazardous land use increases by 1 kilometer, the odds of an increase in vulnerability reduce by 0.63. In the case of a fire hazard, since the odds ratio is more than 1, it indicates a positive relationship. Hence, it is interpreted that a rise in frequency of fire hazards increases the vulnerability of the local population by a factor of 1.19, or 20%.

More interesting observations about the variables of accessibility are made based on the choice preferences of the population of lower income and upper middle-income groups. The choice preferences were sorted based on the income levels recorded in the first part of the survey format. These preferences were analyzed following the above explained procedure to obtain the odds ratios for the variables of accessibility for respective income levels and are as shown in Table 16.3. It can be seen that the effect of change in predictors of time distance to a hospital and open space on the odds of levels of accessibility varies by a considerable level between low income and upper middle-income groups of a population. As observed, the effect on the odds of accessibility from an increase in the time distance to the hospital is more negative for the upper middle-income group than the same for lower income group. Similarly, an increase in time distance to open spaces much lowers the odds of improving accessibility for the upper middle-income group compared to the same for the population belonging to lower income group.

Similiarly, an analysis of the change in the odds ratio of amenities for the slum population and those living in developed areas is done based on values, as shown in Table 16.4. It can be observed that the nature of the relationship between variables

Table 16.3 Odds Ratios for Lower and Upper Middle Income Group Population

Variables	Lower	Upper Middle
Retail market	0.818452	0.8220641
Hospital	0.885574	0.8173403
Open space	0.918434	0.888371
Emergency services	0.945695	0.9224908

Table 16.4 Odds Ratios for Slums and Developed Areas

Variables	Slum	Developed Area
Water supply	1.0505615	1.1647782
Electric supply	0.9914546	0.9668207
Garbage	1.1506968	1.2226568
Drainage	0.9905227	0.9913274

and the odds of levels of amenity are similar across both the categories. However, a closer look reveals that one unit increase in the duration of the availability of water enhances the odds of improving amenities in slums by around 5% as against 16% in developed areas.

This can be interpreted in two ways. First, that the variable of municipal water supply is more preferred by the population living in developed areas as against those in slums. Second, that an increase by one unit is not sufficient enough to bring about a significant change in the odds of improving amenities in slums. Of the two possible interpretations, the second seems to be more valid because the primary observations in the case study city show that the municipal water supply in slums is critical. A similar interpretation can be done for the variable of garbage disposal efficiency in slums versus developed areas.

16.10.2 Spatial Analysis

It is difficult to generalize geographically which parts of the city (the fringe, core or others) are more critical in terms of levels of accessibility because variations are observed in the spatial analysis (see Figure 16.3a). However, it is observed that the locations in closer proximity to services such as a fire station and a hospital show better levels of accessibility compared to distantly located settlements. Therefore, accessibility to emergency services, measured in terms of time proximity, is seemingly influenced by the physical proximity of such services. This poses a question for the efficiency or the role of other supporting infrastructures such as road networks and transportation systems, which can significantly minimize the time factor for accessing these services on the peripheral regions of the city.

The availability of municipal services for water supply, sanitation, electricity, and solid waste indicates lower levels in the fringe as well as the city core areas (see Figure 16.3b). It is noted that there is a considerable portion of the case study city that is not served by infrastructure networks and is mainly located on the east and southeast part of the city. The spatial analysis of levels of the community measures indicates that parts of the city periphery and the old city have lower levels of preparedness (see Figure 16.3c).

The study also makes an interesting observation that in areas with fewer water supplies, the community mitigates the shortage of supply by storing water on own premises. However, at many places this storage is not sufficient enough to avoid the scarcity in summers. In localities that receive a fairly regular water supply, people have lesser contingency water storage capacities. Such areas are susceptible to scarcities caused by technological failures at the city level water treatment and supply point; these failures can take nearly a week to be repaired. The situation further gets aggravated if the failure occurs in the summer.

In case of levels of vulnerabilities, it is observed that they exist more in a pocketed form but have a gradual change in intensity with the adjoining areas (see Figure 16.3d). The analysis of various variables considered in this research on the city of Nagpur suggests that different communities and parts of the same urban area are subjected to different vulnerabilities and intensities. This is evident from the fact that the old

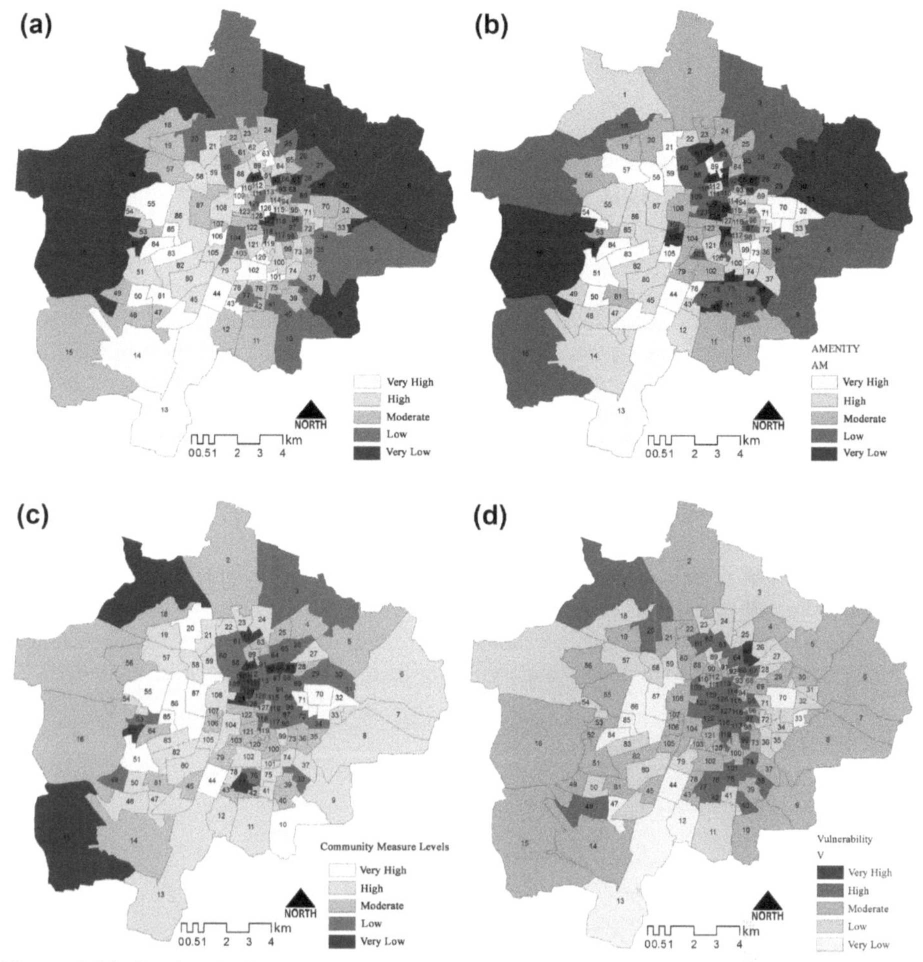

Figure 16.3 Spatial distribution of coefficients of selected components of Infrastructure-Based Community Resiliency; (a) measure of "accessibility"; (b) measure of "amenity"; (c) measure of "community measures"; (d) measure of "vulnerability".

city areas have serious issues of water scarcity while the newer developments on the western side have vulnerabilities owing to complex and mixed developments. While settlements along the river courses are comparatively more vulnerable to floods, at the same time their position with regards to water scarcity is not very positive. A reason for this could be the high unsuitability of river water for any utilization due to serious levels of pollution.

Overall, it is seen that, in the case study city, the high resiliency is achieved only in selected wards, which are either falling under the administrative zone of the city or are aligned with the arterial national highway in the southern end. It is inferred that the resiliency of the community at a place is influenced by its socio-economic profile, characteristics of physical developments such as slums, population density, and the land tenure.

16.11 Prioritizing Infrastructure Development for Gaining Resiliency

Because they are the end beneficiaries of infrastructure provisions and at the same time the first respondents in the case of an adversity, the local population's evaluation of the infrastructure systems are considered to be of paramount significance in this research. Considering this community-centric approach, framing the alternative policy framework for prioritization also needs to be oriented so as to accommodate these evaluations. Secondly, a need has been identified not just to prioritise various components of services but also their locations for implementations. For the purpose of prioritizing infrastructure components, this study adopted the approach of the Kano model, which is one of the scholarly approaches that takes into account the evaluations of end beneficiaries or consumers. It is popularly used for strategizing the product or business developments and in recent times it has been applied to various other fields that require prioritization of policies based on the user's conditions. It defines the attributes of a system as "must be," "one-dimensional," and "delighters," depending upon the predicted impacts on levels of satisfaction of the consumers.

In order to deliberate on the prioritization of selected components for gaining resiliency, a comparative analysis of response values is done. The average response values of accessibility, amenity, community measures, and vulnerability for each category of level of resiliency were plotted to understand their significance for resiliency (see Figure 16.4).

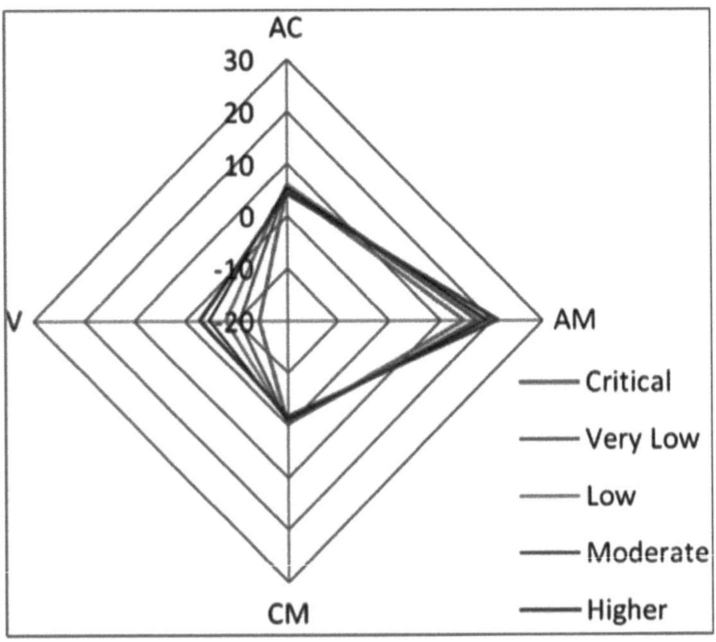

Figure 16.4 Response values of selected components of infrastructure-based community resilience.

It can be prominently seen that as the measure of vulnerability (V) reduces, the resiliency levels increase. A significant variation in the values of vulnerability also hints at the inconsistent profile of physical developments in the city. It can be seen from Table 16.5 that the standard deviation values are higher for the component of vulnerability. In the field of risk management, aspects with significant variations or higher standard deviations are considered to be of a volatile nature and are a first priority for negating. In the language of the Kano model, this is the "must be" attribute and thus vulnerability reduction should be the first priority of the local authorities of Nagpur city.

The component of amenity (AM) also shares a comparatively higher level of standard deviation. The variations in the values of AM indicate a discrepant distribution of services in the city. It can be interpreted that people prefer to have basic levels of availability of infrastructure and anything above it is appreciated but anything below the basic levels is not at all preferred. The component of amenity therefore is the "one-dimensional" attribute in Kano's model, which acts as the element of competition for people to decide their preference between cities and parts of cities.

The ambiguity in the values of accessibility can be interpreted in another way as a mixed and unclear preference amongst the local population for having better accessibility. It is well established and understood that greater levels of accessibility contribute to a higher resiliency of the local population. For people to appreciate the physical environments and at the same time have better levels of accessibility, innovations in generating newer models of mixed land use development are important. If properly planned and implemented, such developments can be the "delighters" for urban population. Similarly, community measures for building up preparedness would need to innovate through raising community awareness, capacity building, innovation in resource conservation and management, etc. These are the future directions and act as the "attractive attribute" for the community. The framework for infrastructure prioritization for community resilience can thus be represented as Figure 16.5.

At the spatial level, the prioritization of the measures for basic need, development needs, and delighters can be done using the distribution analysis of levels of components, as explained in Figure 16.6. The estimates of the community resilience are used in this research to indicate the levels of criticality of life support services and key infrastructures for disaster resiliency in different parts of the case study city. The analysis based on the Kano model suggests that a higher priority for vulnerability reduction is necessary in the case study city for achieving disaster resiliency. A detailed analysis carried out in the case study city of Nagpur indicated that the higher levels

Table 16.5 Standard Deviation Values for Resiliency Components

Components	Mean	Std. Dev.
Accessibility	5.4957	1.6203918
Amenity	17.9048	3.0323394
Community measures	−0.70266	0.7069656
Vulnerability	−8.26918	3.8498265

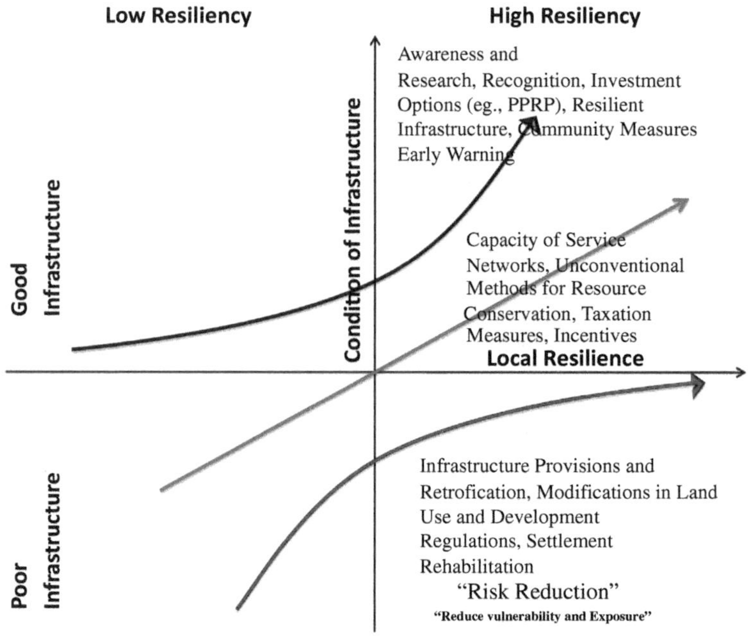

Figure 16.5 Framework for prioritizing action for resilience (based on Kano model for QFD).

of vulnerability are associated with the physical characteristics of the settlement, the lower availability of basic infrastructures, and a poor accessibility to life support services. Therefore, areas with critical levels of resiliency should focus on the provisioning of basic needs and critical infrastructures as a priority measure. However, the option of relocation of settlements to remove the vulnerabilities needs more specific analysis, which is not covered in the scope of this study.

A comaparison of estimated levels of resilience and those of accessibility and availability of basic infrastructures over the case study city indicate that a moderate level of resilience corresponds to an inadequacy of infrastructures and services. As commonly understood, the inadequacy of infrastructures exerts more pressure on the existing systems and therefor makes them prone to failure. At the same time, it is observed that a lower level of infrastructure availability also reduces the capacities of the local community to take measures against the criticalities arising from such non-availability. Therefore, in city areas with moderate levels of resiliency, measures should be initiated to build the capacities of existing infrastructures as well as the local communities so as to enhance their resiliency against any adverse scenario. In the same way, areas with higher levels of resiliency, which are assumed to have a better level of infrastructures, should look to further develop their resiliency. This situation also offers an opportunity for the city developers to adopt innovative approaches against the conventional developmental trends, which only pressurizes the available infrastructure systems and services in the long term by increasing densities and growing in a haphazard way, and thereby generating newer vulnerabilities in urban systems. The innovations or alternate paths can become a benchmark for all other city areas to adopt for gaining resiliency against any disaster.

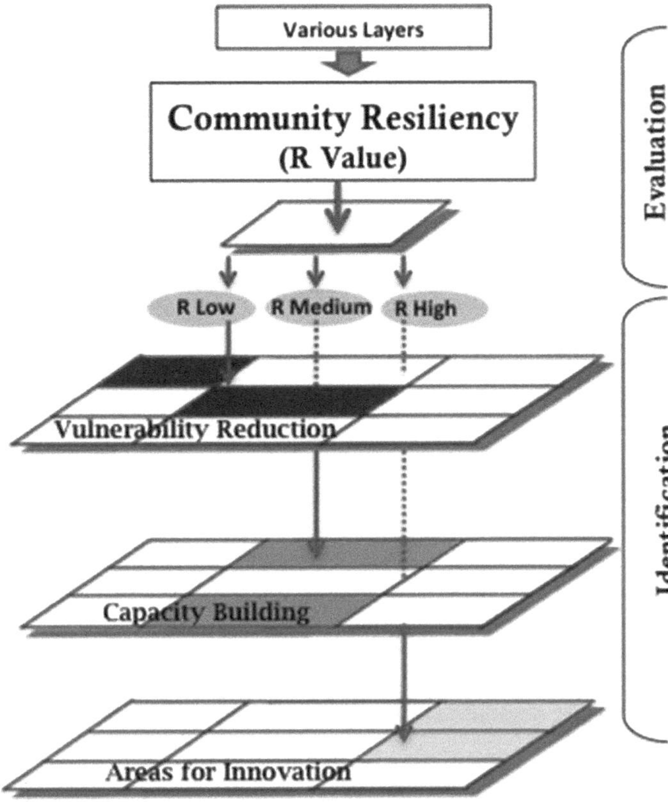

Figure 16.6 Framework for ranking priority regions.

16.12 Discussion

Deliberations on the resiliency of cities and communities under various adverse conditions are a recent phenomenon, and the approaches for addressing it through development policies are being explored. This chapter puts forth an attempt to orient the decision making for urban infrastructures from a sociological perspective. The analysis based on a community's evaluation brought out the fact that vulnerable communities in cities not only exist in the slums but that citizens at large are exposed to various levels of vulnerabilities owing to the conditions of urban infrastructures and the physical development.

However, the study noted a limitated literature available on community resilience with a focus on urban infrastructures, and those available mostly provide theoretical construes. Studies on resilient infrastructures focus on improving engineering capacities and levels for resiliency. Some other estimation methods on resiliency available in literature, such as CDRI, consider the positions of local urban governments and do not accommodate the perception of resiliency of local community. The present study, therefore, is unique in a way as it tries to integrate the evaluations of the local community about the access and availability of basic life support services and infrastructures for evaluating levels of resiliency.

Risks are generated within cities and are reflected through rising urban population, accumulation of economic and physical assets, highly pressurized infrastructures, developments in hazardous areas, and rising urban poor. It is commonly observed through various case studies that the failure of conventional tools of urban planning such as land use distributions and development control regulations are giving rise to newer urban risks. The proposed framework in this research for prioritizing areas and actions for achieving resiliency can potentially give directions to use these conventional tools effectively and generate resiliency in urban communities.

The cities in India are undergoing tremendous transformations owing to the rapid social, economical, and environmental transitions. Approaches for managing urban growth are yet dominated by the top-to-bottom approach, the success of which is being questioned due to rising urban issues. While efforts to develop a bottom-up approach are also being taken in managing urban environments, a similar approach is absent in managing urban infrastructures. The alternative framework for achieving disaster risk resiliency developed in this research using community's evaluation demonstrates the possibility of evolving such a bottom-up approach in planning and managing local resources and infrastructures in urban areas.

References

Abley, J. (2000). Stated preference techniques and consumer decision making: New challenges to old assumptions. *Working Paper Series*. University of Cranfield, UK Accessed 13th November 2015, from https://dspace.lib.cranfield.ac.uk/bitstream/1826/664/2/SWP0200.pdf.

Benoît, R., Jean, P. S., Mathias, G., Frederic, P., & Marie, S. (2003). A new structural approach for the study of domino effects between life support networks. In K. Alcira, A. Margaret, & C. Anne (Eds.), *Building Safer Cities: The Future of Disaster Risk, Disaster Management Series No. 3* (pp. 245–272). The World Bank.

Bruneau, M., et al. (2003). A framework to quantitatively assess and enhance the seismic resilience of communities. *Earthquake Spectra*, *19*(4), 733–752.

Chang, S. E. (2009). Infrastructure resilience to disasters. *The Bridge*, *Vol. 39*(No. 4) Journal of the National Academy of Engineering, Washington, DC.

Deshkar, S., Hayashi, Y., & Mori, Y. (2011). An alternative approach for planning the resilient cities in developing countries. *International Journal of Urban Sciences*, *15*(1), 1–14.

Godshalk, D. R. (2003). Urban hazard mitigation: Creating resilient cities. *Natural Hazards Review*, *4*(3), 136–143.

Government of India/Ministry of Housing and Urban Poverty Alleviation (GoI/MHUPA). (2012). *Report of the Technical Group on Urban Housing Shortage (TG-12)*. National Buildings Organisation. Accessed November 13, 2015, from http://nbo.nic.in/Images/PDF/urban-housing-shortage.pdf.

Government of India/Ministry of Urban Development (GoI/MUD). (2011). *Report on Indian urban infrastructure and services for estimating the investment requirements for urban infrastructure services*. High Power Committee, New Delhi.

Government of India. National Disaster Management Division (GoI/NDM). (2002). Disaster management: The development perspective. In *Tenth five year plan document 2002–2007*. Ministry of Home Affairs, Government of India, New Delhi.

Government of India/National Disaster Management Division (GoI/NDMP). (2004). *Disaster management in India: A status report*. Ministry of Home Affairs, Government of India, New Delhi.
Government of India/Reserve Bank of India (GoI/RBI). (2015). *Handbook of Statistics on Indian Economy*. New Delhi.
Herman, R., & Ausubel, J. (1988). *Cities and their vital systems: Infrastructure past, present and future*. Washington, DC: National Academy Press.
International Federation of Red Cross. (2004). *World disasters report: Focus on community resilience*. Geneva: Switzerland.
International Federation of Red Cross. (2012). *World disasters report: Focus on forced migration and displacement*. Geneva: Switzerland.
IPCC Working Group II. (2007). *Impacts, adaptation and vulnerability*. Contribution of IPCC WG-II to the *IPCC fourth assessment report*. Cambridge, UK: Cambridge University Press.
Krishnan, M. (1999). Disaster management action plans for the state of Maharashtra. *Shelter, Housing, and Urban Development Corporation*. Accessed November 13, 2015 from http://cidbimena.desastres.hn/pdf/eng/doc15213/doc15213-contenido.pdf.
Kroes, E. P., & Sheldon, R. J. (1988). Stated preference methods: An introduction. *Journal of Transport Economics and Policy*, *9*, 110–121.
Morris, S., & Pandey, A. (2009). Land Markets in India: Distortion and Issues. In N. Mohanty, R. Sarkar, & A. Pandey (Eds.), *India Infrastructure Report: Land—A Critical Resource for Infrastructure* (pp. 13–19), New Delhi: Oxford University Press.
Nagpur Municipal Corporation. (2012). *Environmental status report*. Nagpur: India.
Nagpur Municipal Corporation. (2011). *Draft city sanitation plan*. Nagpur: India.
Nagpur Municipal Corporation. (2000). *Flood control action plan*. Nagpur: India.
O'Rourke, T. (2007). Critical infrastructure, interdependencies, and resilience. *The Bridge*, *Vol. 37*(no. 1) National Academy of Engineering. Washington DC.
Pearmain, D., Swanson, J., Kroes, E., & Bradley, M. (1991). *Stated preference techniques: A guide to practice* 2nd ed. Steer Davies Gleave and Hague Consulting Group.
Pelling, M. (2003). *The vulnerability of cities: Natural disasters and social resilience*. London: Earthscan.
Potter, R.B., & Lloyd-Evans, S. (1998). *The city in the developing world*. Harlow: Addison Wesley Longman.
Revi, A. (2008). Climate change risk: An adaptation and mitigation agenda for Indian cities. *Environment & Urbanization*, *20*(1), 207–229.
Sen, A. (1981). *Poverty and famines: An essay on entitlement and deprivation*. Oxford, UK: Clarendon Press.
Twigg, J. (2007). *Characteristics of a disaster-resilient community: A guidance note*. DFID Disaster Risk Reduction Interagency Coordination Group, London.
United Nations Department of Economic and Social Affairs (UNDESA). (2014). *World urbanization prospects: The 2014 revision*. New York: United Nations.
United Nations Environment Programme (UNEP). (2002). *Global environment outlook—3: Past, present, and future perspectives*. London: Earthscan Publications Ltd.
World Bank, India. (2003). *Financing rapid onset natural disaster losses in India: A risk management approach* Report No. 26844-IN, New Delhi.

Vernacular Built Environments in India: An Indigenous Approach for Resilience

17

Priya Choudhary
Smt. Manoramabai Mundle College of Architecture, Seminary Hills, Maharashtra, India

17.1 Introduction

India is a vast country with diversified climatic, topological, and sociocultural conditions. Yet each region has a vast knowledge base about resilient built environments, in the form of its splendid yet mostly unmapped vernacular built environments. Vernacular built environments in India are centuries old and have sustained many natural disasters, thus proving to be resilient. This embedded knowledge in the existing vernacular settlements has evolved over a period of time with a good deal of experimentation by the locals and thus has stood the test of time. Considering the inherent resiliency, the vernacular built environments can be adapted appropriately in the changed sociocultural scenario. The study of vernacular built environments can generate an indigenous approach for the resilient design of contemporary built environments. Vernacular settlements at times were regarded as a romantic and engaging heritage from the past, but they have turned out to be remarkably enduring and thus resilient. Vernacular traditions that have been recognized and encouraged, appropriately supported and serviced, augmented, and complemented may prove to be the only realistic and sustainable solution to the housing and building demands of the future. (Oliver, 2006) Formal design training and the trained designers after modern period have mostly ignored it. (Choudhary, 2014)

In this chapter, the vernacular built environments of two regions with different social, climatic, and topographic conditions are discussed with respect to their inherent resilient aspects of spatial planning, construction methods, and sociocultural aspects. Adoption of the responsive planning principles gleaned from vernacular built environments can contribute to the development of an indigenous approach for the design of resilient built environments in India. Hence, there is a need for and the possibility of transcending them in the present context. Resiliency cannot only be discussed in an urban context; it is equally important in a semiurban or rural context, especially in India. Understanding developed from vernacular built environments is of utmost importance for dealing with rural developments that are in a rapid phase of urbanization. The indigenous approach will always be socially acceptable and culturally appropriate.

Urban Disasters and Resilience in Asia.

17.2 Resilient Built Environments

The term *built environment* basically means everything that is humanely created, modified, arranged, or maintained. Thus, collectively, the products and processes of human creation are classified under that term (McClure et al., 2007). Another definition, by Clarke (2003), states that *built environment* is actually a simplistic title of credible utility for a complex assemblage of various constructions, each produced in specific conditions and various regulatory, financial structures. It is as old as humanity itself. Yet the concept of a built environment is relatively recent and very inclusive. Architects and planners have been studying and analyzing built environment in various areas, such as architecture, urban design, and urban planning. The focus is on the interrelationships between humans and the environment. Resilience as a concept should not be referred to with respect to physical aspects only; the human component is equally important. Hence, while discussing resiliency, it is relevant to discuss with regard to the built environment rather than a single building or settlement. Understanding that the idea of the built environment is all inclusive makes a lot of difference.

The word *resilience* was first used in psychology in the 1950s. It has been used to describe the tolerance abilities of children (Ozel et al., 2015) The term was also used to describe a conglomerate of qualities that allow people to remain psychologically balanced and mentally healthy in the face of negative life circumstances and crises (Petzold & Muller, 2002). Thus, if this logic is applied to a built environment, *resilience* can be defined as the ability to absorb excessive natural forces (i.e., shocks) without losing stability, balance, and any major alterations in its physical, functional, economic, and social organization. Furthering this thought, Ozel and colleagues (Ozel et al., 2015) mentioned that resilient built environments not only can survive potential risks and threats, but also can catch the positive outcomes that the changes and transformations might bring. Flexibility and adaptability have become two fundamental qualities of any built environment that is called resilient. Walker and colleagues (Walker et al., 2004) described resilience as a capacity of a system to absorb disturbance and reorganize itself while undergoing change, so as to retain essentially the same function, structure, identity, and feedbacks. The possibility of constant change and the need to reform and balance in adaptation to change comprise the premise on which the concept of resiliency rests.

Vulnerability refers to the conditions determined by physical, social, economic, and environmental factors or processes, which increase the susceptibility of a community to the impact of hazards (UNISDR, 2004). Resilience and vulnerability are two different yet related concepts which have an overriding concern for response in the case of natural hazards or environmental change. *Vulnerability* refers to the susceptibility to damage from a hazard, whereas *resiliency* emphasizes something regaining its original form. Vernacular architecture settlements, which are referred to as *vernacular built environments,* are those that have evolved, transformed, and adapted to the local physical, climatic, and social changes. Hence, they are less vulnerable and inherently resilient. It is not only the physical aspects of spatial planning and construction methods in these environments that are flexible or adaptable, but also the social systems in vernacular communities that are contributing to resiliency. There are inherent rules of community bonding and participation, and thus the community relationships nurtured are equally responsible for resiliency. There is an intrinsic awareness of the rights and duties of the members of the community, which certainly contributes to reducing vulnerability in the wake of any natural disasters.

17.3 Vernacular Built Environments

The word *vernacular* has its origin in the Latin *vernaculus,* which means "domestic, native, indigenous" (Choudhary, 2014). So a built environment that is indigenous to a specific time or place can be called a *vernacular built environment.* This term is most often applied to residential buildings. In 1964, Rudofsky (2002) used the term *vernacular* in an architectural context and brought the concept to the attention of the public and the mainstream architecture.

Vernacular built environment refers to existing vernacular buildings/settlements designed and built by locals to satisfy their needs comfortably and conveniently. As described by Oliver (2006), the availability of building materials, the capacity of the land to support a population, and the kind of autonomy and trading relationships that evolve have a direct bearing on the form, location, and density of vernacular buildings. Oliver (1997) defines vernacular architecture as comprising dwellings and all other buildings of the people related to their environmental contexts and available resources; they are customarily owner- or community-built and utilize traditional technologies. All forms of vernacular built environments are built to meet specific needs, accommodating the values, economies, and ways of life of the cultures that produce them. Thus, these are inherently less vulnerable. The vernacular knowledge base has evolved over time. Instead of working on high-tech, Westernized solutions to vulnerability issues, it is important to introspect, study, and apply indigenous vernacular principles. This approach certainly can help people cope with immediate crises with more cultural appropriateness and social acceptance. Tradition by itself is not enough to sustain—rather, it should be examined and brought up to date wherever desired and practicable (Oliver, 2006) Hence there is a need to study existing vernacular built environments; understand vernacular design principles and work out the possibilities of extending these principles in contemporary times. Here it's important to address 'vernacular built environments' not just as a typology frozen in time but as a dynamic process of continuous evolution to respond to changing social-cultural and environmental needs.

17.4 Methodology

The study described in this chapter is based on the premise that vernacular built environments are inherently resilient. To test this premise, it is essential to assess its resiliency on the basis of selected parameters. If it is understood why and how these vernacular built environments are resilient, then one can discover ways and means to transformed appropriately that vernacular knowledge base to contemporary times. The parameters for resiliency listed by various researchers are given as well.

The resilience of an ecosystem has been described by the Resilience Alliance (2002), and the characteristics given can very well be applied to a built environment to quantify its resiliency. The assessment is based on the following parameters:

- The amount of change the system can undergo and still retain the same controls on function and structure
- The degree to which the system is capable of self-organization
- The ability to build and increase the capacity for adaptation

Resilient design principles given by the Resilient Design Institute (Langeveld, 2013) are as follows:

- Simple, flexible modular systems have a greater capacity to adapt in case of changing situations.
- In the case of hazards, local and renewable resources and building materials will be easily available to rebuild the damaged portions.
- Diverse and multifunctional spaces can cater to changing uses easily and hence are better able to adapt.
- Resilient cities require the capacity to predict natural shocks and interruptions and respond to continuous changes with a dynamic architecture (Jha et al., 2013).
- The know-how of the living cultures increases their ability to recover and rebuild in the case of hazards. If local people are aware of the practical construction techniques to build houses/residences, which is the case for vernacular communities, they can start rebuilding more quickly.
- Durability increases the life span of buildings and reduces maintenance, thus making buildings resilient.

Vernacular buildings are durable, as is evident since they have existed for hundreds of years.

Ozel and colleagues (2015) had proposed a comprehensive model for better understanding of the resilience potential of vernacular architecture in terms of local traditions, spatial arrangements, land use, and building culture. The elements are as follows:

- Resilience through design—Vernacular architecture is climate-adapted design, so it is resilient. Even vernacular settlements integrate themselves with their binaural contexts in a harmonious fashion.
- Resilience through mixed-use spaces and collectiveness—Vernacular communities imply collective living and shared facilities through the optimum and flexible use of space.
- Resilience by using appropriate materials and construction systems—In the context of resilience, local materials provide buildings with the capacity to adapt to the changing climate conditions as they keep evolving in the terms of time according to the changing environment.
- Resilience by promoting local production and autonomy—Vernacular settlements encourage autonomy and self-sufficiency by several design strategies, including the integration of dwellings with food production areas (namely, farms).

Indigenous (vernacular) earthquake-resistant construction technologies are categorized into four types (Ravi et al., 2004):

- Construction technologies using ductile construction materials such as timber and bamboo—Though these materials offer flexibility to the joints, timber and bamboo are both prone to decay. The earthquake resistance of buildings made from these materials may be severely compromised due to weathering. Such materials are used for buildings that are not required to be as durable. Measures to avoid weathering and decay have made some of the vernacular constructions using timber and bamboo more long-lasting.
- Construction technologies using robust architectural forms—Buildings with symmetrical plan forms, regular geometric shapes, and uniform elevation generally have greater seismic resistance than irregular buildings.
- Construction technologies using resilient structural configuration—Buildings with bands and braces have well-distributed and redundant load-resisting systems, so they perform very well under the force of earthquakes. Earthquake forces can be resisted through walls with lateral ties or through frame systems.
- Construction technologies reducing seismic forces—Various nonstructural building elements, such as partition walls, can be made out of lightweight materials so as to reduce the impact due to heavy mass in case of earthquake.

17.5 Parameters

Based on the studies discussed previously, a set of parameters is listed here, which will form the framework for assessment of vernacular built environments of two different regions in India.

The parameters are divided into three categories:

- Spatial planning
- Construction materials and methods
- Sociocultural aspects

Spatial planning parameters include siting, built form and orientation, spatial flexibility, and climate adaptability. Understanding topography, geology, and placement of settlement accordingly is a very important aspect of vernacular built environments.

Construction materials and methods include the use of local/renewable and durable materials, local knowhow, and local construction technology. As stated by the Resilient Design Institute (Langeveld, 2013), knowledge of the living cultures increases the ability to recover and rebuild after a disaster strikes. If local residents are aware of the construction techniques, which is the case for vernacular communities, they can start rebuilding more easily and efficiently.

Sociocultural aspects include community bonding, local production, and autonomy. Vernacular communities imply collective living and encourage autonomy and self-sufficiency. Planner/designers do not play much of a role in local production and autonomy, but spaces fostering community bonding are important.

Spatial planning parameters such as siting and built forms are of primary importance, as they contribute substantially to the vulnerability of a built environment.

Appropriate siting can prevent the built environment from getting damaged after a disaster. Local materials and local technology play significant roles in imparting resiliency to the built environment. Yet these are of secondary importance, as they will contribute to resiliency significantly if spatial planning parameters are appropriately considered. The sociocultural aspects of a society need to be addressed appropriately through spatial design. The provision of multifunctional community spaces appropriate to the culture of a society encourages community bonding, which in turn will help substantially in rebuilding activities.

17.6 Vernacular Built Environment in Harsil

17.6.1 Harsil Valley

Harsil Valley is located in the mountain district of Uttarkashi, 2620 m above sea level. Most of the terrain is mountainous, consisting of high-rise ridges and hills. The region has only two seasons—winter and summer. Summers are moderate, but winters are extremely cold. There is significant rainfall throughout the year, and heavy snowfall usually occurs from November to January. The maximum temperature in the region is around 15° C and the minimum temperature is around 0°C. The winter season in Harsil runs from October to January. Gangotri, one of the four important pilgrimage destination for Hindus, is located near the valley. Due to the terrain, topography, and Himalayan geology, this region is very ecosensitive and vulnerable to natural hazards. Heavy rainfall, landslides, and earthquakes often devastate the region, which is in seismic zone IV. Large amounts of moraines (i.e., any accumulation of unconsolidated glacial debris) can be seen along the riverbeds. This debris may have been plucked off the valley floor as a glacier, or it may have fallen off the valley walls as a result of frost wedging or landslides. The valley has large deposits of moraines on all the riverbanks.

17.6.2 Spatial Planning

In this section, the vernacular built environment of Dharali, a village in Harsil, is explored.

Siting: As seen in Figure 17.1, residential growth started in Dharali next to a rivulet that meets the Ganges River. There was no planned road network, and the rivulet acted as the guideline for development. The rocky land near the rivulet was not suitable for cultivation, so this piece of hilly land was chosen for building houses. The residences were planned in an organic manner along the length of the rivulet. The placement of the units depended on the gradient of the land. Compact planning was done due to the climatic conditions. Courtyards and streets act as places for multifunctional activities. The houses are constructed not on the riverbanks, but at considerably higher levels. After 1955, as connectivity increased between Gangotri and Uttarkashi, a vehicular road was developed between the two. Dharali gained importance as a transition space. Commercial development subsequently took place adjoining the road (Figure 17.1). This development was haphazard and did not emulate the earlier pattern. The buildings were placed with the longer sides facing the

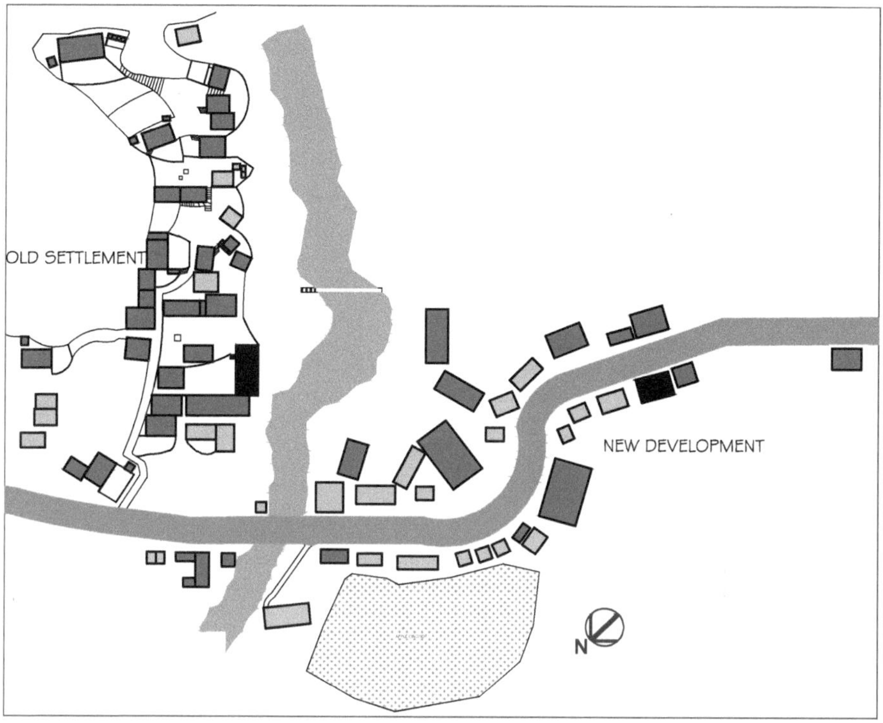

Figure 17.1 Settlement plan of Dharali.

road. As tourism developed, the commercial development along the road included hotels and general stores. There is a clear distinction between residential and the commercial areas, with the rivulet acting as a divider. All these commercial developments are built using reinforced cement concrete (RCC) without considering the topographical or climatic conditions.

When a natural calamity occurred in June 2013 in the form of a cloudburst, the old settlement was hardly damaged, but the new developments were devastated. Figure 17.2 shows that the siting of the old settlement was an important factor in its protection. After the cloudburst, a large amount of mud and debris flowed down along the rivulet. Due to the siting of the old settlement, however, it was hardly affected even though the rivulet overflowed. But the overflowing of the rivulet damaged new developments located along the road. As shown in Figure 17.3, the other commercial development along the road was filled with mud and boulders carried by the rivulet. Thus, the vernacular built environments in the valley are usually placed away from water bodies and on higher firm rocks. But the new developments were built along the riverbeds. As shown in Figure 17.4, in the case of the cloudburst, the road built near the riverbank got washed away. This is an important lesson that one can learn from these old settlements with their well-planned siting.

Built-form and orientation: The houses in the village are mostly two-story wooden houses. The plan-forms are rectilinear and thus symmetrical. Due to the steep, sloping topography, the ground coverage is small and the spaces are stacked one above

Figure 17.2 Siting of Dharali—old and new settlements.

Figure 17.3 New developments were filled with mud and boulders brought by the rivulet.

Figure 17.4 The roads near the riverbanks were washed away.

the other. Many of the houses are 150 to 300 years old and have withstood many natural disasters. The locals strongly believe that their housing should face the river. The longer axis of each house lies from east to west so that it faces south for maximum heat penetration (Choudhary et al., 2014). A minimum cut and fill is seen in the vernacular houses, with rooms being built on different levels. The vertical zoning in the houses is very peculiar, in that the lower part of the house is reserved for animals and fodder and the upper floors are used by the people living there. This planning was done considering the heavy snowfall that happens in winters. The kitchen is usually located on the top floor, for the purpose of venting out the smoke directly to the sky. This helps to keep the living environment free of smoke. The middle floors are sandwiched between the two floors, and hence the area is warmer and suitable for living.

Spatial flexibility: As seen in Figure 17.5, south-facing balconies receive much-needed sunlight and are multifunctional spaces. The open spaces between houses also serve many functions. Spaces enclosed within the cluster of houses boost social interaction, helping to strengthen community bonding. The women use such spaces in the afternoon for household activities such as drying grains and apple pieces and chatting with each other. Such spaces become vibrant living spaces when community worshipping and processions of local gods and goddesses take place.

Climate adaptability: The openings are kept to a minimum size to prevent heat loss. The height of the habitable spaces is kept less than 3 m to conserve heat. Verandas and balconies are provided on the south, which act as buffer zones to protect the

Figure 17.5 A south-facing balcony that gets plenty of sunlight.

interiors from the cold winds. The walls are kept thick for good insulation, and the roof is built sloping with deep overhangs to deal with the heavy rainfall and snowfall.

The aspects of spatial planning of vernacular built environments in the Harsil Valley are in accordance with the physical and social context. As a result, they are less vulnerable to natural hazards, and thus resilient.

17.7 Construction Materials and Methods

The vernacular houses in the valley were built using stone, timber, and mud plaster. The materials were locally available, and locals were aware of the construction methods. Although the use of brick and concrete is increasing in the area, the locals realize

Figure 17.6 Example of a construction system of a vernacular house.

that vernacular houses are stronger, stable, and warmer in winter. As seen in Figure 17.6, the vernacular method of construction includes provision of horizontal wooden ties. This special type of construction method is called as 'Chinai wall construction'. The walls are thick (0.3–0.5 m) and load bearing, and have alternate layers of stone and wood. The walls are mostly unplastered, showing attractive composition. The wooden joints are all carefully crafted, with notches improving their flexibility. The RCC construction lacks this flexibility and thus fails severely under sudden impacts caused by natural hazards.

Figure 17.7 shows a severely damaged RCC building after a cloudburst in 2013. Thus, the construction methods adopted in vernacular built environments in Harsil Valley create resilient structures due to their inherent flexibility and rigidity. The local knowledge base about the construction methods increases the ability of recoverability

Figure 17.7 RCC building damaged due to a cloudburst occurring in 2013.

and rebuilding in the case of hazards. Precast RCC construction with flexible joinery details are elements of a better construction system. Wooden ties can be replaced with precast members.

17.7.1 Sociocultural Aspects

In Harsil Valley, the remoteness of the villages compels people to develop strong community bonds and an active social life. There is a lot of interdependence among the members of the community, which further strengthens the bonds. The built environment is responding well to the spatial needs of the community. Thus, the vernacular communities can cope better with natural hazards due to their strong community bonding. Since villages are not very well connected by road due to the difficult terrain, they usually foster autonomy and self-sufficiency in terms of acquiring food, especially grains. The role of spatial design in fostering internal community bonds is significant.

17.8 Vernacular Built Environment in Vidarbha

17.8.1 Vidarbha

The Vidarbha region in Maharashtra state is an agroclimatic zone. Due to climatic peculiarity, the zones tend to have a particular crop pattern, which is responsible for the evolution of typical sociocultural conditions. Vidarbha lies on the northern part of Deccan Plateau, and it has no major hilly areas. Its primary economic activity is agriculture, and it is rich in forests and mineral wealth. In Vidardha, rains are mostly received from the southwest monsoon, from June to September. Annual rainfall varies from 700 to 1250 mm (PDKV, 2013). Agriculture in the region is mainly dependent on rain, which is scarce and unpredictable. Climate change has made the situation more difficult. There is a droughtlike situation after 2–3 years. The region experiences subhumid to humid conditions in monsoon season, semiarid conditions in winter season, and arid conditions in summer season. Hot summers and moderately cool winters characterize the Vidarbha region. The maximum temperature in summer is usually 48°C and in winters, the minimum temperature is 6°C. Climate change has contributed to vulnerability of the region in the recent past. In the last decade, more than 2 million farmers have committed suicide in Maharashtra, more than 70% of which lived in the Vidarbha region (Anonymous, 2013). Farmer suicide is a socioeconomic phenomenon, and there are many complex reasons for it. Yet climate change, which has caused an increase in average temperature and a reduction in rainfall, is one major reason. The vernacular built environment in this area contains the houses of local farmers and landlords in rural areas built during the last 200 years. These houses of landlords/military officials (subhedars) with a limited autonomy over the immediate hinterland; are called *Wadas*. A representative village, Tarsawanga, located in the Wardha district of Vidarbha, is described in the next section (SMMCA B.Arch VII Sem. 2009, 2009).

17.8.2 Spatial Planning

Siting: Figure 17.8 shows the settlement plan of a village named Tarsawanga. Its siting, on the bend of the river, helps the flow of cool winds from the river to enter the village, and yet it also protects the village from occasional floods. The narrow, north-south-oriented streets are shaded in the afternoon due to North-South orientation and narrow width.

Built form and orientation: The built forms in the village are compact and rectilinear (SMMCA B.Arch VII Sem. 2009, 2009). These wadas are usually closed entities with introvert planning, and the central core comprises living spaces that are protected from heat increases on all sides. The walled enclosure had built, semiopen, and open spaces. The ratio of built spaces to open spaces usually varied from 1:2 to 1:3. Entry to these houses is usually through a long narrow corridor that acted as a well-defined transition from the exterior public space and the inside private courtyard. A semiopen veranda surrounding an open courtyard is displayed in Figure 17.9. This is the most public space, where official guests are entertained and in the past, usually social, political, and business matters were discussed.

Spatial flexibility: The interiors and most of the open and semiopen spaces are multifunctional. Their use changed as per the time of day, or the season.

Climate adaptability: Courtyards are usually square and rectilinear. They act as binding spaces encircled by semiopen verandas and are major sources of light and ventilation (Figure 17.9). Transition spaces such as semiopen verandas are extremely useful in this climate. They prevent direct exposure to sunlight in summers yet in

Figure 17.8 Village siting in Vidharba.

winters, sunlight enters these spaces due to the low altitude. The compact planning of the built spaces around private courtyards helped shade the courtyards and walls. There are usually two to three courtyards, arranged in a hierarchical fashion. They vary in size based on their use. The inner courtyard usually had a Tulsi Vrindavan (planter for Basel plant), which helped get oxygen-rich air in the courtyard. There were at times service courtyards which connect and segregate the toilet from the main residential areas. The height-to-width ratio, which ranges from 1:1 to 1:2, improved the wind flow in the courtyard. The windows, which are about 9% to 12% of the wall area, were on interior walls of the rooms abutting courtyards. There were very few windows (0–6% of the area) on the exterior walls. All the windows had high sills (1.5.m) which were even higher on the exterior (1.6 m). Smaller openings reduced heat increases and also avoided the entry of dry, hot winds in summer. Also, when the air enters through such openings, it gets compressed. The slant walls of the windows allow the entry of compressed air and sudden expansion after entry, producing a cooling effect called the *Venturi effect* (Choudhary, 2014).

Figure 17.9 Semi-open veranda.

17.8.3 Construction Materials and Methods

Structural stability is usually the most important aspect of vernacular built environments. The construction methods in Vidarbha were based on resilient structural configurations. The footings were stone stepped, protecting the superstructure made out of 1.2-m-thick mud walls. These walls were strengthened with teakwood twin columns, as shown in Figure 17.10. The stone base was placed with wooden columns to protect it from moisture and ingress of water through capillary action. The floors were made of timber and mud. The mud gave an insulating effect. There was a double roof system, which provided additional storage space, and this space between the two roofs was ventilated, which helped minimize the heat gain through the roofs.

17.8.4 Sociocultural Aspects

Community spaces around a temple or a peepal tree are common. Strong community bonding, which is often observed in vernacular communities, is gradually decreasing. The Vidarbha region is prone to droughtlike situations due to uncertain and scant rainfall. The weakening of community bonds is one reason that farmers in the area have been taking extreme steps like committing suicide.

The vernacular built environments in Vidarbha, though not found to be resilient in themselves, are not capable of dealing with the vulnerability of the region, as these issues have socioeconomical causes as well. Yet understanding about vernacular design and

Figure 17.10 Wooden twin columns and beams forming a frame for mud walls.

construction principles, if transcended appropriately, can lead to climate-responsive architecture. This development can have a positive impact on climate change in the long run.

17.9 Conclusions

The study of vernacular built environments of two identified regions suggests that these are in response to the physical and social context. Thus, they are less vulnerable to environmental hazards and changes. Siting of the settlements, flexibility of construction methods, and community spaces that help strengthen community bonds are a few of the important aspects of the vernacular built environment in Harsil Valley, which makes it less vulnerable and enhances resiliency. In the case of vernacular built environments in the Vidarbha region, the spatial and built form designs help reduce heat gain and improve ventilation. These are the requirements in the hot and dry climatic conditions of the region, which are aggravated due to climate change. Vernacular built environments are studied in this chapter, with an emphasis on the relationship between humans and their physical environment. Vernacular built environments may not be completely resilient against very-high-risk natural hazards, but they are certainly in much more harmony with their natural surroundings, and hence are less vulnerable. Also, vernacular built environments foster positive social bonding between their residents and between the residents and built forms, which helps communities rebuild after any kind of calamity. This is an important lesson. While discussing resiliency, it is important to understand embedded social systems/cultural aspects, along with the physical aspects of settlement planning and design.

The tradition can be seen as a creative process through which people as active agents interpret past knowledge and experience to face the challenges and demands of the future (Asquith & Vellinga, 2006). The resilient design approach can emerge from the active process of the transmission, interpretation, negotiation, and adaptation of vernacular knowledge, skills, and experiences. As stated by Al Sayyad (2006), the idea of modern knowledge as being different from and possibly opposite to vernacular knowledge should be abandoned, as the vernacular in some instances may in fact be the most modern. Here, one can use vernacular systems in vernacular built environments as the model. The vernacular as a model system for contemporary design may help one identify general principles and mechanisms that characterize the relationship between humans and their environment (Rapoport, 2006). Thus, very important lessons learned from the use of vernacular built environments need to be extended with the help of a modern technological knowledge base. There is another important issue regarding resiliency and rapid urbanization—namely, cultural sustainability and social acceptance. If contemporary built forms do not continue the cultural values and vernacular visual vocabularies of a community, their acceptance and thus success are not ensured. The human capacity to form mental models is genetically programmed (Lewcock, 2006). Hence, every society with its typical genetic and cultural background would develop mental images using the emotional connections of memory. The new developments should resonate with these mental images to foster acceptance by a particular group or society (Choudhary, 2014). So it is not just about technology; the role of the reinterpretation of vernacular environments is much more significant to make resilient and contemporary built environments socially, culturally appropriate, and acceptable. Once culture has been destroyed due to lack of consideration, action, or knowledge, it is almost impossible to reconstruct (Jaulin, 1971). Awareness, knowledge, and options for positive action can empower individuals to create resilient and sustainable built environments.

It is impossible to continue to build and design settlements in contemporary times as they were built hundreds of years ago. Change is inevitable. It is essential to think about the resiliency of many rural and semiurban areas, which are getting urbanized along with urban areas. The twenty-first century is referred to as an "urban" century and that too of "global south." Many Southeast Asian countries like India are urbanizing in a rapid way. All these countries have lots of rural areas with an immense traditional knowledge base in the form of vernacular built environments. The physical and sociocultural aspects of the vernacular built environments have many inherent qualities that foster resilience, either directly or indirectly. Hence, it is essential to develop an indigenous approach to achieving resiliency based on vernacular knowledge when dealing with rapidly urbanizing rural areas.

References

Al Sayyad, N. (2006). Foreword. In L. Asquith, & M. Vellinga (Eds.), *Vernacular architecture in the twenty-first century: Theory, education, and practice*. New York: Taylor & Francis.

Anonymous. (2013). *Vidarbha*. Accessed 26 October 2015, from http://en.wikipedia.org/w/index.php?title=Vidarbha&oldid=576716725.

Asquith, L., & Vellinga, M. (2006). *Vernacular Architecture in the twenty-first century: Theory, education, and practice*. New York: Taylor & Francis. http://www.ecologyandsociety.org/vol19/iss2/arl5.

Choudhary, P. (2014). *Vernacular architecture in Vidarbha*. Nagpur, India: VNIT.

Choudhary, P., Jathar, V., & Chakradeo, U. (2014). *Eco-friendly architecture of Harsil Valley: Anubhuti II*. Nagpur, India: Swaprakashan.

Clarke, P. W. (2003). The Economic Currency of Architectural Aesthetics. In C. A. R (Ed.), *Designing Cities: Critical Readings in Urban Design* (pp. 29). Oxford: Blackwell Publishers.

Jaulin, R. (1971). Ethnocide: The theory and practice of cultural murder. *The Ecologist, 1*, 12–15.

Jha, A. K., Geddes, Z. Z., & Miner, T. W. (2013). *Building urban resilience: Principles, tools and practice*. Washington, DC: World Bank.

Langeveld, M. (2013). The Resilient Design Principles. *The Resilient Design Institute* Accessed November 4, 2014, from http://www.resilientdesign.org/?s=resilient+design+principles&x=0&y=0.

Lewcock, R. (2006). "Generative concepts" in vernacular architecture. In L. Asquith, & M. Vellinga (Eds.), *Verancular architecture in the twenty-first century: Theory, education, and practice* (pp. 199–214). New York: Taylor & Francis.

McClure, W., Bartuska, T., & Bartuska, T. (2007). The Theory, Science, and Practice of Bringing Buildings to Life. In L. Asquith, & M. Vellinga (Eds.), *The Built Environment: A Collaborative Inquiry into Design and Planning*(2nd ed.). Hoboken: John Wiley & Sons.

Oliver, P. (1997). *Encyclopedia of vernacular architecture of the world: Theories and Principles, Volume 1*. Cambridge University Press.

Oliver, P. (2006). *Built to meet needs: Cultural issues in vernacular architecture*. Oxford, UK: Elsevier.

Ozel, B., et al. (2015). Resilience and intangible heritage of vernacular architecture. In C. Mileto, F. Vegas, L. Gracia & V. Cristini (Eds.), *Vernacular architecture: Towards a sustainable future* (pp. 571–576). London: Taylor & Francis.

PDKV. (2013). *Dr. Panjabrao Deshmukh Krishi Vidyapeeth, Akola*. Accessed October 2014, from http://pdkv.ac.in/aicrpAgromet.php.

Petzold, H.G. & Muller, L. (2002). *Resilienz und protektive Faktoren im Alter und ihre Bedeutung für den Social Support und die Psychotherapie bei älteren Menschen*. (Resilience and protective factors at the age and its importance for the social support and psychotherapy in the elderly. Its an online document by FPI-publications, Dusseldorf. http://www.fpi-publikation.de/polyloge/alle-ausgaben/08-2003-mueller-l-petzold-h-g-resilienz-und-protektive-faktoren-im-alter.html) Zurich: s.n.

Rapoport, A. (2006). Vernacular design as a model system. In L. Asquith, & M. Vellinga (Eds.), *Vernacular architecture in the twenty-first century: Theory, education, and practice* (pp. 179–198). New York: Taylor & Francis.

Ravi, S., Brzev, S., & Kharel, G. (2004). Indenous earthquacke resistant technologies—An overview. *Vancouver, Canada: 13th World Conference on Earthquake Engineering*.

Resilience Alliance. (2002). *The characteristics of resilience*. Accessed October 24, 2014, from Available at: http://www.resalliance.org/index.php/resilience.

Rudofsky, B. (2002). *Architecture without architects: A short introduction to non-pedigreed architecture*. Albuquerque: University of New Mexico Press.

SMMCA B.Arch VII Sem. 2009 (2009). *The second renaissance: The vernacular architecture of Vidarbha region*. p. 10 Nagpur, India: SMMCA.

United Nations International Strategy for Disaster Reduction (UNISDR). (2004). *Living with risk: A global review of disaster reduction initiatives*. Geneva, Switzerland: UN Publications.

Walker, B., Holling, C., Carpenter, S., & Kinzig, A. (2004). Resilience, adaptability, and transformability in socioeconomic systems. *Ecology and Society*, *9*(2).

Building Community Resiliency: Linkages Between Individual, Community, and Local Government in the Urban Context

18

Rajib Shaw[1], Gulsan Ara Parvin[2], Akhilesh Surjan[3], Atta-ur-Rahman[4]
[1]Professor, Graduate School of Global Environmental Studies, Kyoto University, Kyoto, Japan; [2]Researcher, Graduate School of Global Environmental Studies, Kyoto University, Kyoto, Japan; [3]Associate Professor, Faculty of Engineering, Health, Science and the Environment, Charles Darwin University, Darwin, Australia; [4]Associate Professor, Institute of Geography, Urban and Regional Planning, University of Peshawar, Peshawar, Pakistan

18.1 Introduction

Community resilience has been discussed in recent years over different literatures (Victoria, 2003, 2009; Shaw & Okazaki, 2003, Delica-Wilson, 2005, Kafle & Murshed, 2006, Shaw, 2012a,b, 2014a). There is no doubt that communities play the most important role in risk reduction, whether in an urban or rural context. Data from the Great Hanshin Awaji Earthquake of 1995 in Kobe, Japan, shows that 98% of the people were rescued by either their family members or their neighbors. The form of a community, its sustained involvement, its form of engagement, and other factors vary from place to place. Community involvement is often linked to household or individual preparedness, on one hand, and neighborhood preparedness, on the other hand. Depending on the scale of the community and its activities, the challenges can vary as well. Urban communities often face a number of issues in terms of disaster risk reduction (DRR) due to such reasons as disaster issues get a relatively lower priority compared to other issues in urban areas, like waste management, sanitation, and traffic. Urban communities have relatively fewer opportunities for direct get togethers due to the busy lifestyles of their residents. Nevertheless, there are good examples of community-based activities in urban areas with different target groups, motivators, and actors.

Shaw (2014b) pointed out a few critical issues that affect community-based activities: the changing nature of disasters, diversity of the community, national local linkages, evidence from past disasters, increasing global awareness of local needs, and sustainability and upscaling. Among these, the changing nature of disasters is quite important because it causes more unexpected damage to urban areas and underlines the increasing importance of urban community resilience. This situation is more significant when dealing with climate-related hazards, where the precipitation patterns change and become more localized. The local drainage, sewage conditions, transport

Urban Disasters and Resilience in Asia.

network, and related infrastructures are some of the causes of local flood inundation, and therefore, localized actions are needed within a megacity. Daily small-scale activities, which may not be directly related to the disaster in question, can bring a community together and help residents cope with larger disasters.

In this chapter, a few examples of urban communities are provided, with specific focuses and implications for handling and recovering from disasters. An analysis is made that links individual preparedness and a community's collective actions, followed by a focused discussion of institutionalization.

18.2 Example of Urban Community Resilience

18.2.1 Urban Communities in Kobe, Japan

After the Great Hanshin Awaji Earthquake of 1995, a different, innovative model of civil society involvement in enhancing community resilience was developed. Shaw and Goda (2004) created the future civil society model, which features the collaboration of specialized nongovernment organizations (NGOs), nonprofit organizations (NPOs), and community-based organizations (CBOs) with the resident association and provide specific services to communities. A trilateral collaboration of city governments, resident associations, and CBOs has created communities where CBOs provide services related to social welfare, health issues, and other elements, local governments provide funding, and resident associations offer opportunities for community participation and forging consensus.

The BOKOMI scheme of the Kobe city government (Shaw, 2014b) is another unique approach, wherein the city government provides grants to elementary school–based communities (191 in total) to start and sustain social welfare, safety, and disaster preparedness activities of the local communities. This initiative, which began in 1995 in 11 pilot school districts, spread to all 191 schools by 2008. In this program, the local community creates a committee, usually including 30-35 members from the resident association, youth association, parent-teacher association (PTA), fire volunteers, women's association, and other local groups. The committee works on disaster prevention and risk reduction activities such as disaster drills, DRR education, and public awareness programs, as well as social welfare activities such as health care training, town watching, and communication with communities. Thus, the unique combination of regular social welfare–related activities and disaster preparedness keep urban communities active, and daily preparedness becomes linked to disaster preparedness.

18.2.2 Environment Protection Communities (EPCs) in Da Nang, Vietnam

Da Nang, a port city in the south central coast of Vietnam, is one of the country's five major cities, with very high economic growth. The city is prone to a growing risk from the effects of climate change, as well as an increased frequency and severity of typhoons and typhoon-induced flooding. A study by Hishida and Shaw (2009) pointed out that an environmental entry point could be a better choice for Da Nang to develop community resilience. Social capital analysis has led to the recommendation that environmental protection communities (EPCs) should be established as a system at the community level, with a specific focus on normal environmental issues like solid waste disposal and sanitation. This has been linked to elementary school education, where students used to observe the environmental conditions of the local lake every month, under the guidance of teachers and community members. EPCs meet periodically to discuss key environmental issues in the community, and they have been very effective during typhoons and other disasters, since they have regular interactions with the community. EPCs make a strong link to the community-level governments, and therefore were very important links between the community and local government.

18.2.3 Advanced Locality Management (ALM) in Mumbai, India

Advanced locality management (ALM) is a unique plan started in 1998 by the municipal corporation of Mumbai, India, to address solid waste management issues in local communities. It urges the formation of committees, registers them with the local government, and receives small grants to initiate activities. The core of the concept behind ALM is locality management, which includes both middle- to high-income residents and slum dwellers. Women's groups in slums and informal settlements are organized into self-help groups to separate waste and sell recyclable waste, which was also a source of income to them. The middle-income group used waste to develop compost and enhance the greenery of the area.

The regular interaction of the group enhanced the community's connectivity, which was reflected during a flood that hit the city in 2005 (Surjan et al., 2009). In the areas where ALMs were very active, there were less of a drainage problem, and thus they were less inundated with water. Also, regular interactions with the communities helped them rescue the neediest residents (like the elderly, small children, pregnant women, and other vulnerable people), and therefore the number of casualties was smaller. The key lesson of using ALMs is that solid waste management provides a good entry point for collective activities, which met the daily needs of urban communities. Through this entry point, communities have been reorganized and community resilience enhanced through participation by various social and economic groups.

18.3 Challenges of Individual and Community Actions

The three examples described here show that to enhance urban community resilience, there needs to be an appropriate entry point, whether it be solid waste management or other environmental issues, social welfare, or health-related issues. Urban communities are more complex and diverse in terms of their social and economic status and needs. Thus, to generate and integrate communities, one must find common issues and problems to focus on, and disaster issues need to be linked to these common problems.

To do that, an urban community needs a change agent, which brings about the appropriate changes within communities. An analysis by Mulyasari and Shaw (2013) in Bandung, Indonesia, found that women's groups such as women's welfare associations (WWAs) can be very effective at communicating risk information in urban areas. When their perceptions and ongoing activities are compared, activities such as dissemination of disaster risk information, conveying early warnings to their peers, and involvement of the local government have been confirmed to work well with the risk communication plans and implementations of WWAs. The results of this study confirmed that women, through their social networks, can become active agents of change, thus acting beyond their usual domestic roles and responsibilities in order to contribute to the overall enhancement of community resilience.

Joerin et al. (2014), from the results of a household survey in Chennai, India, that assessed the physical, social, and economic resilience of individuals through a Climate-related Disaster Community Resilience Framework (CDCRF), revealed that people living in the vicinity of rivers and canals are at higher risk from the impacts of flood (i.e., damages sustained to houses or diseases) compared to others. However, despite their experience with past flood-related disasters, they have not been able to enhance their coping abilities due to their limited adaptive capacity. Thus, their resilience is limited to absorbing, managing, and recovering from climate-related disasters (particularly floods). In collaboration with other stakeholders, mainly the Corporation of Chennai (Municipality), community-driven participatory solutions were concluded to have beneficial effects in enhancing community resilience to climate-related disasters. There, local women's self-help groups, veterans' associations, and other related community-based organizations played important roles in communicating information about risk to the community.

Fernandez and Shaw (2014) have pointed out that youth councils have played a very important role in enhancing community resilience in Makati, the core urban area of metropolitan Manila in the Phillippines. The structured system of having a barangay-level disaster management committee, as well as involvement of a youth council with a specific budget has given the specific provision of community-based activities, which can be linked to seasonal actions before, during, and after typhoons. The youth council is entitled to take some environmental and disaster-related activities (so-called green activities), which have a specific budgetary provision and therefore can be considered as regular events in the community.

One of the key challenges facing urban communities is determining how their actions are linked to individual or household actions. In Japan, there is a higher level of disaster education imparted in elementary and junior high schools. However, with high school, it becomes a difficult issue. The survey by Shaw et al. (2004) showed that while school education programs can enhance the level of perception of disaster in high school students, their actual preparedness depends more on community and family education. For high school students, who spend more time outside the home, it is difficult to define the community, especially in an urban context. It is said that the age group between 15 to 25 (or 15 to 30) is a gap area in the urban context in Japan, especially concerning community participation. The period from the age of 15 onward (i.e., when children reach high school age) to 25 or 30 (when people start having their own families) is a time when people usually live alone, in many cases away from their family and hometown. How to link this age group to their community's activities is a real challenge. Even if they are aware of the need for individual or community actions, the modern lifestyle does not provide enough opportunity to bring this age group into community-based activities (especially in urban settings). These connections can possibly be strengthened by an institution-based approach, either through a high school's or university's community linkage or office's corporate community approach

18.4 Institutionalization: The Missing Link

From these case studies, a few issues can be considered, which are important for urban community resilience:

- There needs to be an *entry point* for community-based DRR in the urban context, where disaster issues need to be linked with daily priority issues.
- There needs to be specific *change agents*, who can motivate actions and lead efforts to make changes.
- Although there can be community-based activities, the key challenge is to reflect these into *individual and family behaviors*. Sometimes incentive schemes work to accomplish this, and other times penalty schemes are more effective.

In this context, the key point is to institutionalize community-based activities for a community's sustainability and upscaling. It is often found that there are proactive, community-based activities after a major event or initiated by an external factor (like project intervention), which often dies up over time. To sustain urban community actions, it is essential to link any community plans with that of local governments, and thereby institutionalize them so they can be sustained. All three examples described in this chapter have a common factor: namely, that they are linked to city government schemes. While in Kobe and Mumbai, the schemes were initiated by the city government and monitored and maintained by the local government, the Da Nang case featured monitoring at the local community level, which is under the purview of the city governments. Thus, the institutionalization of the efforts of local government is the key to the sustainability of an urban community. The relevance of these issues changes over time, so it is important to institute periodic monitoring and updates based on any changing circumstances.

The other dimension of institutionalization is the participation of informal institutions at the community level. Although informal institutions are few and far between in urban areas, they do exist in the old part of cities, in the form of specific groups like women's groups, veterans' groups, youth groups, faith-based organizations like church groups, mosques, and temples, and self-help groups. Community activities, which are linked to these types of formal, informal and semiformal institutions, can also be effective in linking individual and family preparedness, and therefore enhancing community resilience in urban settings.

In conclusion, the formal, semiformal, and informal institutionalization of community activities can be considered as an effective approach for achieving urban resilience. For local governments, the key responsibility is to develop systems, platforms, and mechanisms to monitor their activities. The level of penetration of social media is high in urban areas, even in developing countries, and that can be an effective monitoring tool for sharing information and activities. The key goal should be to create responsible citizens, who either serve their community's needs for disaster preparedness or at least are aware of the needs and convey them widely to family members and other people they know. Thus, in the long run, a citizen-based approach can have a positive influence on the resilience of urban communities.

References

Delica-Willson, Z. (2005). Community-based disaster risk management: Local level solutions to disaster risks. *Tropical Coasts*, *12*(1), 66–73.

Fernandez, G., & Shaw, R. (2014). Participation of youth councils in local-level HFA implementation in Infanta and Makati, Philippines and its policy implications. *Risk Hazards and Crisis in Public Policy*, *5*(3), 259–278.

Hishida, N., & Shaw, R. (2009). Participatory urban environmental management and social capital: Case of Danang, Central Vietnam. In R. Shaw, H. Srinivas, & A. Sharma (Eds.), *Urban risk: An Asian perspective* (pp. 295–318). London, UK: Emerald Group Publishing.

Joerin, J., Shaw, R., Takeuchi, Y., & Krishnamurthy, R. (2014). The adoption of a climate disaster resilience index in Chennai, India. *Disaster*, *38*(3), 540–561.

Kafle, S. K., & Murshed, Z. (2006). *Community-based disaster risk management for local authorities: Participant's workbook*. Pathum Thani, Thailand: Asian Disaster Preparedness Center (ADPC).

Mulyasari, F., & Shaw, R. (2013). Role of women as risk communicators to enhance disaster resilience in Bandung, Indonesia. *Natural Hazards*, *69*(3), 2137–2160. http://dx.doi.org/10.1007/s11069-013-0798-4.

Shaw, R. (2012a). *Community-based disaster risk reduction*. London, UK: Emerald Group Publishing.

Shaw, R. (2012b). Overview of community based disaster risk reduction. In *Community-based disaster risk Reduction* (pp. 3–17). London, UK: Emerald Group Publishing.

Shaw, R., & Goda, K. (2004). From disaster to sustainable civil society: The Kobe experience. *Disasters*, *28*(1), 16–40.

Shaw R. (2014a): Community Practices for Disaster Risk Reduction in Japan, Springer Publisher, 239 pages.

Shaw R. (2014b): Disaster risk reduction and community approaches. In R. Shaw (Ed.), *Community practices for disaster risk reduction in Japan* (pp. 3–20). Japan: Springer.

Shaw, R., & Okazaki, K. (2003). *Sustainability in grass-roots initiatives: Focus on community-based disaster management*. Kobe, Japan: UNCRD Publication.

Shaw, R., Shiwaku, K., Kobayashi, H., & Kobayashi, M. (2004). Linking experience, knowledge, perception, and earthquake preparedness. *Disaster Prevention and Management*, *13*(1), 39–49.

Surjan, A., Redkar, S., & Shaw, R. (2009). Community-based urban risk reduction: Case of Mumbai. In R. Shaw, H. Srinivas, & A. Sharma (Eds.), *Urban risk: An Asian perspective* (pp. 339–354). London, UK: Emerald Group Publishing.

Victoria, L. (2009). Community capacity and disaster resilience. In R. Shaw, & R. Krishnamurthy (Eds.), *Disaster management: Global challenges and local solutions* (pp. 338–351) Hyderabad, India, United Press.

Victoria, L. P. (2003). *Community-based disaster management in the Philippines: Making a difference in people's lives*. Bangkok: ADPC.

Climate Migration and Urban Changes in Bangladesh

19

Reazul Ahsan[1], *Jon Kellett*[2], *Sadasivam Karuppannan*[3]
[1]Research Fellow, MIT-UTM Sustainable Cities Program, Massachusetts Institute of Technology, Cambridge, MA, USA; [2]University of Adelaide, Adelaide, Australia; [3]University of South Australia, Australia

19.1 Introduction

The concept of climate change is defined as a gradual change of global physical conditions, which encompass change of air and sea temperatures, rise of sea level, changes in precipitation levels, and seasonal variation (IPCC, 2007). Such changes are considered as long-term processes and are barely noticeable in everyday life, however the impacts of these changes are likely to undermine the essentials of human survival. These essentials include food, fresh water, health, home, land, and employment. Human activities are largely responsible for the current global climate change trend, and the onus is on humans to identify the likely changes and deal with the consequences. Some expected climate changes and associated impacts will likely be higher average temperature that will affect seasonal variation and crop production, as well as changes in precipitation level, which will increase drought in certain regions. In addition, the rising sea level will increase the risk from extreme climate events, notably cyclones, frequent and prolonged floods, tidal surges, and salinity intrusion (Ahmed, 2011). Climate change impacts manifest in distinct ways in different parts of the world as a result of factors such as geographic location, economic and industrial development, and social structure. For example, the annual mean rainfall trend has increased in southeastern China, the Arabian Peninsula, and the western coast of the Philippines, along with south Asian regions (IPCC, 2001). With the change of mean temperature and sea surface temperature, the frequency and intensity of tropical cyclones in the Pacific has increased over the past few decades. While the frequency of the cyclones originating in the Bay of Bengal and Arabian Sea have decreased, the intensity has increased at least five-fold (Lal et al., 2001). Across the world, hundreds of millions of people are exposed and vulnerable to different extreme events, which may displace more than 200 million people from their land in the near future. Such displacement will form a new social group called "climate migrants." Rising sea level and temperature change will increase the number of climate migrants, particularly in small island states, low-lying deltas, and developing countries. This migration will increase poverty and provoke a crisis for fresh water supply, shelter, food, and energy, leading to socioeconomic disparity.

Half of the world's population lives within 60 km from a coast while three-quarters of the world's cities are located along a coast. In developing countries, much of the

Urban Disasters and Resilience in Asia.

urban population resides in areas exposed to climate change with systems poorly equipped for adaptation. About one billion people live in urban slums, and another 50 million are likely to be displaced by climate factors and will contribute to urban slum populations in Asia, Africa, and Latin America (UNHABITAT, 2008). Urban slums are usually the first destination for displaced people. In these slums, they will have limited access to employment, water, shelter, sanitation, and basic amenities. Often the refugees trade physical safety for the basic human rights they previously enjoyed, namely the right to development and the right to live in their own society and culture. Most of those migrants are internally displaced (rural to urban areas), and only a minority crosses an international border.

Vulnerable and exposed communities across the world are already suffering climate change and adopting different adaptation solutions as a matter of survival and to support their livelihoods. Amongst different social and economic adaptation approaches, involuntary migration is notable, especially for the vulnerable coastal communities in the developing world (Gray and Mueller, 2006). When their local administrative authority fails to cope with environmental hazards, vulnerable households select their own adaptation strategies and often choose to migrate to urban areas. Nevertheless, climate migrants need support when they move to new locations. Their demands are increasingly challenging the status quo of towns and cities that attract a large share of climate migrants. Parvin and Shaw (2011) note that climate migrants are usually poor and usually settle in congested inner city areas for easy access to jobs or on the urban fringe for low rent. Such unplanned and illegal land occupancy changes the urban land use, expanding cities and increasing urban poverty. Socioeconomic inequalities are increasing in such urban areas. Migrants who have lost almost everything due to climate change can often become trapped in a downward spiral of urban poverty when they reach their migration destination. Climate migration is now documented by various international studies and organizations, but very few of these focus on how the migration destination's urban system is impacted. The question of how to address this new migration type and provide housing, services, and economic opportunities is proving to be a major challenge. The prevailing urban system in poor and developing countries like Bangladesh is already vulnerable to climate change, but is often unprepared to deal with unplanned and unpredicted urbanization due to a lack of resources.

This study explores climate change as a potential trigger of population displacement using Bangladesh as a case study. Bangladesh is a low-lying coastal zone where internal migration has long served as an adaptive strategy but where climate adaptation is seldom addressed in development policy. Almost two-thirds of the total population lives in coastal areas and their livelihoods depend on the local climate. We consider aspects of forced displacement that suggest that the aggregate of thousands of individual adaptation decisions has long-term consequences for the migration destination's society; in this case for the cities of Bangladesh (McLeman and Smith, 2006). We use a poststructuralist approach to investigate climate migration and its impact on the urban system, examining the processes and conditions by which the human identities of climate migrants are constructed, using both quantitative and qualitative methods.

19.2 Research Design

The southern coast of Bangladesh is home to 35 million inhabitants at a density of 738 people per square kilometer. This population is projected to increase to between 40 and 50 million by 2050 (Hossain, 2009). These coastal communities mainly depend on subsistence agriculture and fishing from the fertile plains along rivers and the coast, and are the prime victims of extreme climate events. To understand climate migration and urban changes in Bangladesh, this study conducted a field investigation in one of the country's southern coastal cities, Khulna, which is the third largest city and divisional headquarters for the southern coast with a jurisdiction of 4394.46 km^2. The total population in Khulna is 2.5 million, and is growing 3.8% per annum (BBS, 2011). Climate migrants coming from the poor communities along the coastal areas have settled in the urban fringe and slums, where they have cheap accommodation and access to low-paid employment. With the help of local and national nongovernment organizations (NGOs), namely Pothikkrreith and Sheba, a mass scale dialogue program (i.e., an audio visual presentation on climate change and displacement) was conducted with the residents (approximately 700 people) in Khulna to inform them of the research aim and to identify targeted people who had migrated due to varying climatic hazards. From that mass-scale dialogue and participant list, 200 respondents were chosen to be the targeted community. The targeted people had all migrated or been displaced due to cyclones, flash floods, river erosion, or drought. The selected stratum was chosen from different Thanas (a Thana is a local government administrative boundary, mainly the area under a police station jurisdiction) so that diversity could be reflected in the sample. Of the 200 selected people, 100 respondents were chosen through systematic random sampling, in which every alternate person was chosen to provide answers for the questionnaire survey.

Over the past two decades frequent disasters, increases in soil salinity, river erosion, floods, tropical cyclones, and storm surges have forced marginal coastal communities to migrate to other destinations in search of alternative livelihoods. Bangladesh has lost thousands of lives due to catastrophic cyclones in recent history, notably Cyclone Sidr in November 2007, which caused nearly 4000 deaths. Moreover the frequency of cyclones in Bangladesh has increased more than five-fold when compared with the past three decades. Following Sidr, Cyclone Aila hit on May 25, 2009. Approximately half a million people were forced to leave their homes. Large areas of productive farmland were rendered useless as a result of inundation and subsequent salinity, and others were permanently flooded, forcing residents to migrate. Such migration has taken place mostly into the southern coastal cities in Bangladesh, such as Khulna and Barisal. In their search for alternative livelihoods displaced communities generally opt for urban destinations where employment opportunities and livelihoods are perceived to be available; thus, those displaced can be labeled climate migrants.

19.3 Defining Climate Migration and Migrants

Traditional migration theory suggests that economics drives migrant decision-making. However, natural disasters around the world provide a further category of environmentally stimulated migration; which is not a new phenomenon in human mobility, but is becoming more widely understood in recent years. For example Cyclone Nargis (2008) in Myanmar and recent earthquakes in Sichuan, China temporarily displaced almost 17.4 million people combined (IFRC, 2008). Approximately 1.5 million people were temporarily displaced by Hurricane Katrina in 2005, and an estimated 300,000 were displaced permanently (2011). Among those temporarily displaced people, approximately 107,000 were illegal immigrants who experienced secondary displacement due to Katrina; these people were already migrants when natural disasters forced them to move again (Castillo, 2005). The increasing rapid onset of naturally hazardous events is likely to result in substantial human displacement from different parts of the world by 2050. The IPCC assessment report (2007) addresses natural hazards and the resulting forced migration, suggesting that such a development could increase migration to a new magnitude. The Office of the United Nations High Commissioner for Refugees (UNHCR, 2002) estimated approximately 24 million people have fled because of floods, famine, and other environmental degradation processes. A widely cited report on human mobility by Brown (2007), projects that, by 2050, the number of climate-induced migrants could be 200 million. Furthermore, another 700 million will remain under threat because one in every 11 people at that time will be affected by environmental factors (Christian Aid, 2007). Brown (2007) suggests that four decades from now, environmentally induced migration could be an issue for 3% of the current world population. The German Advisory Council on Global Change (WBGU) addressed environmental degradation as drought, desertification, flood, cyclones, and other forms of water scarcity. WBGU estimates that these environmental issues will affect as much as one-third of the world's population, forcing vulnerable populations to secure their livelihoods (Kolmannskog, 2008). The research on migration and environmental degradation stresses that certain parts of the world, such as parts of the Middle and Far East including India, Pakistan, and Bangladesh could experience severe impacts from environmentally induced migration; the vulnerability of these nations is enhanced by their large populations and relative poverty (Fernando et al., 2010).

In addition to extreme events such as hurricanes and cyclones, climatic factors such as desertification, drought, and irregular rainfall pose a threat for people who are already vulnerable to socioeconomic factors such as poverty and malnutrition. These "gradual onset" climatic factors represent a steady worsening of local conditions and can act as an indirect force to migrate (Edwards, 2008). The common factor in all these examples is climate. Displacement due to climate stress is well understood, but there is very little consensus when it comes to defining "climate" or "environmental migrants." Researchers try to draw a line to distinguish "economic migrants" and "climate migrants." A common argument suggests that economic migrants respond to both push and pull factors, whilst climate migrants have no choice but to flee, and are responding only to push factors. These climate migrants often have limited options or choices in terms of destinations in the face of an immediate emergency and may come to represent a new type of migrant because they could become serial, multi-stage migrants.

Environment and migration issues gained some traction in 1985 with the report entitled "Environmental Refugees," published by the United Nations Environmental Program (UNEP). Since the publication of the report, social scientists, demographers, and advocacy groups have produced burgeoning literature regarding this category of migrants and how it might fit within traditional migration theory. The UNEP (2007) definition of environmental migration focuses on causality, reflecting at least part of the neoclassical definition in respect of push factors. It defines environmental refugees as:

> *those people who have been forced to leave their traditional habitat, temporarily or permanently, because of a marked environmental disruption (natural and/or triggered by people) that jeopardized their existence and/or seriously affected the quality of their life. By 'environmental disruption' in this definition is meant any physical, chemical, and/or biological changes in the ecosystem (or resource base) that render it, temporarily or permanently, unsuitable to support human life."*
>
> *p. 312*

The UNEP definition does not provide generic criteria distinguishing environmental refugees from other types of migrants, nor does it specify differences between environmental migrants and refugees. The United Nations Refugee Convention of 1951 did not cover environmental degradation as a motivation to make a migration decision, so it has become a challenge for theorists to define such migration under mainstream migration theory (Bates, 2002). Dun and Gemenne (2008) suggest that the victims of climate events are climate migrants. They argue that communities that are displaced solely because of climate change and hydrometeorological disasters are most likely forced to migrate and often suffer hardship and economic marginalization at their migration destination, which should entitle them to some protection under the UN 1951 Refugee Convention. These migrants have very limited options when selecting their destination because their date of departure is immediate and they are often completely uncertain about their stay duration; as it could be a short, long, or even permanent migration. Often their migration occurs within their country of origin. Baird (2008) shows that, in many cases, the impact of climate change is felt most acutely by the minority and indigenous communities, who often live in the most marginal areas and are already vulnerable to other social conflicts. Many scholars recommend different approaches to define climate refugees and achieve a durable solution that can fit within the UN 1951 Refugee Convention and 1967 protocol and place climate change-induced forced migration under a legal framework (Romer, 2006). Renaud et al. (2007) proposed to address these vulnerable communities as "environmental displaced person(s)," or in short "EDPs" and encourage governments to recognize their plight and provide them with support. A factor that may distinguish these EDPs from refugees under the 1951 Convention is that they may move only within their country of origin. Thus, their putative refugee status may become an issue for internal migration policy.

Social scientists and demographers classify the environmental migration process in three different categories. First, in places affected by a gradual environmental degradation process, people may have time to make a decision to either cope with these environmental changes with some external assistance or move away. Those who move away from the affected area could be referred to as "environmentally motivated migrants." Over long periods of time, increases in drought or floods would be the prime motivations for semi-arid and low-lying coastal communities to migrate. Such migration is considered as regional migration and is difficult to predict (Piguet, 2008).

Second, if the migration decision is supported and triggered by environmental degradation or by frequent natural and hydrometeorological disasters and there is not enough support available to cope with or to resolve these threats, then that migration decision may be considered “forced migration.” Third, if there are any sudden direct or indirect environmental changes, victims do not have time to react and if not enough help is available, then migration is their only option to survive. These people can be labeled as “environmental refugees” (Renaud et al., 2007). The obvious example is the case of low-lying island communities, which are increasingly susceptible to inundation as sea levels rise and storm severity increases.

The fourth IPCC report addressed migration due to climate change factors through two different channels: drought and cyclone. However the IPCC report does not address forced or climate migration as a direct consequence of sea-level rise (IPCC, 2007). Despite the debates on climate refugees or migrants due to climate change actions, a working definition of climate change-related migration was designed by Kniveton et al (2008) and supported by the International Organization of Migration (IOM) (2008):

> *persons or groups of persons who, for compelling reasons of sudden or progressive changes in the environment that adversely affect their lives or living conditions, are obliged to leave their habitual homes, or choose to do so, either temporarily or permanently, and who move either within their country or abroad.*
>
> *p. 31*

There are no complete definitions of climate migrants or environmental migrants although there are a number of studies that have been conducted to document historical evidence to link climate change and migration. Recent work outlined the role of the environment in migration related to Hurricane Katrina in the United States (Reuveny, 2007) and the coastal cyclone in Bangladesh (Kniveton et al., 2008). It is clear that the conventional economic interpretation of migration is stretched by the increasing severity of environmental push factors, which in extreme cases leave populations no choice but to move. Most of these people will move within national borders and may move more than once, usually to larger scale urban settlements, as they perceive better opportunities there. Thus, individuals and families which began as climate refugees may shift their status and become economic migrants, either within the borders of their home country or perhaps internationally, as they make successive moves. These incremental movements will inevitably impact the location to which these populations migrate, particularly as the scale, frequency, and severity of climate-driven events increases and the numbers of migrants proportionally increase.

19.4 Climate Change and Bangladesh

In this chapter, Bangladesh is presented as a case study in order to illustrate the previous discussion. The average elevation of coastal lands in Bangladesh is less than 1.5 m above Mean Sea Level (MSL). According to UNFCC (2007) with 1 m of Sea Level Rise (SLR), approximately 15 million people, or 11% of the total coastal region population, and 17,000 km^2, or 18% of arable land, will be permanently submerged. Associated with this loss of land, 60% of the total population will be affected by flash floods. Furthermore, with 1 m of SLR, 8000 km of roads and two major cities could be affected (Islam, 2001). Much of the population in these vulnerable coastal regions will have no choice but to migrate to other parts of the country as their homes and land are destroyed. There is also the risk of serious health impacts due to shortages of water and the polluted water supply in their original settlements, where these areas survive.

Population, settlement, and hydrology share a long and tense history in Bangladesh. The topography, high rainfall, and frequently shifting pattern of water courses mean that, over time, the people and settlements have moved as natural conditions have changed. There is a historical record of settlements that have become depopulated as rivers have changed their courses and the local population has sought more favorable territory. As Siddiqui (2008, p.6) points out:

> *In Bangladesh, migration is considered as a livelihood strategy for both the poor and better-off. Choice of migration destination, migration cost, level of benefits and risks vary significantly, based on the economic and social network of the migrant.*

While the constant adaptation to changing conditions appears to be the historic norm in Bangladesh, the onset of rapid climate change is accelerating and intensifying this process, to the point that existing urban areas are struggling to cope.

Geographically, Bangladesh is located between 20°34'–26°-38' N and 88°01'–92°41' E. Bangladesh is bordered on the west, north, and east by India, on the south-east by Myanmar, and on the south by the Bay of Bengal (Agrawala et al., 2003). The country is located at the unique juxtaposition of the composite, sprawling, second-largest river system in the world, the Ganges- Brahmaputra-Meghna. This river system drains an area of 1,086,000 km^2 from China, Nepal, India, and Bangladesh. Due to its unique geographical location, the nation has rich biological diversity, hosting a variety of species which populate the ecosystems of the country (Aminuzzaman, 2010). However, much of the terrain is low-lying, with more than 80% of the country occupied by floodplains. Mean elevations range from less than 1 m on the tidal floodplains to the northeast Sylhet basin, which is up to 6 m above MSL. There are only elevations greater than 30 m above MSL in the northwest of the country (Rashid, 1991). The nation is rapidly urbanizing, although by international standards, it has a high rural population. More than a quarter of the total population (28%) lives in major urban areas in Bangladesh. Dhaka, the capital, has a very high population density at 1,502 people per km^2. Uneven development, natural hazards, low economic growth, and lack of employment in rural areas, along with rapid urbanization make the country vulnerable to climate change (Islam, 2007). From 1974 to 1981, Bangladesh experienced a high urban population growth rate of 10.3%, due to economic push and pull factors (BBS, 2001). The pull factors included both the economic opportunities of jobs and the social benefits of better education and health in the cities. The push factors included poverty, natural disasters, and ethnic discrimination. Migrants moved both from rural areas and from small towns to adjacent large cities. Each year, between 300,000 and 400,000 migrants migrate to Dhaka to improve their economic prospects.

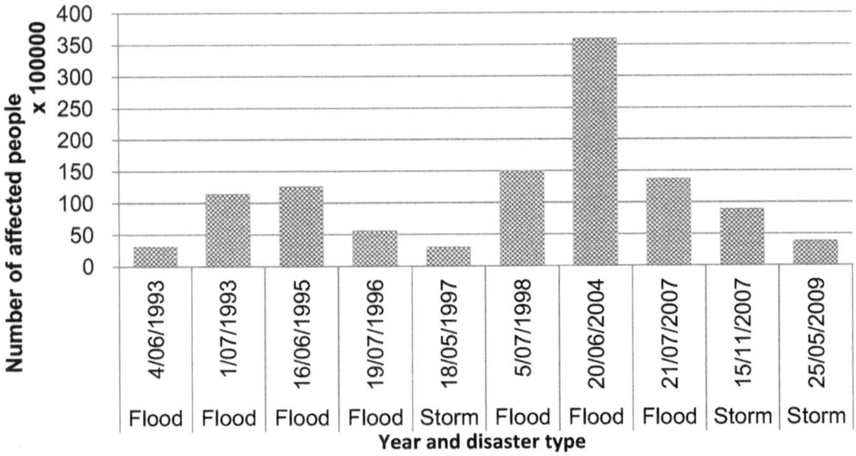

Figure 19.1 Number of people affected in a number of natural disasters in Bangladesh since 1993.

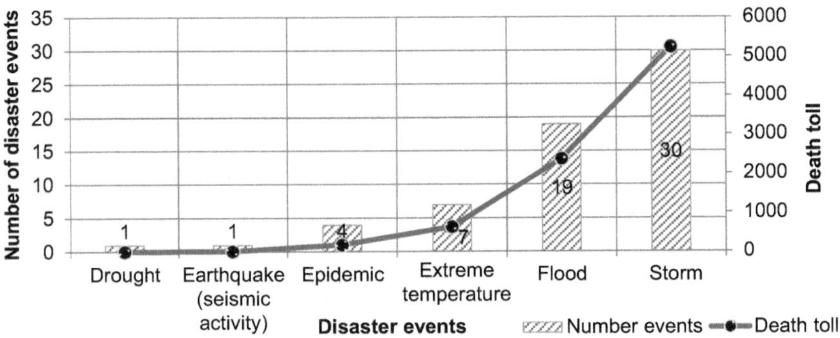

Figure 19.2 Loss of human lives in different natural disasters in Bangladesh (2000–2010). *Source:* Adapted from EM-DAT (2010).

Therefore, internal migration is the dominant factor driving rapid urbanization in Bangladesh. This scale of migration presents a challenge for balanced development (Rana, 2010). Figure 19.1 shows the number of natural disasters that have affected Bangladesh since 1993 and the total number of affected people. Concurrent natural disasters at high intensity forced many victims to consider migration as their only viable adaptive response (EM-DAT 2010).

Over the past decade, Bangladesh has been affected by 63 major natural disasters which have caused significant monetary loss, as well as loss of human lives. Figure 19.2 shows the death toll in Bangladesh as a result of different natural disasters up to 2010 (EM-DAT 2010).

Over the past two decades, extreme climate events including flash floods and cyclones have increased the rate and volume of adaptive migration. This is particularly the case after 2006. The sample survey of climate migrants conducted in 2010 for

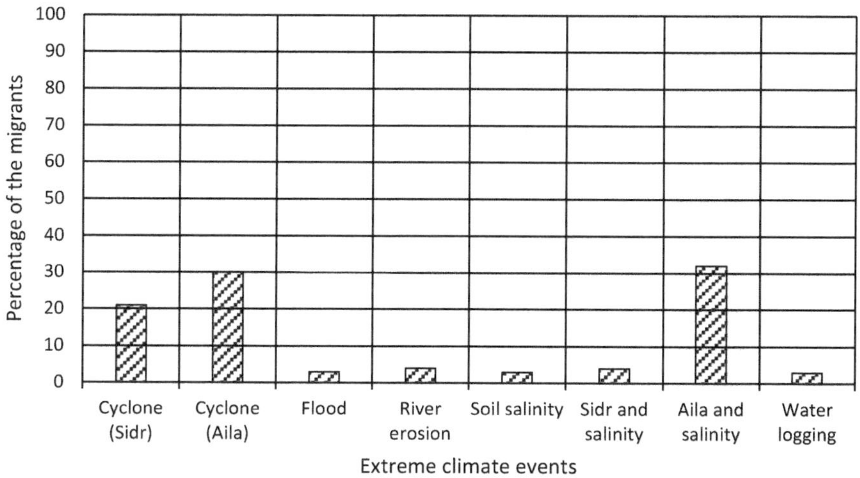

Figure 19.3 Migration trend due to extreme climate events from 1980–2010.
Source: Ahsan et al. (2014).

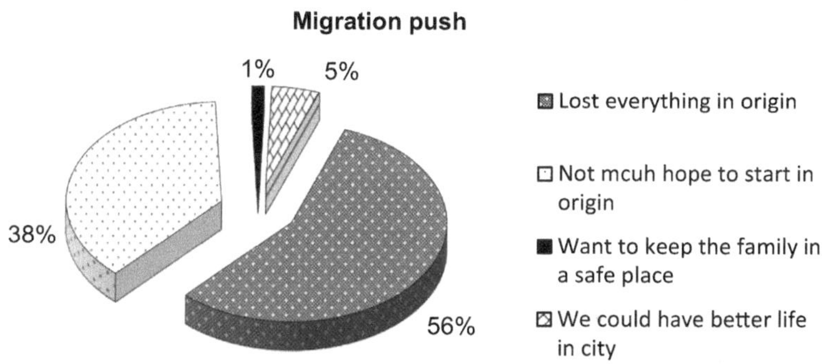

Figure 19.4 Push factors forcing the coastal community to migrate to urban areas.
Source: Ahsan et al. (2014).

this study suggests that the volume of migrant flow is clearly influenced by extreme climatic events. Figure 19.3 shows the migration trend based on field survey of 100 heads of households, who were identified as climate migrants (Ahsan et al., 2014).

Field research identified migrant households that became residents in the nearest urban center, with some migrating to the capital in search of better livelihood and safety from 1970, following a coastal flash flood. The largest number of migrants identified themselves as driven out by Cyclone Aila in May 2009. Estimates suggest about half a million people were forced to leave their homes, more than 200,000 people were trapped in water, more than 300 people died, and 1120 people were missing as a result of this cyclone (Land, 2010). The field research also demonstrates that forced migration may be considered an adaptive measure to a range of different climate-driven factors as addressed in Siddiqui and also demonstrated in Figure 19.4.

It is not simply the cause that is a matter for concern, but the treatment and conditions these people face when they arrive at their destinations. The migrant communities studied faced natural hardships and hazards at their point of origin, however these hardships did not end when they arrived in their urban destination. In the urban slums, which are usually the first destinations for these climate-displaced communities, the migrants experience multiple aspects of deprivation, which restrict their efforts to obtain a decent standard of living. For example, more than 65% residents living in urban slums in Khulna and Dhaka do not have any sanitation facilities, 45% of slum dwellers do not have a fixed place to dispose of their garbage, and 35% are outside of the garbage collection system (CUS, 2006). This increased influx of migrants, driven by extreme climate events, poses new challenges for the existing urban system. Next, we focus on urban conditions and the problems posed by increased rural to urban migration based on our fieldwork conducted in Khulna.

19.4.1 Climate Migration and Urban Changes

The combination of long-term gradual degradation, such as soil salinization and extreme climate events, can eventually force rural coastal farmers and fishermen to leave their traditional lands and move to the nearest city. When forced to leave hurriedly, they may bring few if any possessions with them. When more time is available to make the decision to leave, they may be able to collect some household goods, but their real wealth—their land and established way of life— is inevitably left behind.

19.4.2 Climate Migration and Social Changes

The first destination for most of the climate migrants to Khulna was its slums and squatters settlements to which they had easy access, but the migration destination offered them little or no improvement in their housing because they could not afford the cost of formal housing. Not all the migrant households could gain access to established urban slums; only a few who had relatives or social networks could access or rent a place in the established urban slums. Most of these urban slums are maintained or owned by "musclemen" (local community leaders whose power and influence stems from the political leaders and current ruling party of the country), who did not allow just anyone to live in their slums. Migrant households who could not afford or gain access to the urban slums usually established themselves as squatters in urban fringe areas, on marginal agricultural land, along rail corridors, next to the highway, or even in the natural drainage network, as well as in low-lying flood-prone areas and on river banks. These migrants settled wherever they could get free or cheap land. In most cases, they built their own informal shelters with dry leaves, plastic bags, bamboo, and mud on whatever land they could access. In some cases, migrant households chose to live on public construction sites (bridges or culverts). Sometimes, private landowners allowed these migrant families to live on their agricultural land or construction sites in order to protect their land and provide security for the assets on the site. These informal settlements provide immediate shelter for the migrant communities, but impact the existing land use pattern. Figure 19.5 shows the location of the climate migrants in Khulna's urban area (Ahsan, 2013).

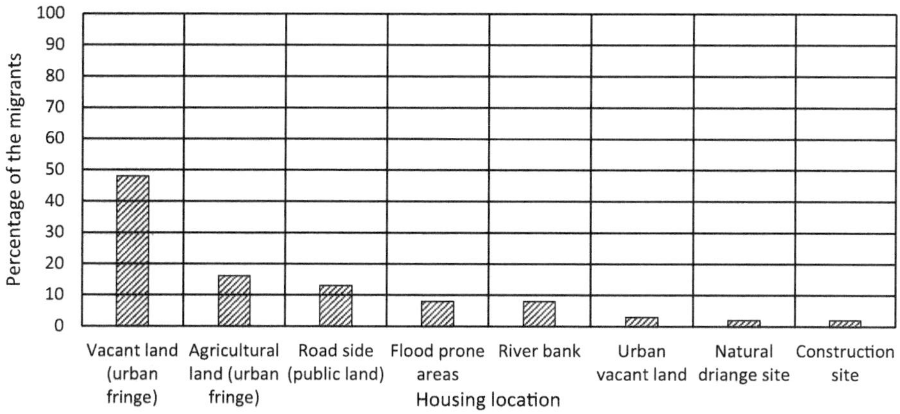

Figure 19.5 Location of climate migrants in urban areas.
Source: Ahsan (2013).

Approximately 47% of the total respondents in Khulna settled in urban fringe areas where they found cheap or free access to land. This urban fringe land has been preserved for further urban expansion or to support landless or homeless communities. Local leaders (i.e. locally elected members representing local government authorities and local political leaders) rent this vacant land to the urban poor, especially migrant communities, and allow them to build their houses and use the land as a source of income. Sometimes, local leaders build informal housing and rent to the urban poor and migrant communities. Usually, housing on these vacant lands is available to all newcomers and does not require any previous tenancy records; therefore, most of these migrant communities along with the urban poor prefer to build informal structures to live there. In Khulna, 18% of the sample of climate migrants lived on agricultural land in urban fringe areas, mostly with the permission of private owners. The most marginal 14% of respondents in Khulna had no place to live, but accommodated themselves by the side of highways on the land preserved by the government for road safety. Others lived on riverbanks (8%) and on flood-prone, low-lying land (8%).

Rapid access to housing is one of the key concerns for migrants, which forces them to take whatever they can get. Finding a place to live for free or at a very low rent is a priority. As a result of the constraints placed upon them by adverse circumstances, they tend to have little regard for the impacts they impose on the urban system. For example, those who occupy low-lying areas, culverts, and drainage systems can block the flow of water with their waste and makeshift structures, affecting the drainage of surrounding areas. Access to sanitation and waste disposal also directly affects surrounding neighborhoods. Figure 19.6 depicts existing informal settlements and service facilities, such as sanitation facilities of migrant communities along roadside urban vacant land and low-lying flood plains preserved for urban protection and natural drainage.

Many of these squatter settlements are illegal and present hazards to the safety and health of their occupants in addition to creating problems for their surrounding areas. A survey of local government policy documents and interviews with government

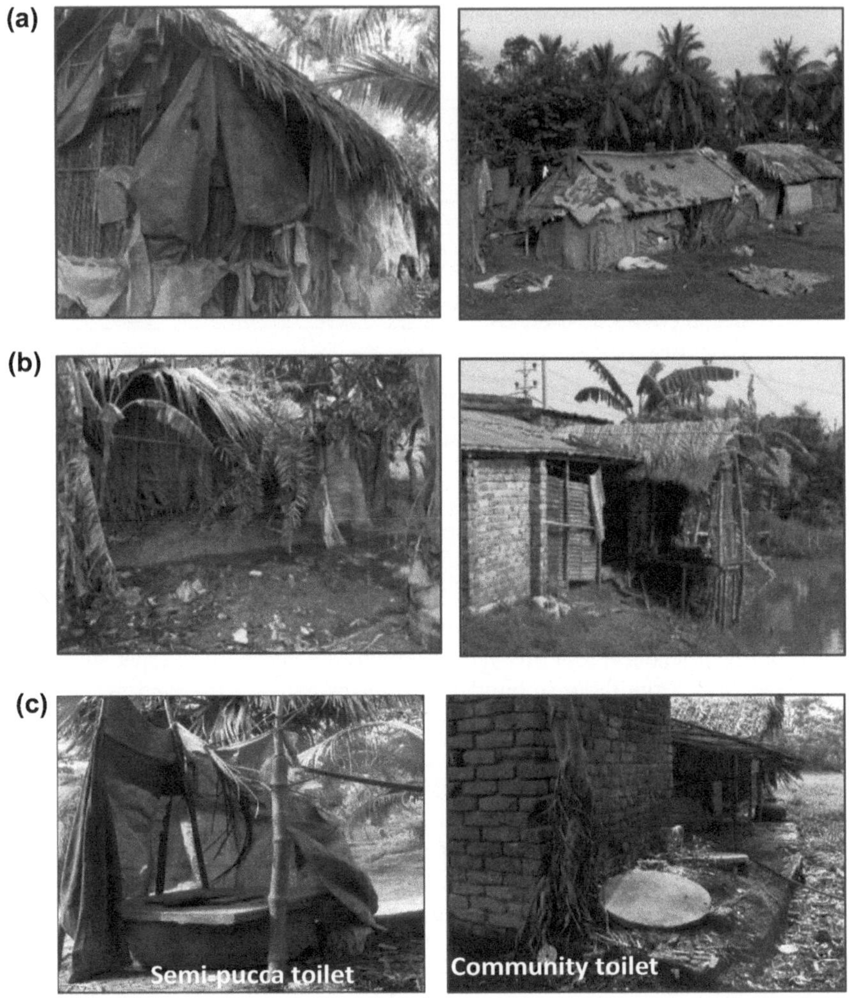

Figure 19.6 Informal Settlement and available service facilities for forced migrants. (a) Informal settlements in urban agricultural land and along the highways in Khulna. (b) Informal settlements in urban fringe and low-lying areas in Khulna. (c) Sanitation facilities for urban poor and the migrant residents in fringe and urban areas.

officials revealed that there has been little official action to address the impact on urban areas of internal climate refugee settlements. This issue has not yet been addressed in the National Adaptation Program of Action (NAPA), 2005. The Ministry of Environment and Forests, which prepared the NAPA 2005, agrees that some of the effects of climate change are linked to migration but it does not prescribe any adaptation program or policies specifically related to the issue of climate migration (Displacement Solutions, 2012). Unsurprisingly, given this policy void at the national level, local governments have also failed to address migrants as a consequence of climate change, but treat them merely as part of the urban poor with no specific guidelines to support or accommodate them.

19.4.3 Climate Migration and Economic Changes

The survey demonstrated that climate actions not only forced migrants to change their place of residence, but also to change their occupation in order to survive in an urban environment. Migrant communities were forced to become involved with different economic activities to support themselves and since they often have no experience beyond farming or fishing, the city offers them few options. An analysis of the change of occupation of migrant household heads was performed, which demonstrated the financial and social impact of the occupation change on the migrant households. For example, most of the jobs available in urban areas are daily basis jobs. Thus, the migrants lack job security and, as they might not get a job every day, they are increasingly vulnerable. Migrants often need to work longer hours to meet the cost of living in the city. Table 19.1 shows the range of employment of the migrant households surveyed and demonstrates the economic solutions available to them to survive in the urban environment as unskilled laborers.

Table 19.1 indicates that nearly half (49%) of the total respondents who were working as day laborers in the city had different occupations in their place of origin. For example, they had been farmers, seasonal laborers, and small businessmen, but now were forced to be day laborers. Women were also forced to become day laborers to maximize the family income. Approximately 24% of respondents were rickshaw pullers and almost none of them had had previous experience in labor-intensive jobs. Among the rickshaw pullers, one was a college student; he had no other option but to work to support his family whose land had been waterlogged in their place of origin. Housewives were forced to work as day laborers and domestic helpers to support their families in the city. This change in economic status could also impact social relations and self-esteem. The following cases demonstrate how climate migration and occupational change can affect the social and family structures of migrant households.

19.4.3.1 Case 1

Mariam, a 32-year-old woman, migrated with her husband and two sons from Kyora due to Cyclone Aila. Mariam's family decided to migrate because they had lost everything including their home and agricultural land. Her husband worked as a day laborer, her 8-year-old son worked in a shop, and her 4-year-old son stayed at home with her. Her husband and son's income combined were not enough to pay rent and buy food in the city because they did not get work every day. Therefore, her husband forced her to earn extra income, but she failed as she had never been to the city and was not skilled enough to fit into the urban job market. Due to her inability to support her family, her husband divorced her. Now, Mariam is living on the urban fringe with her 4-year-old son and is serving as a housemaid. Forced migration has made her earn an income but has taken her family.

Table 19.1 Occupation of Migrant Household Heads in Urban Areas Compare to Their Previous Occupation at Their Place of Origin

Occupation at Place of Origin	Occupation at Migration Destination									
	Day Laborer	**Rickshaw Puller**	**Petty Business**	**Construction Worker**	**Office Work**	**Van Puller**	**Jobless**	**Housemaid**	**Total**	**(%)**
Agriculture	11	4		1			2		18	18
Fisherman	8	2	2	1					13	13
Fishing business person	8	7	4			1			20	20
Petty business person	5	3	2	1	3				14	14
Land laborer	13	4		1		4			22	22
Housewife	1			1			1	1	4	4
Day laborer	3	3		2					8	8
Student		1							1	1
Total	**49**	**24**	**8**	**7**	**3**	**5**	**3**	**1**	**100**	**100**

Source: Ahsan (2013).

Climate migrants' lack of job security and opportunity for regular income in the informal sector has direct impacts on their household income. The next case is an example of a forced migrant who was well off at his place of origin and forced to leave his land and all his possessions due to Cyclone Aila and find an illegal accommodation next to a highway.

19.4.3.2 Case 2

Mohammed Pahar was well-off farmer back in Kayra Thana. He was forced to migrate due to Cyclone Aila in 2009. Pahar migrated with his two wives and a son because they lost everything and have no hope of returning to resume their former life. They are now staying on government vacant land next to the highway, which is reserved for road safety. Pahar is working as a carpenter in the city, although he is in his late sixties, and his wives are afraid of urban culture and stay at home. His son also works as a salesman in a small shop. They built their own dwelling with bamboo, mud, and palm leaves, which is very temporary and cannot provide any safety or protection from rain or sun. They do not have any toilet facilities; they use local primary school toilets and are only allowed to use them when the school is closed. This is the best that Pahar can arrange and afford for his family after migrating from the coast.

According to the field research, the average income of migrant household heads in Khulna was 4000–5000 taka[1]: only a few earned more than 8000 taka per month and they had regular jobs working long hours. Fewer than 5% of respondents earn more than 8000 taka per month and they are generally office workers or small businessmen. Monthly income varied depending on job availability and a person's working ability. Those who could work six or seven days a week and had regular work were able to earn up to 6000 taka a month. On the other hand, those who worked for one to three days a week might earn as little as 1000–3000 taka a month. Only those who were involved with petty business and those who had formal work could earn 8000 taka or more per month. Rickshaw pullers earned from 2000–7000 taka per month; their income also varied based on their working hours, days, and availability. Paying rent for houses in the slums or a toll for a squatter settlement is new for most respondents, as all of them had had their own homestead prior to displacement. Living expenses in the city were also much higher than at their place of origin. Therefore, to sustain their lives in the city, every household member was involved with some kind of earning activity, except for children younger than 8 and the elderly older than 65 years of age.

The housing conditions of migrant communities in urban areas are highly related to income. Most of the housing structures are informal in nature. These informal structures are mostly made of mud, hard plastic, cardboard, dry leaves, and bamboo; none of these materials are formally used as construction materials (Figure 19.6). Migrant communities have no other option but to build their houses or shelters with the available materials they can obtain for free or at low prices. Houses or structures in the existing slums are constructed with similar materials and provide shelter for the urban poor and migrant communities, but they have very limited facilities such as access to clean water and toilets: the typical size of a slum house was 14 m^2 for a family of four

[1] The Bangladeshi currency is called taka. In May–July 2013, US$1 = 72 taka; therefore, the average income of a migrant household was US$65–75 per month.

persons. The average rent for such informal structures in a slum was 500–650 taka per month, which equates to one-seventh of the average household income. On the other hand, the average size of an individual squatter dwelling on urban public land and in fringe areas was 10 m^2 and the migrants were always under the threat of eviction, either by the government or by local musclemen if they failed to pay the required tolls.

In Khulna, climate migrants clustered on the urban fringe are highly exposed and vulnerable to floods and the soil becoming waterlogged. At the same time, complex urban governance systems have no options or policy strategies to recognize and support these migrants with basic service facilities beyond their service jurisdiction. For example, the study found that the urban fringe where the migrants lived had very poor or no drainage or sanitation facilities. Some of these settlements were established 10–15 years ago, but they still they do not have any basic service facilities, such as drainage provided by the local authorities. According to the field survey, the main drainage facilities available in urban fringe areas are open drains and half of the residents (50%) are using these. Open drains are not covered and frequently overflow, especially during rainy weather. At these times, drain water even flows into houses. Only 8% of respondents have proper concrete drainage facilities. On the other hand, 31% of respondents do not have any drainage facilities because they live next to the roadside and along the drainage networks. Others use natural drainage systems. Since most urban services do not extend to urban fringe areas, migrants in these areas are deprived of access to these facilities and consequently risk an unhygienic and unhealthy life.

In addition to inadequate or nonexistent drainage facilities, migrants in the survey had very limited access to sanitation and water supply. Most migrant households who lived as squatters made their own latrines and those who could not afford the construction cost used common and open facilities such as those in local public schools or Madrasas (colleges for Islamic instruction). Those who lived in slum areas used the common latrines made or constructed by the slum owners. Common latrine complexes in slum areas usually had four toilets and a bathroom shared by at least 60 households (approximately 350 people). In fringe and riverbank areas, some migrant households had constructed personal hanging latrines over the water, which was neither sanitary nor permitted by the local municipal authority. However, because they had no facilities, they created their own. Overflow was a very frequent problem of common latrines, due to excessive use and lack of maintenance. Overflow channels were linked to the nearest ponds and water bodies. Thus, raw sewage escaped into the water and caused another health hazard, especially in monsoons. Most of these latrines were made of slabs and rings; although sometimes there were no slabs on the rings and the bad smell was a constant problem. Those who could not afford ring and slab latrines either used open drains or made some temporary arrangement with bamboo and dry leaves. Figure 6 presents the existing sanitation facilities for urban poor and the migrant residents in fringe and urban areas (Ahsan, 2013).

Migrant communities also had very limited access to a safe drinking water supply for domestic use. Most migrant households were living beyond service jurisdiction; at the same time, the service provider, Khulna Water Supply and Sewerage Authority (Khulna WASA), had a limited capacity to meet the growing water demand. According to *The Daily Star* (2011), Khulna WASA provided 90 million l of water daily to

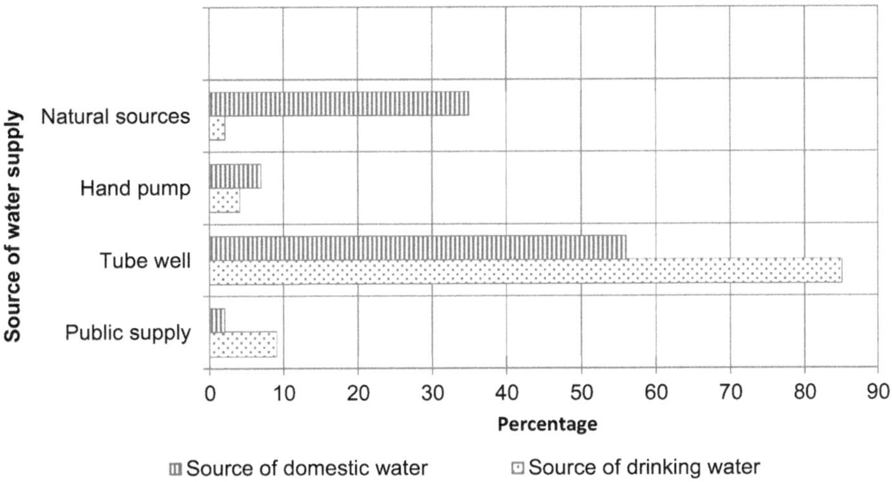

Figure 19.7 Water-supply facilities for respondents for drinking and domestic use.
Source: Ahsan (2013).

meet the demand of 240 million l per day, a shortage of about 150 million l per day. Therefore, the existing services are limited to the city residents, unevenly in the existing slum areas, and not at all for the fringe areas. Some slums within urban areas have piped water supply, but the service is very limited and only available for three hours a day. Tube wells are the main sources of fresh water supply for migrant residents. Some of them also use surface water, such as water from ponds, for everyday use and drink water from tube wells. Figure 19.7 presents the sources of water supply used for both drinking and general use by respondents.

Only 2% of the migrants surveyed had access to piped water supply for their daily use and only 9% used treated piped water for drinking. Piped water was mostly for common uses and was only available during morning sessions for 2–3 hours. Almost 85% of respondents got their drinking water from tube wells and 56% got water for everyday use from the same source. For domestic use, 35% of respondents also used natural sources such as ponds, rivers, and lakes. The 2% of respondents who lived far from other sources used natural water sources (mostly pond water) for drinking purposes. Some residents also used saline water from shallow tube wells (which use water from the surface water table from approximately 15–20 feet deep) and set up those tube wells as community water sources. Approximately 7% of respondents used saline water for domestic use and 4% used saline water for drinking.

19.5 Policy Gap

Bangladesh is a poor country which is rapidly urbanizing and where poverty and lack of urban services for the urban poor are common phenomena. The national government and the urban local authorities are aware of poverty and urbanization issues, but often fail to address the problem of unpredictable internal migration induced by environmental rather than economic factors. The field research found that internal migration towards urban areas increases after any climate hazards hit the coastal areas. For example, migration towards Khulna increased in 2007 after Cyclone Sidr and after Aila in 2009. The Bangladeshi government has two national level policy documents that address climate adaptation. For example, NAPA (2005) suggests that migration due to climate change is an undesirable outcome. NAPA (2005, p. 17) identifies crime as an outcome of the economic impact of climate change, but fails to explain where and when this crime comes from or occurs. At the same time, NAPA also suggested that climate-affected communities could be deterred from migrating to cities if community-based adaptation for food and agriculture could be practiced locally. Thus, the social consequences of mass-scale migration to cities would, to some extent, be halted (NAPA 2005, pp. 35, 36). NAPA agreed that climate victims are migrating to cities, but does not address these victims in the adaptation process.

There have been several initiatives in recent years that have sought to address problems of landlessness and poverty, which have sometimes involved victims of climate change. The aim of the government sponsored "Adarsha Gram" program was poverty alleviation and land reform. Phases 1 and 2 of the Adarsha Gram ended on December 31, 2008. Following this program, the government approved a similar new project titled "Guchhogram" in May 2009. The aim here was to settle landless, homeless, address-less, and river eroded people on public land in a small village cluster with basic support. It was launched by the Ministry of Land (MOL) (Martin et al., 2013). This project addressed climate victim rehabilitation along with the landless and poor. A typical Guchhogram is constructed on a minimum site size of two acres of land. In each village, a minimum of 30 families is rehabilitated. Each of the families receives a minimum of 0.04 acres of land for a homestead. Each is provided with a 300 ft^2 house, a kitchen, and a sanitary latrine. Adequate provision is made for safe drinking water through suitable types of tube-well (MoL, 2013). However the project has been criticized for its lack of transparency and ineffectiveness. In terms of accommodating the climate victims, the project failed to contribute much as there is no clear process to distinguish the climate migrants from the economic migrants. Recently, the project has tended to accommodate the landless poor rather than climate victims (Displacement Solutions, 2012).

It is clear that climate change and migration are not unknown terms in Bangladesh, but the consequences of forced migration and how it can be addressed are not yet clear at the policy level. The Bangladesh Climate Change Strategy and Action Plan (BCCSAP) 2009, the most updated policy guideline for "Climate Change and Adaptation" in Bangladesh, predicts that "hundreds of thousands of people" will be forced to migrate, especially from coastal zones, because of decreasing livelihood opportunities and decreasing agricultural productivity. It estimates 6-8 million people could be displaced by 2050. The policy guideline acknowledges forced displacement and rapid urbanization, but outlines no clear policy lines for migrants (Martin et al., 2013, MoEF, 2008). The policy paper BCCSAP 2009 notes: "climate change is likely to impact most severely on the poorest and most vulnerable in society...[and that] every effort will be made to ensure that they are protected and that all programmes focus on the needs of this group for food security, safe housing, employment and access to basic services, including health" (MoEF, 2008). With this statement the policy paper concurs with our study findings that climate migration affects migrants who need to have safe housing and all other urban facilities to cushion the often severe social impacts of their migration. It still fails to recognize that the ingress of large numbers of climate migrants also impacts on the existing urban system, placing stress on basic services, exacerbating flooding, and threatening public health. Central and local government need to address these challenges and provide a minimum quality of life for migrants to minimize social and urban conflicts. Unfortunately however, the BCCSAP does not propose any specific projects or policies regarding how exactly the government might ensure safe housing and other facilities for the most vulnerable members of society (Displacement Solutions, 2012, MoL, 2013).

Existing development policy in Bangladesh clearly addresses migration as a livelihood and development strategy. The sixth five-year action plan (FYP 2011–2015) and the National Strategy for Accelerated Poverty Reduction and the Millennium Development Goals give a national perspective, and the Coastal Zone Management Plan provides a regional focus on migration and national development. However all these strategies mainly address migrant workers who cross the international border and contribute to the Bangladeshi national economy through remittance. These are people who migrate to the Middle East and European countries in search of work. These planning and strategy documents do not tend to address internal migration, especially not climate change migration (IMF, 2013). Current development and economic policy in Bangladesh do address migration as a part of the development process, but fail to address forced migration induced by climate change and its impact on the social and economic system.

Climate migrants are considered an isolated community and, in order to understand their perceptions of social vulnerability, this research encouraged migrant households to identify or rank the main problems they face as climate migrants and separate them from local urban societies. The aim of conducting this problem ranking or prioritizing was to verify the questionnaire-based findings and to understand the social vulnerabilities faced by the migrants in different urban contexts. In Khulna, a total of 46 respondents identified five different problems they face as migrants and the study used a five-degree scale to identify the challenges faced by the migrants in a new social context, with the following ranking model. For example, in Khulna, 41 respondents out of 46 identified "eviction threat" as one of the main problems—it is ranked 4.4 under the five-degree scale. To draw the five-degree scale, this research used the following equation:

$$V = \left\{ \left(\frac{Ro}{Rt} \right) \times Di \right\},$$

where V = value under the five-degree scale, Ro = Number of respondents, Rt = Total number of respondents, and Di = 5, a value of common scale.

Figure 19.8 presents the urban problems prioritized by the interviewees.

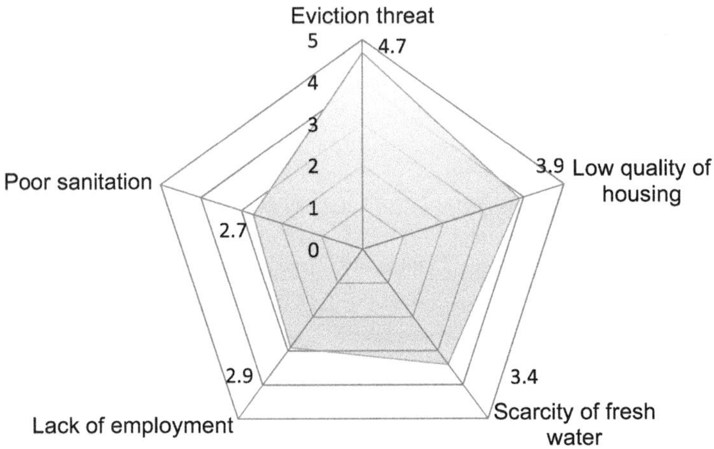

Figure 19.8 Prioritizing urban problems by the migrant households in Khulna.
Source: Ahsan (2013).

In Khulna, the respondents' main concern was always fear of eviction from their home because they were occupying land without any permission or legal approval. Though they were paying rent or protection money, they knew that a landlord, muscleman, or government could evict them without any notice and that they would be on the road again. They also ranked housing quality as another key problem. This relates to the finding that most of the climate migrants in Khulna lived in urban fringe areas and constructed their own housing with locally available materials. Water supply, employment, and sanitation facilities were also ranked third, fourth, and fifth, respectively, on a five-point scale based on the respondents' opinions.

It is important to develop and design innovative and effective rights-based domestic solutions to climate displacement in Bangladesh. Regardless of the lack of understanding of climate migration and policy support, the most truly durable and effective solutions to climate displacement can be found in Bangladesh itself, where there is not only a large amount of public land available for relocation programs, but where common language, cultures, and traditions make relocation more sustainable. The policy gap and lack of understanding of climate migration and its urban impacts is the key limitation of the government's response to climate change, therefore the international community could lend support in developing and implementing domestic solutions, both through technical as well as financial assistance (IMF, 2013).

19.6 Conclusions

Displacement Solutions (2012, p. 34) indicates "climate displacement in Bangladesh will only worsen as climate change increases the frequency of extreme natural hazards that are already leading to displacement across Bangladesh." The field survey findings in this chapter present clear evidence of this statement. Climate events in

Bangladesh induce forced migration from coastal areas towards urban areas. Forced migration due to sudden-onset events is often an immediate decision of those affected. On the other hand, migration due to slow-onset events is a gradual and continuing process. Forced migration due to extreme climate events is not a new phenomenon, but in recent decades it has increased and its consequences for urban areas, which are the usual migration destinations, are becoming more apparent and better understood. Migration due to climate change not only affects migrants' livelihoods, but also affects the broader social and urban system. However, the lack of policy guidelines, both internationally and within the country, fails to distinguish climate migrants from economic or traditional migrants. Although the government of Bangladesh has two national-level policy vehicles which address climate change, neither truly appreciates the scale of the resulting urban problem. They also do not offer any viable policy solutions. Drawing quantitative and qualitative data from surveys of climate migrants in Khulna, we have presented internal migration as an immediate and long-term consequence of climate change and extreme climate events. We have also documented the consequent, knock-on impacts on Bangladesh's cities where these migrants have settled. The condition of Bangladesh's cities is directly and adversely affected by this migration in respect to their amenity, quality of life, access to basic services, and safety. We present the experience of the major social and economic disruption that stems from climate migration in the context of individual stories. This chapter has discussed issues such as arriving at suitable definitions for climate refugees, however, while this represents an important step in the development of policy solutions, it is important not to ignore the devastating impacts on individuals and families that the research discussed in this chapter has revealed. The following quotes from climate refugees interviewed as part of this research summarize the human impact of continuing climate change in Bangladesh:

> *"We lost everything in cyclone Sidr and come to city and now we are moving from one slum to another because the government does not want us to live in city. This is our only option, to move from place to place," said Ibrahim, who is now living in the urban fringe area and is looking for a place where he will not have any threat of eviction.*

> *"We keep on shifting our home because of river erosion, since we do not have any more land to shift, we shift to Dhaka in 2004," said Jhora Begum from Bhola.*

Climate migrants have traditionally been subsumed into the economic migrant category since worsening conditions have prompted populations to move away from a declining natural economic resource base. Their migration continues to be viewed fundamentally as a search for improved quality of life and better economic prospects. In the future, the most demanding concern for the government is to go beyond acknowledging the facts of the case and begin to mainstream climate migration strategies into climate change adaptation policies and programs. A national climate adaptation plan that respects, protects, and fulfills the rights of climate migrants is required. Climate migration is not an impossible problem to address, but it needs to be seen as more than just an increase in the number of urban poor. The Bangladeshi government and indeed the world need to take heed of the climate displacement crisis that is unfolding in this country and act on their findings.

References

Agrawala, S., Ota, T., Ahmed, A. U., Smith, J., & Aalst, M. V. (2003). *Development and Climate Change in Bangladesh: Focus on Coastal Flooding and the Sundarbans*. Paris: Organisation for Economic Co-operation and Development (OECD).

Ahmed, A. F. (2011). *Community-Based Adaptation to Climate Change in Bangladesh: The Global Initiative at Local Level. 5th Annual Community Based Adaptation to Climate Change*. Melbourne, Australia: Monash University.

Ahsan, R. (2013). *Climate migration and urban changes: A study of adaptation in Bangladesh*. Doctor of philosophy thesis (PhD). Adelaide, Australia: University of South Australia.

Ahsan, R., Kellett, J., & Karuppannan, S. (2014). Climate-induced migration: Lessons from Bangladesh. *Climate Change Impacts and Responses*, *5*, 1–14.

Aminuzzaman, S. M. (2010). *Environment Policy of Bangladesh: A Case Study of an Ambitious Policy with Implementation Snag. South Asia Climate Change Forum*. Victoria, Australia: Monash University: Monash Sustainability Institute.

Baird, R. (2008). Climate change and minorities. In I. Matheson (Ed.), *State of the world's minority*. London: Minority Rights Group International.

Bangladesh Bureau of Statistics (BBS). (2001). *Preliminary report on household income and expenditure*. Dhaka, Bangladesh: Ministry of Planning.

Bangladesh Bureau of Statistics (BBS). (2011). *Population & Housing Census 2011*. Dhaka, Bangladesh: Bangladesh Bureau of Statistics, Ministry of Planning.

Bates, D. C. (2002). Environmental refugees? Classifying human migrations caused by environmental change. *Population and Environment*, *23*, 465–477.

Brown, O. (2007). *Climate change and forced migration: Observation, projection and implication. Human Development Series*. Geneva, Switzerland: United Nations Development Programme (UNDP).

Castillo, E. E. (2005). *Illegal immigrants afraid to get storm aid*. Associated Press.

Centre of Urban Studies (CUS). (2006). *Slums of Urban Bangladesh: Mapping and Census, 2005*. Dhaka, Bangladesh: Centre for Urban Studies.

Christian Aid. (2007). *Human tide: The real migration crisis Christian Aid Report*. London.

Displacement Solutions. (2012). *Climate Displacement in Bangladesh*. Geneva, Switzerland, and Portsea, Portsea, Australia: Displacement Solutions.

Edwards, S. (2008). Social breakdown in Darfur. In *Climate change and displacement*. Washington, DC: Amnesty International.

EM-DAT. (The International Disaster Database). (2010). *Emergency Events Database*. US Office of Foreign Disaster Assistance (OFDA) and the Centre for Research into the Epidemiology, Belgium. http://www.emdat.be.

Fernando, N., Warner, K., & Birkmann, J. (2010). Migration and natural hazards: Is relocation a secondary disaster or an opportunity for vulnerability reduction. In T. Afifi, & J. Jager (Eds.), *Environment, forced migration, and social vulnerability*. London: Springer.

Gray, C. L., & Mueller, V. (2006). Natural disasters and population mobility in Bangladesh. *Proceedings of the National Academy of Sciences of the United States of America (PAN)*, *1*, 1–6.

Hossain, R. (2009). *Climate change environmental forced migrants and rural-urban migration: Are big cities prepared in Bangladesh? Urban Poverty and Climate Change Infrastructures of Development Conference*. BRAC University, Dhaka, Bangladesh: BRAC University.

Intergovernmental Panel of Climate Change (IPCC). (2001). Climate change 2001: Impact, adaptation, and vulnerability. In J. J. McCarthy, et al. (Ed.), *Third Assessment Report of the Intergovernmental Panel on Climate change*. Cambridge, UK, and New York: Cambridge University Press.

Intergovernmental Panel of Climate Change (IPCC). (2007). Climate change 2007: Synthesis Report. In A. Abdelkader, et al. (Ed.), *An Assessment of the Intergovernmental Panel on Climate Change*. UK: Cambridge.

International Federation of Red Cross (IFRC). (2008). *China: Sichuan Earthquake. Emergency appeal*. IFRC, China.

International Monetary Fund (IMF). (2013). *Bangladesh: Poverty Reduction Strategy. International Monetary Fund (IMF) Country Report*. Washington, DC: IMF.

International Organization of Migration (IOM). (2008). *Migration and Climate Change. IOM Migration Research Series*. Geneva, Switzerland: IOM.

Islam, N. (2001). *Urbanization, Migration and Development in Bangladesh: Recent Trends and Emerging Issues*. Dhaka, Bangladesh: Centre for Policy Dialogue and United Nations Population Fund.

Islam, N. (2007). *Towards a National Urban Sector Policy. Policy Dialogue Series*. Dhaka, Bangladesh: Centre for Policy Dialogue and United Nations Population Fund.

Kniveton, D., Schmidt-Verkerk, K., Smith, C., & Black, R. (2008). *Climate change and migration: Improving methodologies to estimate flows*. Sussex, UK: Sussex University.

Kolmannskog, V. O. (2008). *Future floods of refugees: A comment on climate change, conflict, and forced migration*. Oslo: Norwegian Refugee Council.

Lal, M., et al. (2001). Asia. Climate Change 2001: Impacts, Adaptation, and Vulnerability. Contribution of Working Group II. In J. J. McCarthy, et al. (Ed.), *Third Assessment Report of the Intergovernmental Panel on Climate Change*. Cambridge, UK: Cambridge University Press.

Land, G. (2010). *Bangladesh: 100 thousands still stranded from cyclones*. Dhaka, Bangladesh: Greenfudge.

Martin, M., Kang, Y. H., Billah, M., & Siddiqui, T. (2013). *Policy analysis: Climate change and migration in Bangladesh. Migration Working Paper 4*. Dhaka, Bangladesh: Refugee and Migratory Movements Research Unit, Dhaka University Sussex Centre for Migration Research, University of Sussex: Climate and Development Knowledge Network.

McLeman, R., & Smith, B. (2006). Migration as an adaptation to climate change. *Climate Change*, *76*, 31–53.

Ministry of Environment and Forests (MOEF). (2008). *Bangladesh Climate Change Strategy and Action Plan 2008*. Dhaka, Bangladesh: Ministry of Environment and Forests, Government of the People's Republic of Bangladesh.

Ministry of Land (MOL). (2013). *Guchhogram (CVRP) Project [Online]*. Dhaka: Ministry of Land (MOL), Government of the People's Republic of Bangladesh. Accessed January 14, 2015, from http://www.guchhogram.gov.bd/index.php.

National Adaptation Programme of Action (NAPA). (2005). *Final Report: November 2005*. Dhaka, Bangladesh: Government of the People's Republic of Bangladesh.

Parvin, G. A., & Shaw, R. (2011). Climate disaster resilience of Dhaka City Corporation: An empirical asessment at zone level. *Risk, Hazards, and Crisis in Public Policy*, *2*, 2–32.

Piguet, E. (2008). *Climate change and forced migration. Research Paper 153: New Issues in Refugee Research Series.* Geneva, Switzerland: United Nations High Commissioner for Refugees (UNHCR).

Rana, M. P. (2010). Urbanization and sustainability: Challenges and strategies for sustainable urban development in Bangladesh. *Environment Development and Sustainability, 3,* 237–256.

Rashid, H. E. (1991). *Geography of Bangladesh.* Dhaka, Bangladesh: University Press Limited.

Renaud, F., Bogardi, J. J., Dun, O., & Warner, K. (2007). Control, adapt, or flee: How to face environmental migration. In *Interdisciplinary Security Connections.* Bonn, Germany: UNU Institute for Environment and Human Security (UNU-EHS).

Reuveny, R. (2007). Climate change-induced migration and violent conflict. *Political Geography, 26,* 656–673.

Romer, K. (2006). *"Environmental" refugees How should government support those at risk of displacement cliamte change?* Forced Migration Review, Refugee Studies Centre, Department of International Development, University of Oxford, UK.

Siddiqui, T. (2008). Migration and Gender in Asia. United Nations Expert Group Meeting on International Migration and Development in Asia and the Pacific. United Nations Economic and Social Commission for Asia and the Pacific Population Division, Department of Economic and Social Affairs Bangkok, Thailand.

United Nations Environment Programme (UNEP). (2007). *Global Environment Outlook GEO4.* Washington, D.C: UNEP.

United Nations Framework Convention on Climate Change (UNFCC). (2007). *Climate Change: Impacts, Vulnerabilities and Adaptation in Developing Countries.* Bonn, Germany: UNFCC.

United Nations High Commissioner for Refugees (UNHCR). (2002). *A critical time for the environment.* Geneva, Switzerland: UNHCR.

United Nations Human Settlements Programme (UNHABITAT). (2008). *Cities and climate change adaptation. UN-Habitat Donors Meetings.* Seville-Spain: UNHABITAT.

Water Stress in the Megacity of Kolkata, India, and Its Implications for Urban Resilience

20

Bhaswati Ray[1], *Rajib Shaw*[2]
[1]Assistant Professor, Sivanath Sastri College, Kolkata, India; [2]Professor, Graduate School of Global Environmental Studies, Kyoto University, Kyoto, Japan

20.1 Introduction

Around one-fifth of the world's population currently lives in regions affected by physical water scarcity with inadequate water resources to meet the total demand, including the water needed to fulfill the demand of ecosystems to function effectively (Watkins et al., 2006). Symptoms of physical water scarcity include environmental degradation and declining groundwater resources. Economic water scarcity is caused by a lack of investment in infrastructure or technology to draw water from rivers, aquifers, or other sources and affects another quarter of the world's population. Symptoms of economic water scarcity include a lack of infrastructure, causing the people without reliable access to water to travel long distances to fetch water. A global water crisis is occurring because the capacity of the hydrological cycle to supply water is being outstripped by the volume of human demand, pollution of water resources, and poor management (Fitzroy, 1998). The crisis is apparent from the global to local scale, including third-world cities, and is accelerated by climate change.

Changes in global and regional climates are projected to influence both the total amount and variation of water throughout the globe (Easterling et al., 2000; Tudhope et al., 2001). Future climate scenarios suggest spatial variation in precipitation responses (Giorgi & Bi, 2005). Many regions are expected to have substantially altered water availability. Due to increased variability in climatic conditions, reduced water availability associated with short-term drought will be more likely to occur in all regions (Royer et al., 2002; Arnell, 2004; Thomson et al., 2006).

Increasing populations, enhanced per capita water use, climate change, and allocations for water conservation are potential threats to adequate water availability. Cities are especially vulnerable to water shortage because their population is highly concentrated and requires reliable water supplies to maintain their intense pace of human and economic activity. With climate change producing variations in rainfall, these challenges could intensify in nearly every region of the world.

With projected increases in urbanization, there will likely be similar increases in per-capita urban water consumption for some regions of the world, although the locations of these increases are sensitive to uncertainties in social and physical dynamics (Alcamo & Henrichs, 2002). Asian cities are the worst affected. Expanding urban population and changing lifestyles, without a proportional increase in civic amenities,

Urban Disasters and Resilience in Asia.

is already putting pressure on water resource management in water-stressed urban centers. Roughly 95% of the urban growth will occur in less-developed countries, with more than 60% of the increase in the world's urban population occurring in Asia (McGee, 2001).

In conjunction with increasing demands on water supplies, there has been an increase in the concentration of human habitation. Currently, more than 50% of humans live in cities (United Nations Centre for Human Settlements, 2001) and the growth of urban water uses, combined industrial and domestic sectors, is associated with the rapid increases in urbanization (Falkenmark, 1998; Gleick, 2003; Jury & Vaux, 2005). It is estimated that the proportion of people living in urban areas will be more than 60% of the total global population, totaling 4.9 billion people, by the year 2030 (United Nations Development Programmes, 2006). The percentage of population living in urban areas would increase to 82% for the developed countries and 57% for the less-developed countries. The number of megacities, with more than 5 million residents, is expected to increase globally from 46 to 61 between 2015 and 2030, with disproportionate increases in Asia and Africa (United Nations, 2004). Ensuring the sustainable supply of water for cities is an increasing concern (Lundqvist et al., 2005). The United Nations focuses on issues of governance as the core of the water crisis. There is enough water for everyone. Water insufficiency is often due to mismanagement, corruption, lack of appropriate institutions, bureaucratic inertia, and a shortage of investment in both human capacity and physical infrastructure (Watkins et al., 2006). Sustainable urban-industrial growth across regions needs to minimize the demand-supply gap of the carrying capacity differentials (Bishop, 1974) in urban infrastructure resources, through both supply management with capacity building and demand management with rational pricing of infrastructure and activity allocation strategies (Joardar, 1998).

The challenge facing humanity is to make cities resilient to future urban growth and ecological degradation. With its origin in ecological sciences, resilience is considered to have the capacity to absorb external shocks without significant deformation (Cutter et al., 2008). The broader socioecological notion of resilience focuses on learning, innovation, adaptive capacity, and transformation (Folke, 2006). The term *resilience* refers to urban systems bouncing back to equilibrium in the aftermath of shocks such as natural disasters (drawing from the classical literature), as well as the adaptive capacities of systems to long-term stressors or shifts such as climate change (Leichenko, 2011; Romero-Lankao & Qin, 2011). One of the major challenges is to build resilient cities, particularly in less-developed nations where cities face multiple stressors. Growing population, polluted ecosystems, and climate change are increasing the vulnerability of cities to inefficient and unsustainable water supply systems, which makes cities less resilient. Increasing vulnerability indicates less resilient cities.

Sustainability development indicators fail to consider resilience or the ability of the system to maintain (or improve) upon its current state over time. Thus, the Water Provision Resilience Index was developed for urban water systems. Built on the "percent of the population with access to safe water," the indicators evaluate the ability of an urban water system to maintain or improve the current level and quality of access over the next 50 years (Milman & Short, 2008). Indicators must take into consideration the risks and uncertainty that affect the system (Loucks, 1997), the assimilative capacity of the system (Alberti, 1996; Azar et al. 1996), and the interlinkages between system variables (Gallopin, 1996). Resilience is also commonly understood as the capacity of the system to absorb disturbance and re-organize while undergoing change so as to still retain essentially the same function, structure, identity, and feedbacks (Folke, 2006). Defined as such, resilience includes more than maintaining given system characteristics; it includes the adaptive capacity of the system—its ability to adapt to stresses and changes and to transform into more desirable states. The study focuses on the ability of a municipal corporation to maintain or improve upon the existing water supply mechanism. The most commonly used mechanism for evaluating resilience in water provision is the one put forth by Walker et al. (2002) that seeks to identify potential stresses and areas of intervention to increase urban resilience. Resilience has been addressed as a function of the types and quantities of pressures on the system (stresses) and the ability of the system to accommodate those pressures (response). The long-term functioning of the water supply system is influenced by the accountability of the water supply provider (Stephen, 2000), stakeholder involvement (Loucks, 1997; Sahely et al., 2005) and the durability of the water supply system (Foxon et al., 2002). Six critical aspects of urban water supply systems that impact water system resilience include supply, finances, infrastructure, service provision, water quality, and governance (Milman and Short, 2008).

The scope of the study discussed in this chapter is limited to Kolkata's drinking water supply system, in keeping with the National Water Policy of 2012, which has made adequate supply of drinking water the highest priority.

20.2 Water Stress in Urban India

Urban centers in India are characterized by a rapid and ever increasing growth of population. 377 million Indians, comprising approximately 31% of the country's population, live in urban areas with an average annual increase of 8 million (Census of India, 2011). The rate of growth for the urban population was 3.4% in the decade from 2001 to 2011, which had increased from the 2.1% level of growth from 1991 to 2001. There is a definite concentration of the urban population in large cities and existing urban agglomerations. According to the 2011 census, there are more than 53 million cities that account for about 43% of India's urban population. This includes three cities with a population of more than 10 million. Recent projections show that by 2030, about 600 million Indians will reside in urban areas, an increase of more than 200 million in just 20 years (Twelfth Five-Year Plan, 2012–2017). The ranking of world cities by population size places Mumbai and Delhi among the top 10 cities, while Kolkata is among the top 15 (United Nations Human Settlements Programme, 2011), indicating the massive size of the megacities in India.

Under the impact of rapid urbanization and expanding city size, supplying water to urban centers is a major cause of concern. Most Indian cities face severe water stress. The concerns of the city water supply pertain to quantity, quality, and equity across different segments of the population, inefficient wastewater management, and a lack of long-term vision. At the aggregate level, although nearly 91% of the urban population is reported to have access to safe drinking water supply, there are severe deficiencies with regard to quantity of water available to urban residents. Cities continue to grow at a rapid pace and encroach upon periurban villages and increase pressure on the existing infrastructure to perform at urban standards. The available groundwater is exploited in an unplanned and unregulated manner, resulting in ecological degradation.

Surveys conducted by the Asia Development Bank in 2007 show that in a sizeable number of urban centers, the availability of water is even less than 100 l per capita per day (Table 20.1). The norms and standards range from 45 l per capita per day (Zakaria Committee, 1963) to 80 l per capita per day (Operations Research Group, 1989) for small towns. It is even higher at 95–125 l per capita per day according to the National Institute of Urban Affairs (NIUA), while for medium-sized towns, it is 150 l per capita per day (National Institute of Urban Affairs, 1997). The norm for water supply in cities, according to the National Commission of Integrated Water Resource Management (NCIWRD, 1999), is 220 l per capita per day for Class I cities. It may be noted that, in the absence of a single norm for domestic water supply, 100 l per capita per day has been considered as the benchmark for identifying water stress in urban households. Only 2.7% of sample municipalities reported supplying more than 100 l of water per capita per day (NIPFP, 2000). Almost 28% of the municipalities provide less than 50 l per capita per day.

As early as the 1980s, under the International Drinking Water Supply and Sanitation Decade Program launched in 1981, the aim was to provide adequate drinking water facilities to 90% of the urban population. Now, three decades from its launch, the mission remains unaccomplished. The hours of supply are also inadequate, forcing people to depend on supplementary water sources. Intermittent supply exposes the water supply system to contamination, as well as the wasting of water by consumers. None of the urban centers in India, except Thiruvananthapuram and Kota, have a 24-h water supply. A study on 20 cities in India showed that approximately 80% of the urban population is covered by water supply systems. The urban centers had only 4.3 h of water supply on average per day, with 30% of the water being lost in transmission (Asian Development Bank, 2007). The lack of water metering for residential establishments in most Indian cities puts further stress on the urban water supply system.

Drinking water was given the highest priority in the 2012 National Water Policy; however, there have been limited efforts to create new surface water sources and groundwater tables have been steadily declining.

Like most Indian cities, Kolkata, the third-largest city of India, also fails to adhere to the norms set by the National Commission of Integrated Water Resource Management. Supply is also intermittent and peripheral parts of the city remain mostly uncovered.

Table 20.1 Water Supply in Major Cities of India

	Coverage	Availability	Consumption	Demand	Unaccounted-for Water	Connections Metered	Average Tariff
Cities	%	h	L/capita/day	million L	%	%	(Rs/m^3)
Kolkata	79.0	8.3	130.0	125	35.0	0.1	1.1
Mumbai	100.0	4.0	191.0	3200	13.0	75.0	4.6
Chennai	89.3	5.0	87.0	600	17.0	3.5	10.9
Delhi	86.0	4.0	N.A.	3600	57.0		
Bangalore	92.9	4.5	74.0	970	45.0	95.5	20.6

Source: Asia Development Bank (2007).

20.3 Historical Background to Current Water Supply System of Kolkata

Kolkata, known as Calcutta until 2001, has one of the oldest water supply systems in India. In its earliest days, Kolkata was dependent on the Hugli River for its drinking water supply. Devout Hindus preferred to use water from the Ganges River (here referred to as Hugli River) for drinking. Contamination in the water supply system has been a concern since the city was founded as a trading port by the East India Company in 1690. The city's European residents preferred rainwater stored in large jars, having an aversion to river water polluted by corpses and other upriver contaminants. However, more substantial means of storage were needed as the city grew in the 18th century. In 1773, the city of Kolkata was declared as the capital of British-occupied India and remained so until 1912. The rise of Calcutta has been phenomenal since 1850. Between 1857 and 1900, Calcutta was connected to all-important destinations in India through the railways. During this time, coal mines northwest of Calcutta began production, the first modern wet dock went into action, the first iron and steel factory in West Bengal started producing iron, and the jute industry recorded phenomenal growth. A rail and road bridge connecting the two banks of the river was constructed in the second half of the nineteenth century.

Many public and private tanks were excavated to conserve monsoon rain for drinking water supply, although their common uses for washing and bathing soon made many unfit as sources of drinking water. The Great Tank, known as Lal Dighi, in the city's heritage hub at Dalhousie Square, was deepened and extended in 1709 to ensure a good supply of water for the garrison of old Fort William. Colonial Kolkata later developed around that area. Within the Kolkata Metropolitan Area, between 1805 and 1836, the excavation of large water tanks in Cornwallis Square, College Square, Wellington Square, and Wellesley Square kept the city hydrated during the days of British rule.

It was only in 1820 that a small pumping station and a system of open masonry aqueducts was constructed (Dasgupta, 1991). In 1822, the pumping station constructed at Chandpal Ghat lifted water from River Hugli into an open masonry aqueduct to distribute water to small portions of town like Old Court House Street, Dhurrumtollah, Chowringhee, Park Street, parts of Chitpur Road, LalBazar, and BowBazar, including water for washing the streets and for fire fighting. Pumping Stations operated for 7 h a day, between 6 a.m. and 10.00 a.m. and between 4 p.m. and 7 p.m. Apart from individual connections, stand posts along the streets provided water for the common citizen. The population of Kolkata was only 200,000 in 1822. By 1870, most principal streets of Kolkata had access to piped water through 500 standposts. The only purification process the water underwent was filtration through a bed of sand and charcoal in their own houses.

In 1870, a 6 million-gallon-per-day (MGD) plant was constructed at Palta, along with a 1-MGD reservoir and pumping station at Tallah, with another pumping station at Wellington Square for distribution. The residents obtained access to filtered water supply, though the per capita availability of water was only 60 l per capita per day. The site of the water works was 30 km upstream at Palta, which not only reduced the risk of contamination of abstracted water by Kolkata's wastes, but also diminished the presence of tidal seawater.

In 1888, three underground reservoirs were constructed at Wellington Square, Halliday Street (Md. Ali Park), and Auckland Square (Bhowanipur), to distribute water to consumers. However, the supply was only intermittent. From 1909 to 1914, production capacity increased at Palta and an elevated reservoir was constructed at Tallah with an aim to create an uninterrupted supply of water. Continuous supply could not, however, be maintained due to unusual usage and wastage by consumers. Although the supply was being supplemented by groundwater sources by that time, the latter hardly compensated for water waste of approximately 30% in the Palta network (Calcutta Environmental Management Strategy and Action Plan, 1995). As worn distribution pipes began to leak, they consequently delivered a much lower volume of water to consumers in the worst-affected areas of parts of south Calcutta, towards the limits of the supply system emanating southwards from Palta.

After independence, production capacity of Palta was augmented to 120 MGD, and further to 160 MGD in 1952. This capacity was sufficient to cater to the population of 2.7 million with a per capita availability of more than 130 liters per day. The production capacity increased from 6 MGD to 160 MGD for a six-fold increase in the city's population. For distribution of filtered water into different parts of the city, underground reservoirs at Wellington Square, Bhawanipur, and Md. Ali Park were revived and converted to booster pumping stations to increase pressure available from the Tallah pumping station.

20.4 Current Status of Water Supply in the Megacity of Kolkata

Until the 74th constitutional amendment, India had a federal structure in the water supply system and the Indian constitution empowered the states to govern water resources. However, with the introduction of the 74th constitutional amendment in 1992, devolution of authority towards the urban local bodies/municipalities was preferred. Urban local bodies have the constitutional right to plan, design, implement, operate, and maintain water and sanitation projects (Nallathiga, 2008). The federal government only plays an advisory role, along with the national and state water policies that provide a broad guideline for the sector. Thus, for the megacity of Kolkata it is the urban local body or the Kolkata Municipal Corporation that is responsible for water supply within the city.

The Water Supply Department in one of the largest urban local bodies, the Kolkata Municipal Corporation, has the objective to provide uninterrupted water supply to the residents of Kolkata. In spite of its favorable hydrological setting, with the Hugli River on one side and an extensive groundwater reserve, a large part of Kolkata suffers from inadequate supply or a total lack of municipal supply.

The Hugli River is the main source of water supply for Kolkata. The Kolkata Municipal Corporation treats water drawn from the river at three main water treatment plants. The oldest of these is the Indira Gandhi Water Treatment Plant, with a capacity of 260 MGD at Palta Waterworks. Water from this waterworks is sent to the Tallah

Table 20.2 Water Supply to the Megacity of Kolkata

Location of Water Treatment Plant	Plant Capacity (MGD)	Supply (MGD)	Remarks
Indira Gandhi Water Treatment Plant, at Palta	260	200	Treated surface water supply
Garden Reach Water Treatment Plant	120 + 15*	82 + 15*	
Jorebagan	8	5	
Wattgange	5	3	
Dhapa	30	10	
Total	438	315	
Groundwater through power-driven tube wells	30	30	Groundwater
Total	**468**	**345**	

*For Budge Budge, Maheshtala and Pujali municipalities.
Source: Kolkata Municipal Corporation (2015).

pumping station for temporary storage and onward distribution to the city. The other two water treatment plants include a 120 + 15*-MGD-capacity treatment plant at Garden Reach and another of 120-MGD capacity at Dhapa in the eastern fringe of Kolkata, where the groundwater is contaminated by arsenic. Dhapa Water Treatment Plant now functions at 30 MGD. It was targeted to shift from groundwater to treated surface water supply by 2014. Jorabagan and Wattgunge have smaller plants of capacities 8 MGD and 5 MGD, respectively. According to the Kolkata Municipal Corporation, the total capacity of the water treatment plants as of 2015 is 438 MGD, while the supply to the city is 315 MGD. A total of 30 MGD of water is also supplied with the help of power-driven tube wells in areas not covered by surface water supplies (Table 20.2). The average per capita availability of water rests at 130 liters per capita per day. 35% of the total urban population, however, remains uncovered by municipal water supply.

Hours of supply vary between 6.5 h per day in the city proper, to 4.5 h in the peripheral areas under Dhapa and Garden Reach Water Works (Table 20.3). However, in most parts of the city other than the main city, the municipal water supply is supplemented by individual groundwater sources. Groundwater continues to be the main source of water in areas planned to be covered by Dhapa Waterworks.

It is believed that the inability of the Kolkata Municipal Corporation to provide piped water to 100% of the population for 24 h is not caused by limitations on production capacity, but rather by production and delivery inefficiencies. Major deficiencies of water service delivery include intermittent supply, inadequate pressure, high leakage loss, high energy consumption, and dependence on groundwater (Majumdar & Gupta, 2007). The city also suffers from inadequate zonal mains and old networks in an advanced state of dilapidation (Sivaramakrishnan, 1993). This also leads to an additional problem of high transmission and distribution losses of 30%. To minimize loss due to leakage, the water pressure is kept low.

Table 20.3 **Water Supply in the Megacity of Kolkata**

Area	Name of the Treatment Plant	Hours of Supply	Duration of Supply/Day
Main city	Indira Gandhi Water Treatment Plant, at Palta, Jorebagan, and Wattgange	6 a.m. to 9 a.m. 11:30 a.m. to 12 p.m. 4:30 p.m. to 6 p.m. 7:30 p.m. to 9 p.m.	6.5 h
Southern fringe areas	Garden Reach Water Treatment Plant	6 a.m. to 9 a.m. 4:30 p.m. to 6 p.m.	4.5 h
Eastern part of the city	Dhapa		
Not covered by piped water supply	Groundwater		

Source: Kolkata Municipal Corporation (2015).

The intermittent supply and low pressure expose the water to contaminants. It is easy for sewage and other contaminants to enter the pipelines when the water pressure is low. Over dependence on groundwater exposes the population to arsenic contamination. The saline content is also high in some parts of the city. A drop in the piezometric level also increases chances of land subsidence.

Improving water supply for the city has been a major focus for Kolkata Municipal Corporation during the past few years. This has involved augmenting piped water supply through commissioning of water treatment plants, booster pumping stations, reservoirs, and head works, as well as the sinking of new tube wells as a short term measure in areas uncovered by piped water supply. The distribution system is also being revamped continuously so that it can meet the burgeoning water demands of the city.

20.5 Spatial Variation of Water Stress Within the City

The average performance of water supply subsector in the administrative area of Kolkata Municipal Corporation is comparatively high. Service level indicators for (i) water supply and (ii) water supply amount and hours exceed the national average, and some are not far from the national targets. A closer look, however, reveals that the result is the average of very high service levels in central areas of Kolkata and poor service levels in the periphery (Table 20.4).

A questionnaire survey was also conducted by the authors in 100 households selected both from the peripheral wards and the main city or the central parts of Kolkata. The survey revealed that the hours of supply are much less than claimed by the civic authority. In the peripheral wards served by Garden Reach and Dhapa Water Works, the duration of service is not more than 4 h a day, 2.5 h in the morning, and approximately 1.5 h in the afternoon. The supply is highly erratic and is available for a very limited duration. In the survey, 32% and 27% of respondents claimed that the

Table 20.4 **Urban Service Delivery Status in Different Areas of Kolkata Municipal Corporation**

	Population	Coverage	Per Head Amount	Hours per Day	Nonrevenue	Metered Connections
	%	%	(liters per capita per day)		%	%
National targets		100a	135b–150c	24b	20	100b
National average		81	123	4	N/A	25
All KMC areas		92	134	8	93	0.1
Central area	69	100	146	9	N/A	N/A
East peripheral area	6	76	109	6	N/A	N/A
West peripheral area	7	45	120	6	N/A	N/A
South peripheral area	18	83	120	6	N/A	N/A

[a]The 11th Five-Year Plan 2007–2012.
[b]*A Handbook on Service Level Benchmarking,* Ministry of Urban Development, 2010 (June).
[c]Manual on Water Supply and Treatment, Central Public Health and Environmental Engineering Organization, 1999.
Source: Kolkata Municipal Corporation (2011). Kolkata Urban Sector Investment Plan, Kolkata.

municipal supply was insufficient and provided water for less than 2 h twice a day and once a day, respectively (Figure 20.1).

The households thus depend on alternative sources of water supply, particularly in the peripheral wards (Figure 20.2). Nearly 63% of the nonslum population have their own devices to pump groundwater and supplement the municipal supply. Nearly 50% of these households even use this water for drinking, with some water treatment at the household level and without regard to potability standards. Another 47% depend on the "water man" to fetch drinking water. Another interesting observation during the household survey was that of the 47% who buy water from the "water man," more than 65% buy the municipal supply, which is available at stand posts and not as individual connections. The core city is better served with nearly all households depending on municipal water supply. Earlier, the supply and pressure was enough to lift water to individual overhead tanks to ensure 24 h of supply. Now, the pressure is reduced and is difficult to lift for storage as a result. The reduction in pressure is also an attempt to minimize leakage from the supply mains. Only 18% of the surveyed households in the main city reported the supply to be sufficient for 24 h when stored in overhead tanks during hours of supply.

Large parts of the city, particularly the peripheral wards, are currently dependent on groundwater (Figure 20.3). This water travels through tube wells and into a network of

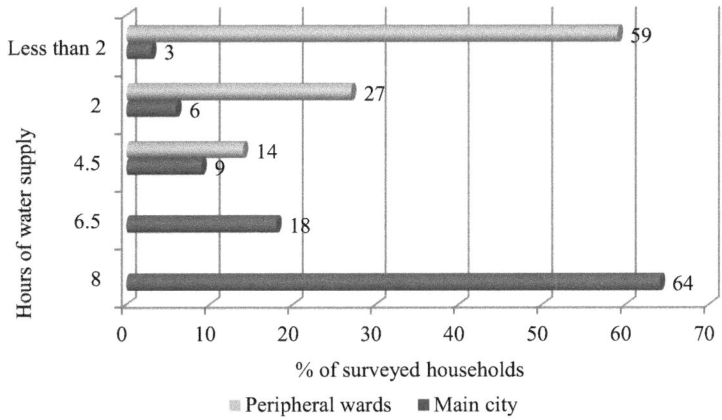

Figure 20.1 Hours of municipal water supply in the main city and in the periphery.

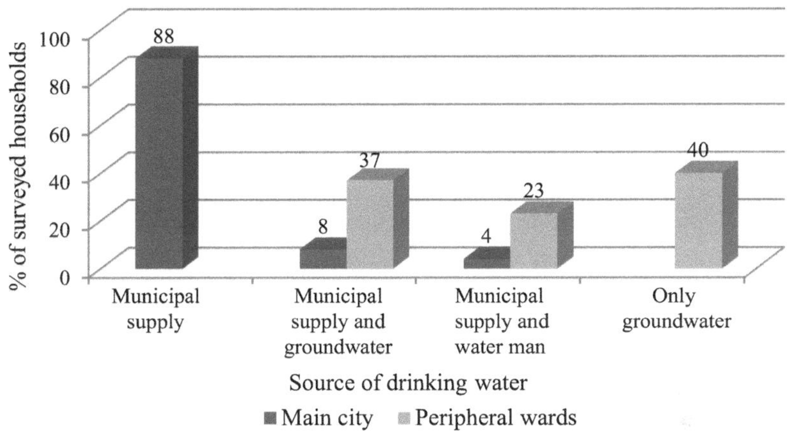

Figure 20.2 Source of drinking water supply in Kolkata.

pipelines and reservoirs before reaching the city's taps. A recent study by the Geological Survey of India and Geo-Scientific Studies, however, found that the water levels in these groundwater reserves are sinking.

The water supply in Kolkata is insufficient. This is evident from the fact that the greater part of the city enjoys a dual water usage system where municipal water supply is supplemented by underground water. Dependence on alternate sources also raises doubts about the water quality. While the Kolkata Municipal Corporation supplies potable water to the residents of Kolkata, individual households that pump groundwater use it untreated. This is a cause of grave concern, especially because the groundwater in the peripheral parts of the city has reported cases of arsenic contamination. The treatment measures followed at the household level are ineffective to make the water arsenic free.

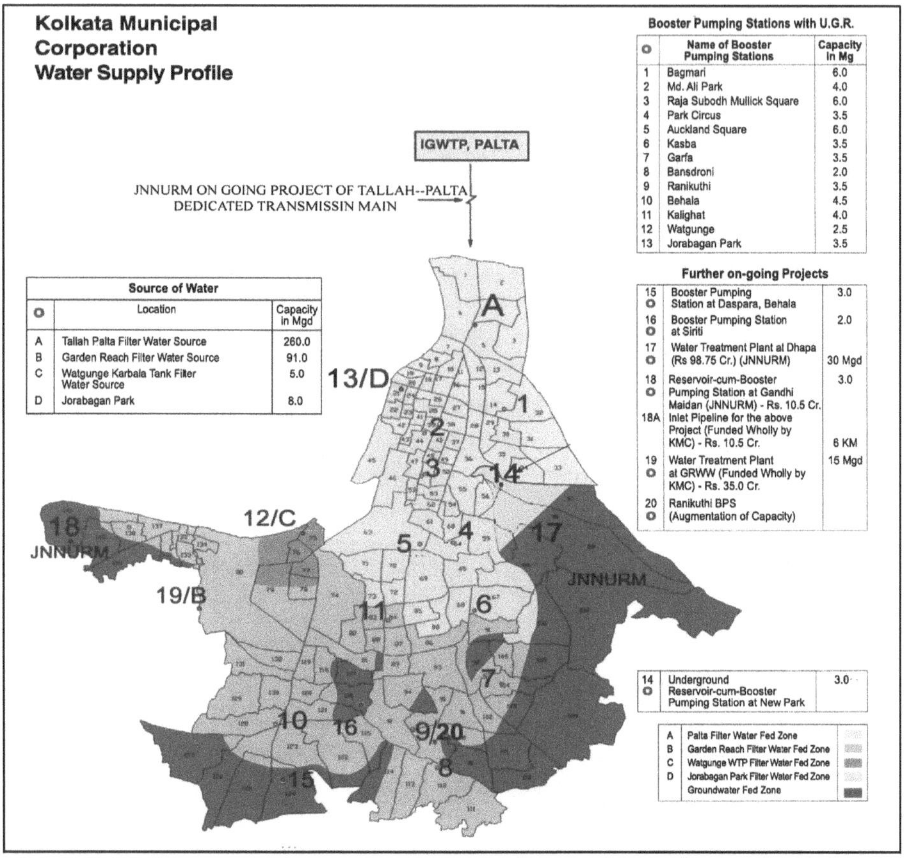

Figure 20.3 Drinking water supply in Kolkata.

20.6 Vulnerability and Challenges to Water Stress Within the City

The water supply system in the megacity of Kolkata faces innumerable challenges. In spite of its favorable hydrological settings, the city of Kolkata still faces acute water shortages. The peripheral wards remain uncovered by the surface water system of the Kolkata Municipal Corporation. These regions may also suffer from arsenic contamination, making the population vulnerable to diseases caused by excessive arsenic in groundwater. No part of the city has 24 h of supply, which exposes the water supply lines to sewage and other contaminants that easily enter the pipelines through cracks when the pressure is low. The inadequacy of the supply and the short duration of that supply forces the population to store water that will be thrown out when the municipal supply is restored, leading to waste and adding to the supply burden. The per capita supply is well below the desired norm. The old, worn-out supply lines also result in leakage loss to an extent of 30%. Backflow of water through the stand posts during hours of no supply in the service main also amounts to wastage. An attempt has been made to identify vulnerability and challenges for the water supply system that spans issues of water supply, infrastructure, finance, water quality, and governance (Table 20.5).

Thus, the water supply system is vulnerable to water contamination, intermittent flow, leakage, and loss due to unregulated flow, insufficient coverage in the slum pockets, excessive dependence on the Tallah tank (which also supplies water to Salt Lake City in the eastern periphery), lack of metering and low-cost recovery, dual water supply (including both surface water and groundwater), lack of awareness about judicious water use, the nonparticipatory nature of governance, and a supply-driven approach. Vulnerability in the water supply system increases with increasing distance from the central city. Peripheral wards are the most vulnerable, especially the eastern periphery.

20.7 Ensure Resilience in the Water Supply System in Kolkata

Despite some uncertainties, there is little doubt that the frequency and amplitude of water-related shocks—droughts, floods, and dry spells—will increase as a result of climate change (IPCC, 2007). An early and very significant negative impact of climate change will be reduced water availability and increased water shocks in many regions of the world. Building resilience to global environmental change, therefore, will have to focus on adapting to increasing water shocks. The pursuit of sustainability is aimed at initiatives for protecting and conserving natural resources and promoting lifestyles, and their supporting infrastructure, that can endure indefinitely because they are neither depleting resources nor degrading environmental quality (Wong & Eadie, 2000). A critical challenge to urban communities are their designs for resilience to the impact of climate change and population growth, particularly in regards to sustainable management of water resources and the protection of water environments (Wong & Brown, 2009).

It is accepted that the conventional urban water management approach is highly unsuited to addressing current and future sustainability issues (Butler & Maksimovic, 1999; Newman, 2001; Ashley et al., 2003, 2005). With the widespread realization of the significance of global warming and climate change, urban communities are seeking resilience to existing and future uncertainties in urban water supplies brought on by a combination of climate variability, population growth, and climate change (Wong & Brown, 2009).

The challenge to building resilience in the urban water supply system arises from the fact that water is the most valuable natural resource and is continuously degenerated under urban pressure, as well as the fact that its availability is threatened by climatic variability.

The implications of excessive water utilization including groundwater, stressed further by climate change, would have far-reaching consequences for the functioning of ecosystems that could adversely affect mankind. Unlike natural disasters, the implications of urban water stress become evident only when the damage is irreparable. The increased arsenic content in groundwater became evident only when people were affected by it. No warning system works, and resilience building remains a neglected issue. Thus, it is important to make the urban water system resilient to increasing water demand, climate change, reduced water availability, increased water pollution, and inefficient delivery systems (Table 20.6).

Table 20.5 Stresses in the Water Supply System of Kolkata

Issues	Indicators	Central City	East Peripheral	West Peripheral	South Peripheral	Comments
Supply	Total withdrawal	1657.83 million liters				Augmentation possible
	Capacity of treatment plants	1771.38 million liters				Augmentation needs investment
	Percentage of population served	100*	76	45	83	Inadequate zonal main. Augmentation through service main and booster pumping stations
	Per capita supply	146	109	120	120	Below norm
	Hours of supply	6.5	4.5	4.5	4.5	24 h supply in Thiruvananthapuram and Kota
		Intermittent supply				
Infrastucture	Percentage of population covered by surface water supply	100	Groundwater supply	More than 60% of the population depends on groundwater as alternate source		Peripheral regions depend on groundwater which has a high arsenic content
	Number of booster pumping stations	9	–	–	9	
	Booster pumping stations under construction	1 (proposed)	4		3 1 (proposed)	
	Stand posts	17,000				
		Backflow of water from reservoir during hours of no supply in the service main leads to wastage				
	Leakage	30% unregulated flow of water				Old dilapidated network

Finance	Water metering for individual connection	No water metering; water is declared as an economic good in the International Conference on Water and Environment in Dublin 1992		100% metering in Coimbatore
Water Quality	Quality of water supplied	Potable	Chances of arsenic contamination in groundwater for individual connections and tube wells used.	
Governance	Supply-driven approach Centralized system with no scope for private-public partnership, people's participation, community involvement. Unaccountability due to lack of water meter.			

*The slum population, which varies between 15% and 40% of the total population in the central city, is covered by standposts.

Table 20.6 Suggested Areas of Intervention for Resilient Water Supply System in Kolkata

		Beneficial		
Intervention	**Explanation**	**Water stress**	**Water quality**	**Resilience**
Increase capacity of treatment plant	Through resource mobilization and private-public partnerships	√		√
Increase population served and area coverage	Increase booster pumping stations and zonal mains to maintain flow of water	√		√
Reduce wastage	Control backflow of water from service main to stand post	√		√
Reduce leakage	Replace existing worn-out pipes	√	√	√
Cost recovery	Water metering at least for individual connections above the optimum norm of 100 liters per capita per day	√		√
Reduce groundwater use	Shift to surface water, artificial recharge of groundwater to dilute chemical contaminants	√	√	√
Rainwater harvesting	Collect and store rainwater to supplement municipal supply at all residential complexes and institutions	√	√	√
Increase surface supply water	Protecting surface water sources	√	√	√
Ensure community participation	Reduction of wastage	√	√	√
	Use grey water for nonpotable use			
Demand based system	Demand management to increase sustainability	√	√	√
Reduce dependence on groundwater	Gradual phasing out and limiting groundwater extraction options in areas served by municipal supply	√	√	√

Hydrologically, the city is endowed with enough freshwater, and yet it is one of the most water-stressed megacities, crippled by poor water quality and inadequate supply.

20.8 Conclusion

The Kolkata Municipal Corporation is planning to implement a few of the suggested areas of intervention that would determine the resilience of urban communities. It is already planning to increase supply and coverage through the seven booster pumping stations under construction and the two proposed ones. In its Vision 2025 proposal, the corporation aims to replace the groundwater supply by surface water for the entire city. However, the corporation should only consider groundwater sources as an option of supply during emergencies, and ensuring groundwater recharge is highly recommended for increasing adaptive capacity to ensure resilience. Replacement of the worn-out pipelines should be done on an emergency basis to stop leakage. Charging for water to the households with individual connections for the water in excess of 100 l per capita per day would increase awareness about the economic and ecological importance of water. Demand management options also need to be explored to make learning and awareness building a part of the increasing urban resilience.

Acknowledgments

The authors acknowledge support from the Kolkata Municipal Corporation and the residents of Kolkata. Financial support for the research in Kyoto University was provided by the Sumitomo Foundation, for which the authors remain grateful.

References

Alberti, M. (1996). Measuring urban sustainability. *Environmental Impact Assessment Review, 16*(4), 381–424.

Alcamo, J., & Henrichs, T. (2002). Critical regions: A model-based estimation of world water resources sensitive to global changes. *Aquatic Sciences, 64*(4), 352–362.

Ashley, R., et al. (2003). Sustainable decision making for the UK water industry. *Proceedings of the ICE-Engineering Sustainability, 156*(1), 41–49.

Ashley, R. M., Balmforth, D. J., Saul, A. J., & Blanksby, J. R. (2005). Flooding in the future. *Water Science and Technology, 52*(5), 265–274.

Arnell, N. W. (2004). Climate change and global water resources: SRES emissions and socio-economic scenarios. *Global Environmental Change: Human and Policy Dimensions, 14*(1), 31–52.

Asian Development Bank. (2007). *Benchmarking and data book of water utilities in India*. Asian Development Bank, Metro Manila, Phillipines.

Azar, C., Holmberg, J., & Lindgren, K. (1996). Socio-ecological indicators for sustainability. *Ecological Economics, 18*(2), 89–112.

Bishop, A. B. (1974). *Carrying capacity in regional environmental management.* U.S. Environmental Protection Agency, Washington D.C.

Butler, D., & Maksimovic, C. (1999). Urban water management-challenges for the third millennium. *Progress in Environmental Science, 1*, 213–236.

Calcutta Environmental Management Strategy and Action Plan. (1995). *Working Paper 1. Situation Report (Revised Version November 1995).* Department of Environment, Government of West Bengal, Calcutta.

Census of India. (2011). *Final population tables.* Government of India, New Delhi.

Central Public Health and Environmental Engineering Organization. (1999). *Manual on Water Supply and Treatment Ministry of Urban Development.* New Delhi: Government of India.

Cutter, S. L., et al. (2008). A place-based model for understanding community resilience to natural disasters. *Global Environmental Change, 18*(4), 598–606.

Dasgupta, B. (1991). *Calcutta's urban future: agonies from the past and prospects for the future.* Government of West Bengal, Calcutta.

Easterling, D. R., et al. (2000). Climate extremes: observations, modeling, and impacts. *Science, 289*(5487), 2068–2074.

Eleventh Five Year Plan. (2007–2012). *Volume II Social Sector. Planning Commission.* New Delhi: Government of India.

Falkenmark, M. (1998). Preparing for the future: water for a growing population. *Journal of Water Services Research and Technology-Aqua, 47*(4), 161–166.

Fitzroy, N. (1998). *Liquid assets: is water privatisation the answer to access?* London: Panos Institute.

Folke, C. (2006). Resilience: The emergence of a perspective for social–ecological systems analyses. *Global Environmental Change, 16*(3), 253–267.

Foxon, T. J., et al. (2002). Sustainability criteria for decision support in the UK water industry. *Journal of Environmental Planning and Management, 45*(2), 285–301.

Gallopín, G. C. (1996). Environmental and sustainability indicators and the concept of situational indicators. A systems approach. *Environmental Modeling & Assessment, 1*(3), 101–117.

Giorgi, F., & Bi, X. (2005). Regional changes in surface climate interannual variability for the 21st century from ensembles of global model simulations. *Geophysical Research Letters, 32*(13), 1–5.

Gleick, P. H. (2003). Water use. *Annual Review of Environment and Resources, 28*, 275–314.

Intergovernmental Panel on Climate Change (IPCC). (2007). Climate change 2007. Appendix to synthesis report. In A. P. M. Baede, P. Van der Linden, & A. Verbruggen (Eds.), *Climate Change 2007: Synthesis report* (pp. 76–89), Contributions of Working Groups I, II, III to the Fourth Assessment Report of the Intergovernmental Panel on Climate Change.

Joardar, S. D. (1998). Carrying capacities and standards as bases towards urban infrastructure planning in India: A case of urban water supply and sanitation. *Habitat International, 22*(3), 327–337.

Jury, W. A., & Vaux, H. (2005). The role of science in solving the world's emerging water problems. *Proceedings of the national academy of sciences of the United States of America, 102*(44), 15715–15720.

Kolkata Municipal Corporation. (2012). *Kolkata Urban Sector Investment Plan 2012–2022.* Kolkata: Volume I- The Urban Service Plan.

Leichenko, R. (2011). Climate change and urban resilience. *Current Opinion in Environmental Sustainability, 3*(3), 164–168.

Lundqvist, J., Tortajada, C., Varis, O., & Biswas, A. (2005). Water management in megacities. *Ambio, 34*(3), 267–268.

Loucks, D. P. (1997). Quantifying trends in system sustainability. *Hydrological Sciences Journal, 42*(4), 513–530.

Majumdar, C., & Gupta, G. (2007). The debate over municipal water pricing: Evidence from Kolkata, India. *Water Resources Development, 23*(4), 571–582.

McGee, T. (2001). Urbanization takes on new dimensions in Asia's population giants. *Population Today, 29*(7), 1–2.

Milman, A., & Short, A. (2008). Incorporating resilience into sustainability indicators: An example for the urban water sector. *Global Environmental Change, 18*(4), 758–767.

Ministry of Urban Development. (2010). *A Handbook on Service Level Benchmarking*. New Delhi: Government of India.

Nallathiga, R. (2008). Contradictions of sustainable urban development: The choice of compact city development approach. *ITPI Journal, 5*(2), 55–59.

National Commission for Integrated Water Resource Development. (1999). *Integrated water resource development: A plan for action* (Vol. I). Ministry of Water Resources, Government of India, 25–50.

National Institute of Public Finance and Policy. (2000). *Option for Closing the Revenue Gap of Municipalities, 2000/01 to 2004/05*. New Delhi: National Institute of Public Finance and Policy.

National Institute of Urban Affairs. (1997). *Financing urban infrastructure in India*. New Delhi: National Institute of Urban Affairs.

Newman, P. (2001). Sustainable urban water systems in rich and poor cities—Steps towards a new approach. *Water Science & Technology, 43*(4), 93–99.

Operations Research Group. (1989). *Delivery and financing of urban services*. New Delhi: Operations Research Group.

Romero-Lankao, P., & Qin, H. (2011). Conceptualizing urban vulnerability to global climate and environmental change. *Current Opinion in Environmental Sustainability, 3*(3), 142–149.

Royer, J. F., et al. (2002). Simulation of climate changes during the 21st century including stratospheric ozone. *Comptes Rendus Geoscience, 334*(3), 147–154.

Sahely, H. R., Kennedy, C. A., & Adams, B. J. (2005). Developing sustainability criteria for urban infrastructure systems. *Canadian Journal of Civil Engineering, 32*(1), 72–85.

Sivaramakrishnan, K. C. (1993). *Managing urban environment in India*. New Delhi: Times Research Foundation.

Stephen, D. (2000). *Performance Indicators for Rural Water Schemes. 26th WEDC Conference: Water, Sanitation and Hygiene*. Dhaka, Bangladesh: Challenges of the Millennium.

Thomson, A. M., Brown, R. A., Rosenberg, N. J., Srinivasan, R., & Izaurralde, R. C. (2006). Climate change impacts for the conterminous USA: An integrated assessment: Part 4. Water resources. *Climatic Change, 69*, 67–88.

Tudhope, A. W., et al. (2001). Variability in the El Niño-Southern oscillation through a glacial-interglacial cycle. *Science, 291*(5508), 1511–1517.

Twelfth Five-Year Plan. (2012–2017). *Volume II: Sectoral Policies and Programmes, Planning Commission*. New Delhi: Government of India.

United Nations. (2004). *United Nations World Population Prospects: The 2004 Revision*. Department of Economic and Social Affairs, United Nations, New York.

United Nations Centre for Human Settlements (Habitat). (2001). Cities in a globalizing world. *Global Report on Human Settlements*. London, UK: Earthscan.

United Nations Development Programmes. (2006). *Human Development Report- beyond scarcity*. New York, USA: power, poverty and the global water crisis.

United Nations Human Settlements Programme (UN - HABITAT). (2011). Cities and climate change. *Global Report on Human Settlements*. Earthscan Ltd., London, UK.

Walker, B., Carpenter, S., Anderies, J. M., Abel, N., Cumming, G. S., Janssen, M., & Pritchard, R. (2002). Resilience management in social-ecological systems: Aworking hypothesis for a participatory approach. *Conservation Ecology*, *6*(1), 14.

Watkins, K., Carvajal, L., Coppard, D., Fuentes, R., & Ghosh, A. (2006). *Human Development Report 2006: Beyond scarcity: Power, poverty, and the global water crisis*. United Nations Development Programme (UNDP).

Wong, T. H. F., & Brown, R. R. (2009). The water-sensitive city: Principles for practice. *Water Science and Technology*, *60*(3), 673–682.

Wong, T. H. F., & Eadie, M. L. (2000). *Water sensitive urban design: A paradigm shift in urban design*. Proceedings of the 10th World Water Congress, Melbourne, Australia: March 12–16.

Zakaria Committee. (1963). *Financial Resources for Urban Local Bodies*. New Delhi: Government of India.

Index

Note: Page numbers followed by "f" indicate figures, "t" indicate tables, and "b" indicate boxes.

B

C

E

H

I

T

V

W

FSC
www.fsc.org
MIX
Papier | Fördert
gute Waldnutzung
FSC® C083411

Kolibri 360 GmbH
Ferdinand-Jühlke-Straße 7
99095 Erfurt, Deutschland
produktsicherheit@kolibri360.de